Rhizosphere Biology

Series Editor

Anil Kumar Sharma
Biological Sciences, CBSH, G.B. Pant University of Agriculture & Technology, Pantnagar, Uttarakhand, India

The Series Rhizosphere Biology, emphasizes on the different aspects of Rhizosphere. Major increase in agricultural productivity, to meet growing food demands of human population is imperative, to survive in the future. Along with methods of crop improvement, an understanding of the rhizosphere biology, and the ways to manipulate it, could be an innovative strategy to deal with this demand of increasing productivity. This Series would provide comprehensive information for researchers, and encompass all aspects in field of rhizosphere biology. It would comprise of topics ranging from the classical studies to the most advanced application being done in the field. Rhizoshpere is a dynamic environment, and a series of processes take place to create a congenial environment for plant to grow and survive. There are factors which might hamper the growth of plants, resulting in productivity loss, but, the mechanisms are not very clear. Understanding the rhizosphere is needed, in order to create opportunities for researchers to come up with robust strategies to exploit the rhizosphere for sustainable agriculture.

There are titles already available in the market in the broad area of rhizosphere biology, but there is a major lack of information as to the functions and future applications of this field. These titles have not given all the up-to-date information required by the today's researchers and therefore, this Series aims to fill out those gaps.

More information about this series at http://www.springer.com/series/15861

Anita Sharma
Editor

Microbes and Signaling Biomolecules Against Plant Stress

Strategies of Plant- Microbe Relationships for Better Survival

Springer

Editor
Anita Sharma
Department of Microbiology
GB Pant University
of Agriculture & Technology
Pantnagar, Uttarakhand, India

ISSN 2523-8442 ISSN 2523-8450 (electronic)
Rhizosphere Biology
ISBN 978-981-15-7093-3 ISBN 978-981-15-7094-0 (eBook)
https://doi.org/10.1007/978-981-15-7094-0

This Springer imprint is published by the registered company Springer Nature Singapore Pte Ltd.
The registered company address is: 152 Beach Road, #21-01/04 Gateway East, Singapore 189721, Singapore

Preface

The ever-increasing human population is constantly putting pressure on agriculture production globally. Plants encounter various biotic and abiotic stresses during their life span which significantly affect the overall agricultural produce. Although various agrochemicals provide instant solutions to the problem and help plants to cope up with major biotic and abiotic challenges, their injudicious use also threatens the sustainability of the environment. The present state of environmental conditions demands eco-friendly approaches that help to restore environmental sustainability by using microorganisms as biofertilizers and biopesticides which can be a good alternative for the same. The plant is directly or indirectly affected by its rhizospheric microbiome, a major constituent of soil–plant interaction. A better understanding of plant microbiome, their interactions, mechanisms, and signaling pathways involved in microbe-driven stress alleviation in plants is necessary to develop such eco-friendly approaches. The content of the book revolves around the plant–microbe interaction and the mechanisms involved in stress management by both the partners. Besides, this book also focuses on communication and signaling between plant and associated microorganisms under stress. The content of the book may help in developing a better knowledge regarding cross-talk between plant and its associated microflora and may also help in developing new approaches and techniques that might help in combating the pressure of extreme environmental conditions on plants. Exploring plant–microbe communication and signaling pathways can pave a path for future research as well as rhizosphere engineering which may help in the maintenance of agricultural sustainability as well as harmony with the environment. The synergistic effect of plants and microbes in phytoremediation is also covered in the book. The book helps to uncover the knowledge of molecular (omics) and physiological aspects of the various processes in improving strategies employed by the plants under stress/extreme environmental conditions and ultimately optimizing plant tolerance to various environmental stresses that pose a major threat in agricultural productivity in an eco-friendly way by microbe-driven methods.

Pantnagar, Uttarakhand, India Anita Sharma

Acknowledgments

The editor thankfully acknowledges all the authors for their contributions in the form of book chapters, Ms. Damini and Ms. Harshita, Ph.D. students of the Department of Microbiology for their help in providing technical support, and Dr. Barkha Sharma, Assistant Professor, Department of Epidemiology, Mathura Veterinary University, Mathura in finalizing the manuscripts and book contents.

Contents

Editor and Contributors

About the Editor

Anita Sharma is presently working as a Professor at the Department of Microbiology, GBPUA&T, Pantnagar. She is an alumna of Pant Nagar University and obtained her M.Sc. degree in 1981 in Microbiology with a minor in Biochemistry. She has been a recipient of merit scholarships throughout her academic career. She obtained her PhD from BHU in 1990. She was awarded a two-year CSIR Fellowship for her project on Nitrogen Fixation by Cyanobacteria, after which she explored plant growth promotory rhizobacteria in a DBT-funded project at the Department of Microbiology, Pantnagar. She joined the Department of Microbiology, Gurukul Kangri University, in 1997, and took on her current position at the Department of Microbiology, Pantnagar, in 1999.

Her main research interests are in plant–microbe interaction and biodegradation; in this regard, she has worked as PI for three projects and Co-PI in more than 10 projects including one INDO SWISS project. Her most recent focus is on soil metagenomics under the influence of nanocompounds and pesticides. She has published more than 60 papers in national and international journals, book chapters, and research articles. She has also edited one book: *Microbial Biotechnology for Environmental Cleanup*. She has presented her work at various national and international conferences and is a life member of the AMI (Association of Microbiologist of India) and a member of several other societies. She has served as Coordinator of the Women Empowerment Cell, Faculty Secretary for the College of Basic Science and Humanities, and as a member of the University's Academic Council.

Contributors

Amit Ahuja ICAR-IARI, New Delhi, India

Sandeep Arora Department of Molecular Biology and Genetic Engineering (CBSH), G. B. Pant University of Agriculture and Technology, Pantnagar, Uttarakhand, India

Asifa Mushtaq Department of Botany and Microbiology, H.N.B. Garhwal (A Central) University, Srinagar, Uttarakhand, India

Geeta Bhandari Sardar Bhagwan Singh University, Dehradun, Uttarakhand, India

Ram Naresh Bharagava Laboratory of Bioremediation and Metagenomics Research, Department of Microbiology, B B Ambedkar University, Lucknow, Uttar Pradesh, India

Deepesh Bhatt Department of Biotechnology, Shree Ramkrishna Institute of Computer Education and Applied Sciences, Veer Narmad South Gujarat University, Surat, Gujarat, India

Pankaj Bhatt State Key Laboratory for Conservation and Utilization of Subtropical Agro-Bioresources, Guangdong Province Key Laboratory of Microbial Signals and Disease Control, Integrative Microbiology Research Centre, South China Agricultural University, Guangzhou, People's Republic of China

Jaideep Kumar Bisht ICAR-Vivekananda Parvatiya Krishi Anusandhan Sansthan, Almora, Uttarakhand, India

Sharini Chaudhary ICAR-NIPB, New Delhi, India

Stanzin Idong Department of Plant Pathology, College of Agriculture, Pantnagar, Uttarakhand, India

B. Jeewan ICAR-Vivekananda Parvatiya Krishi Anusandhan Sansthan, Almora, Uttarakhand, India

Manisha Joshi Department of Microbiology, College of Basic Sciences and Humanities, GBPUA&T, Pantnagar, Uttarakhand, India

Amir Khan Department of Microbiology, College of Basic Sciences and Humanities, GBPUA&T, Pantnagar, Uttarakhand, India

Mohd. Saghir Khan Department of Agricultural Microbiology, Faculty of Agricultural Sciences, Aligarh Muslim University, Aligarh, Uttar Pradesh, India

Arun S. Kharat Laboratory of Applied Microbiology, School of Life Sciences, Jawaharlal Nehru University, New Delhi, India

Priyanka Khati ICAR-Vivekananda Parvatiya Krishi Anusandhan Sansthan, Almora, Uttarakhand, India

Kuldeep Kumar ICAR-Vivekananda Parvatiya Krishi Anusandhan Sansthan, New Delhi, India

Rajesh Kumar Rhizosphere Biology Laboratory, Department of Microbiology, Babasaheb Bhimrao Ambedkar University (A Central University), Lucknow, India

Damini Maithani Department of Microbiology, GB Pant University of Agriculture & Technology, Pantnagar, Uttarakhand, India

M. L. Manoj ICAR-NIPB, New Delhi, India

K. K. Mishra ICAR-Vivekananda Parvatiya Krishi Anusandhan Sansthan, Almora, Uttarakhand, India

Pankaj Kumar Mishra ICAR-Vivekananda Parvatiya Krishi Anusandhan Sansthan, Almora, Uttarakhand, India

Sandhya Mishra Key Laboratory of Microbial Signals and Disease Control, Integrative Microbiology Research Centre, South China Agricultural University, Guangzhou, People's Republic of China

Nandkishor More Department of Environmental Science, B B Ambedkar University, Lucknow, Uttar Pradesh, India

Sikandar I. Mulla Department of Biochemistry, School of Applied Sciences, REVA University, Bangalore, India

Pallavi Department of Botany, J.N.L. College, Patliputra University, Patna, Bihar, India

Manoj Parihar ICAR-Vivekananda Parvatiya Krishi Anusandhan Sansthan, Almora, Uttarakhand, India

Amit U. Paschapur ICAR-Vivekananda Parvatiya Krishi Anusandhan Sansthan, Almora, Uttarakhand, India

Arunava Pattanayak ICAR-Vivekananda Parvatiya Krishi Anusandhan Sansthan, Almora, Uttarakhand, India

Swati Rastogi Rhizosphere Biology Laboratory, Department of Microbiology, Babasaheb Bhimrao Ambedkar University (A Central University), Lucknow, India

Sheel Ratna Rhizosphere Biology Laboratory, Department of Microbiology, Babasaheb Bhimrao Ambedkar University (A Central University), Lucknow, India

P. Rawat Department of Plant Physiology, College of Basic Sciences and Humanities, G. B. Pant University of Agriculture and Technology, Pantnagar, Uttarakhand, India

Seema Rawat School of Life Sciences, Central University of Gujarat, Gandhinagar, Gujarat, India

Sudipta Saha Department of Pharmaceutical Sciences, B B Ambedkar University, Lucknow, Uttar Pradesh, India

Olfa Ben Said Laboratory of Environment Biomonitoring, Coastal Ecology and Ecotoxicology Unit, Faculty of Sciences of Bizerte, University of Carthage, Zarzouna, Tunisia

Saima Saif Department of Agricultural Microbiology, Faculty of Agricultural Sciences, Aligarh Muslim University, Aligarh, Uttar Pradesh, India

Saurabh C. Saxena Department of Biochemistry, Central University of Haryana, Mahendragarh, Haryana, India

D. Shankhdhar Department of Plant Physiology, College of Basic Sciences and Humanities, G. B. Pant University of Agriculture and Technology, Pantnagar, Uttarakhand, India

S. C. Shankhdhar Department of Plant Physiology, College of Basic Sciences and Humanities, G. B. Pant University of Agriculture and Technology, Pantnagar, Uttarakhand, India

Anil Kumar Sharma Department of Biological Sciences, CBSH, G.B. Pant University of Agriculture and Technology, Pantnagar, Uttarakhand, India

Anita Sharma Department of Microbiology, GB Pant University of Agriculture & Technology, Pantnagar, Uttarakhand, India

Ajay Veer Singh Department of Microbiology, College of Basic Sciences and Humanities, GBPUA&T, Pantnagar, Uttarakhand, India

Ashish Kumar Singh ICAR-Vivekananda Parvatiya Krishi Anusandhan Sansthan, Almora, Uttarakhand, India

Harshita Singh Department of Microbiology, GB Pant University of Agriculture & Technology, Pantnagar, Uttarakhand, India

Lal Bahadur Singh ICAR-NIPB, New Delhi, India

N. Sreeshma ICAR-NIPB, New Delhi, India

Almas Zaidi Department of Agricultural Microbiology, Faculty of Agricultural Sciences, Aligarh Muslim University, Aligarh, Uttar Pradesh, India

Abbreviations

2DE	Two-dimensional electrophoresis
AAS	Atomic absorption spectroscopy
ABA	Abscisic acid
ABC	Antigen-binding cassette
ACC	1-amino-cyclopropane-1-carboxylase
ACCD	Acetyl-coenzyme A carboxylase carboxyl transferase subunit beta
AFP	Antifreeze proteins
AHO	Acetonyl hydroxyoindole
AM	Arbuscular mycorrhizae
APX	Ascorbate peroxidase
ATAF	Arabidopsis transcription activation factor
ATSDR	Agency for Toxic Substances and Disease Registry
AzA	Azelaic acid
BR	Brassinosteroid
BTH	Benzothiadiazole carbothioic acid S-methyl ester
CDF	Cumulative distribution function
CDPKs	Calcium-dependent protein kinases/calmodulin-like domain protein kinases
CMC	Critical micelle concentration
COR	Cold-regulated genes
CREs	Cis-regulatory elements
DA	Dehydroabietinal
DAPG	Diacetylphloroglucinol
DHAR	Dehydroascorbate reductase
DREB	Dehydration responsive element binding
EBR	Epibrassinosteroid
EDS	Enhanced disease susceptibility
EEA	European Economic Area
ELISA	Enzyme-linked immunosorbent assay
EPA	Environment protection agency

EPS	Exopolysaccharide
ERD	Early responsive to dehydration
ERM	Extraradical mycelium
ESI-LC-MS/MS	Electrospray ionization liquid chromatography-mass spectrometry
ESTs	Earlier expressed sequence tags
ETC	Electron transport chain
ETI	Effector-triggered immunity
FAO	Food and Agriculture Organization
FT-ICR-MS	Fourier-transform ion cyclotron resonance mass spectrometry
FTIR	Fourier-transform infrared spectroscopy
GC-MS	Gas chromatography-mass spectroscopy
GOGAT	Glutamine synthase
GP	Guaiacol peroxidase
GPx	Glutathione peroxidase
GR	Glutathione reductase
GS	Glutamine synthetase
GSH	Glutathione
GST	Glutathione-s-transferase
HA	Hyperaccumulator
HAPs	Hyperaccumulator plants
HCMV	Human cytomegalovirus
HKT	Histidine kinase transporter
HM	Heavy metals
HNE	Hydroxynonenal
HPLC	High-pressure liquid chromatography
HR	Hypersensitive response
HSP	Heat shock proteins
ICP-AES/OES	Inductively coupled plasma-atomic/optical emission spectrometry
ICP-MS	Inductively coupled plasma-mass spectrometry
ICS	Isochorismate synthase
INA	Isonicotinic acid
IPCC	Intergovernmental Panel on Climate Change
IPLs	Isopyruvate lyases
IP-MS	Immunoprecipitation-mass spectrometry
ISR	Induced systemic resistance
IST	Induced systemic tolerance
ITPS	International Technology Professional Solutions
iTRAQ	Isobaric tag for relative and absolute quantification
JA	Jasmonic acid
LC-MS	Liquid chromatography-mass spectroscopy
LEA Protein	Late embryogenesis abundant protein
LEA	Late embryogenesis abundant protein

LPO	Lipid peroxidation
MALDI-TOF	Matrix-assisted laser desorption/ionization time-of-flight
MAMPs	Microbe-associated molecular pattern
MAPK	Mitogen-activated protein kinase
MDA	Malondialdehyde
MDHAR	Monodehydroascorbate reductase
MG	Methylglyoxal
MT	Metallothionine
NAA	Neutron activation analysis
NAM	No apical meristem
NGS	Next-generation sequencing
NMR	Nuclear magnetic resonance
NPR	Nonexpressor of pathogenesis-related protein
NRAMP	Natural resistance-associated macrophage protein
PAH	Polycyclic aromatic hydrocarbon
PAL	Phenylalanine ammonia-lyase
PAMPs	Pathogen-associated molecular pattern
PC	Phytochelatins
PCBs	Polychlorinated biphenyls
PCD	Programmed cell death
PEG	Polyethylene glycol
PFOA	Perfluorooctanoic acid
PGPR	Plant growth-promoting bacteria
PIP	Plasma membrane intrinsic protein
POPs	Persistent organic pollutants
PPI	Protein–protein interaction
PR	Pathogenesis-related protein
PRX	Peroxiredoxin
PSB	Phosphate-solubilizing bacteria
PUFA	Polyunsaturated fatty acid
PYR	Pyrrolidonyl arylamidase
QTL	Quantitative trait locus
RBOH	Respiratory burst oxidase homologs
RISR	Rhizobacteria-induced systemic resistance
RLKs	Receptor-like kinases
RND	Reflex neurovascular dystrophy
ROPs	Recalcitrant organic pollutants
ROS	Reactive oxygen species
RUBISCO	Ribulose-1,5-bisphosphate carboxylase/oxygenase
RWC	Relative water content
RYMV	Rice yellow mosaic virus
SA	Salicylic acid
SAA	Systemic acquired acclimation
SAR	Systemic acquired resistance

SDS-PAGE	Sodium dodecyl sulfate-polyacrylamide gel electrophoresis
SNPs	Single-nucleotide polymorphisms
SOD	Superoxide dismutase
TCTP	Translationally controlled tumor protein
TEM	Transmission electron microscope
TEM-EDX	Transmission electron microscopy energy-dispersive X-ray spectroscopy
TF	Transcription factor
TMV	Tobacco mosaic virus
TNT	Trinitrotoluene
TRXh	Thioredoxin
VOCs	Volatile organic compound
XRF	X-ray fluorescence
ZFP	Zinc finger protein

Chapter 1
Rhizospheric Engineering for Sustainable Agriculture

Stanzin Idong and Anil Kumar Sharma

Abstract The mammoth soil entails seemingly high number of microbes in order of tens of thousands of species. The intricate plant-microbiome community is of paramount significance for crop health. Understanding, predicting, and controlling the structure and function of the rhizosphere will allow us to tackle plant–microbe interactions precisely and which may act as a means to increase plant ecosystem productivity, improve the mechanism to a wide range of environmental distress, and mitigate effects of climate change by designing ecosystems for long-term soil carbon storage. Modifying the rhizosphere helps to increase beneficial microbes that increases nutrient availability and reduces biotic and abiotic stresses. It is highly connected and interactive therefore engineering needs to be cautious.

1.1 Introduction

For several years components of rhizosphere (plant, soil, and microbes) are studied separately. The rhizosphere is a narrow zone and gets influenced by root secretions that contain up to 10^{11} microbial cells per gram and more than 30,000 prokaryotes. Rhizospheric soil is mesotrophic in nature which means it is a home for many bacteria, archaea, viruses, and fungi. Due to the intimate association with microbes, plants could be considered as meta-organisms or holobionts (between the plants per se and its interacting microbiota) and the genome of the plant microbiome is sometimes referred to as the second genome of the plant. Rhizodeposition may change qualitatively and quantitatively and affect the microbial component of the plant (Hartmann et al. 2009). This can be described as the characteristic rhizosphere feedback loop that maintains the rhizosphere in a dynamic equilibrium.

S. Idong
Department of Plant Pathology, College of Agriculture, Pantnagar, Uttarakhand, India

A. K. Sharma (✉)
Department of Biological Sciences, CBSH, G.B. Pant University of Agriculture and Technology, Pantnagar, Uttarakhand, India

A. Sharma (ed.), *Microbes and Signaling Biomolecules Against Plant Stress*, Rhizosphere Biology, https://doi.org/10.1007/978-981-15-7094-0_1

Rhizosphere has its importance in different activities such as crop production, nutrient uptake, and carbon storage. The rhizosphere ecology is badly affected due to current changes in climate and weather, which ultimately leads to decrease in crop yield (Adl 2016). Abiotic stresses such as drought, salinity, and high temperature have detrimental effect on plant growth and development (Lobell and Field 2007). Additional environmental stresses show negative impact on plant growth and development in a number of ways like disturbing hormone balance and increasing susceptibility to diseases (Ashraf 2003; Glick et al. 2007). The physical and chemical context of rhizosphere are the result of many competing and interacting processes that depend on soil type and water content, the composition of microbial communities, and the physiology of the plant itself.

All the components of the rhizosphere can be engineered or manipulated to improve plant health and productivity or to limit the effect of various stresses of biotic or abiotic origins, which is of paramount interest in the current situation of global climate change and for the need of sustainable agricultural practices. Plants can be engineered through modifying rhizosphere pH or by releasing compounds that improve nutrient availability, protect against biotic and abiotic stresses, or encourage the proliferation of beneficial microorganisms. Rhizobacteria that promote plant growth have been engineered to interfere with the synthesis of stress-induced hormones such as ethylene, which retards root growth, and by producing antibiotics against soil borne root pathogens (Ryan et al. 2009). Type of crop species and the cultivars supporting the antibiotic producing microbes are helpful in decreasing different pathogens.

In order to understand the complex interaction in the rhizosphere, new molecular tools are being employed which ensure whether the strategies to engineer the rhizosphere are safe, beneficial to productivity, and substantially improve the sustainability of agricultural systems. This approach may reduce the use of agrochemicals and improve the quality and quantity of crop plants. Some of the materials are still to be developed while others are being tested in the field. Role of phytohormones should also be considered while dealing with rhizosphere engineering.

1.2 How Rhizospheric Engineering Works?

1.2.1 Engineering the Soil

Since ages soil engineering has been in use and helpful in changing the physical properties of the soil and leading to beneficial growth. Classical amendment practices e.g. biochar improves the water retention capacity of the soil and increases the pH of acidic soil. Another successful amendment practice is with calcium silicate that leads to a partial control of sugarcane borer pest, *Diatraea saccharalis* on rice (Sidhu et al. 2013) and plant residues for the partial control of *Rhizoctonia*. Application of the ammonium-based fertilizers tends to acidify the rhizosphere whereas nitrate-based fertilizers result in an alkaline rhizosphere. Shifts in pH can alter soil chemistry around the roots and influence the growth and composition of microbial communities of the soil.

1.2.2 Engineering the Plants

Plant Engineering has gained attention from over 30 years with the development of genetic engineering. Genetic engineering is an accepted technique for plant rhizospheric engineering. A large part of it deals with resistance to pathogens. Transgenic tobacco (*Nicotiana tabacum*) and potato (*Solanum tuberosum*) plants overexpressing a *Trichoderma harzianum* endochitinase gene were generated and they were tolerant to fungal pathogens such as *Alternaria alternata*, *A. solani*, *Botrytis cinerea*, or *Rhizoctonia solani* most likely because the transgenic plants were more efficient in degrading the fungal cell wall (Nion and Toyota 2015).

Several attempts have been made to change root architecture as it is the main part where absorption of minerals takes place. Root architecture is influenced by temperature, nutrient availability, and internal system of plants. These factors are subject for modification that helps in developing resistance against drought and salinity. The exudates released from the plants are known as rhizodeposition and used as another method. Many genes have been identified controlling the root exudates which can be manipulated via genetic engineering so rhizosphere can be changed. Iron uptake in dicots and non-graminaceous monocots also involves the release of organic compounds that chelate $Fe3^{+}$ and facilitates its reduction to $Fe2^{+}$ for uptake (Hinsinger et al. 2003). Manipulating efflux of H^{+} and organic anions from roots in transgenic plants help in engineering rhizosphere.

1.2.3 Engineering Microbes

Many studies have been dealt with microbes that surround roots specially with PGPR. Plant growth promoting rhizobacteria (PGPR), the soil microbes that influence plant growth have been known for more than 100 years (Hartmann et al. 2009). They stimulate plant growth through various direct and indirect mechanisms. These mechanisms include nutrient solubilization, biological nitrogen fixation, induction of systemic resistance, production of plant growth regulators, organic acids, and volatile organic compounds (VOCs) as well as enzymes like 1-aminocyclopropane-1-carboxylate (ACC)-deaminase, chitinase, and glucanase.

PGPR are used as bioinoculants for many years but now bioinoculants involve microbial community or mixture of bacterial strains. Knowing the genes/proteins/metabolites involved in PGPR–plant interactions, responsible for abiotic stress resistance may allow creating engineered plants harboring genes that prevent stress, and/or microbes that could be used to alleviate stress. *Bacillus amyloliquefaciens* strain KPS46 was investigated for its ability to activate extracellular protein elicitors for enhanced plant growth and induced systemic resistance in soybean plants.

1.3 Conclusion

The Scientists related to agriculture research fields have sensed rhizospheric engineering as a tool that can help to meet out demands of growing population without disturbing the ecosystem. It provides an exciting opportunity to unleash the diverse rhizosphere and fills the gap to find solutions to the problems. Unravelling the plant–microbe interaction will be important to understand and exploit the full potential of a cropping system, understanding the mechanisms behind rhizosphere priming and managing the carbon cycle in the soil under current and future climate. Molecular techniques have allowed manipulation of genes that influence rhizospheric function and development in biotechnology but raised the bars to ensure that it is safe, beneficial, and improves sustainability of the agriculture. Major obstacle that hampers development is the complexity of the rhizosphere including chemistry and biology.

Molecular techniques that allow the direct manipulation of genes influencing rhizosphere functions will be progressive in future. High throughput and "omics" techniques further make it possible to screen and analyze large and complex microbial communities in the soil. Genomics has given rise to metagenomics, an approach that gives remarkable development in mass sequencing procedures and which will enable us to explore the microbial diversity of the rhizosphere more rapidly and in greater detail. The complexity of rhizosphere chemistry and biology continues to present a great challenge for further research. The complex relationships between the structure of microbial communities and their function make attempts to predict and manipulate their ecology very difficult, and will certainly remain for several more years the touch stone of rhizosphere ecology.

Acknowledgements We acknowledge the facilities provided by the GB Pant University, Pantnagar.

References

Adl S (2016) Rhizosphere, food security, and climate change: a critical role for plant-soil research. Rhizosphere 1:1–3

Ashraf M (2003) Relationships between leaf gas exchange characteristics and growth of differently adapted populations of blue panicgrass (Panicum antidotale Retz.) under salinity or waterlogging. Plant Sci 165:69–75

Glick BR, Cheng Z, Czarny J, Duan J (2007) Promotion of plant growth by ACC deaminase-producing soil bacteria. In: New perspectives and approaches in plant growth-promoting Rhizobacteria research. Springer, Dordrecht, pp 329–339

Hartmann A, Schmid M, Van Tuinen D, Berg G (2009) Plant-driven selection of microbes. Plant Soil 321:235–257

Hinsinger P, Plassard C, Tang C, Jaillard B (2003) Origins of root-mediated pH changes in the rhizosphere and their responses to environmental constraints: a review. Plant Soil 248:43–59

Lobell DB, Field CB (2007) Global scale climate–crop yield relationships and the impacts of recent warming. Environ Res Lett 2:014002

Nion YA, Toyota K (2015) Recent trends in control methods for bacterial wilt diseases caused by Ralstonia solanacearum. Microbes Environ 30(1):1–11

Ryan PR, Dessaux Y, Thomashow LS, Weller DM (2009) Rhizosphere engineering and management for sustainable agriculture. Plant Soil 321:363–383

Sidhu JK, Stout MJ, Blouin DC, Datnoff LE (2013) Effect of silicon soil amendment on performance of sugarcane borer, Diatraea saccharalis (Lepidoptera: Crambidae) on rice. Bull Entomol Res 103:656–664

Chapter 2
Stress Alleviation in Plants Using SAR and ISR: Current Views on Stress Signaling Network

Damini Maithani, Harshita Singh, and Anita Sharma

Abstract Plants are armored with several defense mechanisms that are inducible in the presence of stress stimuli to protect them from pathogen attack, insect mediated herbivory or to increase systemic tolerance towards abiotic stimuli such as heat, light, cold, mechanical injuries, etc. These inducible defenses include multifaceted molecular, biochemical, and morphological changes. SAR and ISR are the two forms of induced resistance pathways in which the plant defense system is hiked up against a variety of pathogens as well as abiotic stresses. Plant defenses are usually preconditioned by previous encounter with the pathogens. Molecular and biochemical studies have revealed the elicitors mediating both types of responses. These elicitors are commercially available and commonly used in conventional agriculture practices for disease management. Research on SAR and ISR over the past few years has elaborated the understanding of the mechanism of entire process at molecular level. Use of PGPR as well as exploiting ISR and SAR machineries for disease management proves to be a promising alternative for environmentally malignant chemical pesticides. This chapter summarizes response pathways of plants under stress conditions, role of ROS (Reactive Oxygen Species), metabolic cues including hormones involved in signaling cascade, and mechanisms of systemic protection in plants.

2.1 Introduction

Plants follow sessile lifestyle and are usually exposed to plethora of biotic and abiotic stresses. Thus to cope up with the extreme conditions, plants have inbuilt intricate machineries that help them to minimize, avoid, or tolerate the stress conditions. To elude the effect of various biotic stresses like attack by pathogenic microorganisms, insects, or herbivores, plants are armored with complex constitutive and inducible machinery involving interplay of genes, proteins, hormones, and

D. Maithani · H. Singh · A. Sharma (✉)
Department of Microbiology, GB Pant University of Agriculture & Technology, Pantnagar, Uttarakhand, India

A. Sharma (ed.), *Microbes and Signaling Biomolecules Against Plant Stress*, Rhizosphere Biology, https://doi.org/10.1007/978-981-15-7094-0_2

various other chemicals at molecular, biochemical, and physiological levels. Some defense pathways are continuously operating in plants for maintaining plant immunity but some are activated or induced only under certain conditions. Defense pathway in plants on attack by any pathogen, insect, or herbivore is not linear but rather a complex signaling circuit, and these complex synergistic, cooperative, and antagonistic interactions determine the output response towards the stimuli (Kunkel and Brooks 2002; Rojo et al. 2003). On exposure to biotic stress, plants use a mechanism known as Induced Systemic Resistance (ISR) to elude the consequences of stress (Schenk et al. 2012). Experimental studies conducted on diseases caused by Tobacco Mosaic virus (Ross 1961b) and *Peronospora tabacina* (Cruickshank and Mandryk 1960) on tobacco plants were landmark studies demonstrating that plants could be preconditioned against a variety of diseases. These studies articulated the concept of SAR more clearly. Later, other plants were also studied as model plants to understand SAR. Studies conducted on biological control of soil borne diseases led to the discovery of another form of resistance known as Induced Systemic Resistance (ISR) which is associated with plant growth promoting rhizobacteria (PGPR). On encountering an attack by pathogen, herbivore, or insect at local area, plants produce certain chemical compounds that minimize the effect and may also protect plant from further attack. Response not only occurs at the site of attack (local response) but also at distant parts in plants (systemic response). An enhanced defensive state is shown by plants when appropriately stimulated. Systemic Acquired Resistance (SAR) and Induced Systemic Resistance (ISR) are such induced responses of plants when stimulated by pathogen attack that provide resistance to the host plant against subsequent challenges offered by pathogens. Plant growth promoting rhizobacteria (PGPR) are able to suppress growth of pathogen by either antagonistic activity or by inducing systemic resistance in plant against soil borne and foliar pathogens. Pathogen induced SAR response is similar to ISR response induced by rhizobacteria, as in both cases plant becomes resistant against future pathogenic challenges.

Rhizobacteria can trigger systemic resistance in plants via Salicylic Acid (SA) dependent and independent pathways. SA independent pathway, which depends upon jasmonic acid (JA) and ethylene signals, is known as ISR. Combined effect of ISR and SAR produces an extended protection to plant against a broad spectrum of pathogens as compared to ISR or SAR alone. SAR and ISR can be differentiated on the basis of nature of the elicitor as well as cascade of signaling pathway followed after induction (Choudhary et al. 2007). Induced resistance in plants is a complex signaling cascade and the mechanism has been elucidated in *Arabidopsis*. Three pathways have been proposed for induced resistance in *Arabidopsis*, out of which two pathways involve the role of PR proteins. In one pathway, pathogenesis related (PR) proteins are produced as a result of pathogen attack whereas in other, as a result of mechanical injury, wounding, or necrotic pathogens. Pathogen induced pathway involves SA whereas pathway induced by wounding relies on JA as signaling molecule. JA dependent pathway, known as ISR is used for pathway initiated by root associated bacteria. A plethora of proteins collectively known as PR proteins are involved in both salicylate and jasmonate induced pathways. These proteins are chitinases, glucanases, thaumatins, phytoalexins, and oxidative enzymes viz., peroxidases, polyphenol oxidases, etc. Third pathway evoked by non-pathogenic rhizobacteria, known as rhizobacteria induced systemic resistance (RISR) is different from abovementioned pathways as compounds like PR proteins and antimicrobials like phytoalexins do not accumulate in the absence of pathogen. Local infection elicits salicylic acid dependent signaling cascade that leads to disease resistance against various pathogens in distantly located plant parts. Plants respond to local attack by de novo synthesis of pathogenesis related proteins as chitinase, glucanase, and other defense related compounds. PGPR besides inducing ISR can also protect germinating seedlings against variety of pathogens by inducing plant vigor. They colonize emerging seedlings and provide protection to the plants against a broad spectrum of pathogens. Effectiveness of SAR and ISR against a variety of pathogens has been demonstrated through various lab and field experiments (Murphy et al. 2000; Nandakumar et al. 2001; Zehnder et al. 2001; Niranjan Raj et al. 2003). SAR is generally induced in plants when foliar and root tissues are exposed to biotic and abiotic elicitors. Signaling cascade is mediated via salicylic acid and accumulation of PR proteins whereas ISR is induced on the exposure of plant roots to PGPR and signaling is dependent on the production of jasmonic acid and ethylene (Vallad and Goodman 2004). SA-JA crosstalk works in concentration dependent manner. Low concentration of SA interacts synergistically with JA for activation of various defense related genes. At higher concentrations, an antagonistic interaction occurs between two pathways which may lead to oxidative burst and cell death (Mur et al. 2006). ROS acts as a mediator between the two pathways, in simple terms thiol signaling and redox homeostasis play crucial role in SA-JA crosstalk. Different stress pathways are interconnected and an overlap exists between their signaling cascades (Mullineaux and Baker 2010; Straus et al. 2010). High light intensity is reported to increase immunity of the plants against pathogen infection and tolerance towards oxidative stress, thus indicating cross-link between two stress pathways (Rossel et al. 2007; Mühlenbock et al. 2008). Figure 2.1a, b represents SAR and ISR mechanisms in brief.

2.2 Local Versus Systemic Signaling

Different types of systemic signaling exist in plants based on different stimulus triggering the response. Systemic acquired resistance (SAR) is activated in response to the pathogen attack and provide immunity to entire plant against bacterial, viral, and fungal pathogens. Systemic wound response is activated in plants due to mechanical injury caused by insect attack and provides immunity to entire plant against further insect attack. Systemic acquired acclimation (SAA), stimulated in response to a variety of abiotic stress stimuli makes entire plant tolerant to abiotic stimuli such as heat, salt, light, etc. Systemic metabolic responses are triggered due to fluctuations in the levels of metabolites (Baxter et al. 2013). Defense mechanisms can not only be activated in those tissues which are directly exposed to stress but also

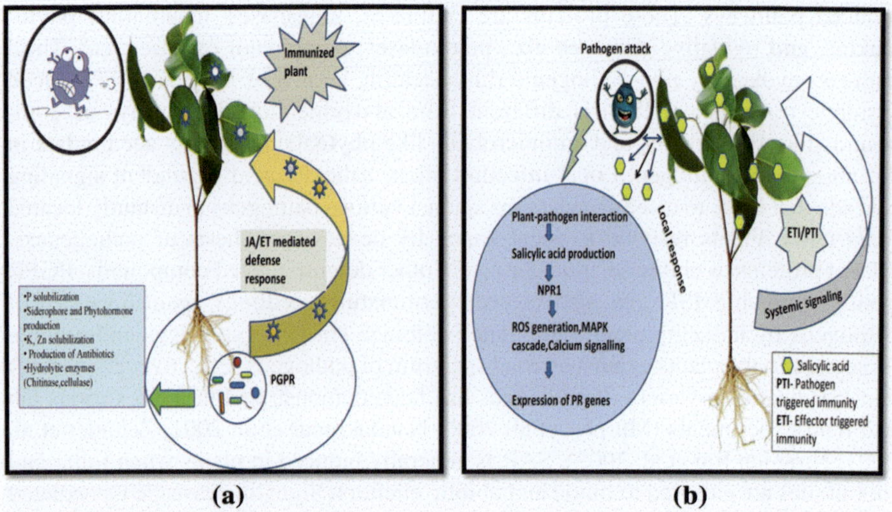

Fig. 2.1 (**a**) Induced systemic resistance via JA/ET mediated signalling renders plant immune to herbivory, wounding, pathogen attack and increases tolerance towards abiotic stress stimui. (**b**) Plant pathogen interaction leads to production of SA that binds with NRP1 and leads to downstream signalling which results in expression of Pathogenesis Related proteins

in distant non-damaged plant parts and thus immunize entire plant systemically from further damage. Most of the ISR triggering rhizobacteria are beneficial and have plant growth promoting properties as compared to SAR, which is associated with accumulation of PRs and negatively affects plant growth (Kim et al. 2004). Long-distance signaling is a concept that shows local exposure of a plant to an environmental stimulus which induces physiological and molecular response systemically in organs that are not exposed to the stimulus (Heil and Ton 2008). It also provides a general mechanism for the coordination of responses in different plant parts and appears to be a more general sensing system for environment and/or nutrient status (Davies et al. 2005; Dempsey and Klessig 2012). Whereas, SA is required for the establishment of SAR. Grafting experiments have shown that SA is not a mobile signal (Vernooij et al. 1994). Putative phloem-mobile systemic SAR signals are reported to include methyl salicylate, glycerol-3-phosphate, lipid-transfer protein DIR1, and JA (Dempsey and Klessig 2012). ROS may also be involved in long-distance mediation of SAR responses (Alvarez et al. 1998). Recognition of virulent pathogens by a plant triggers a localized oxidative burst and cell death in hypersensitive response, which after a short delay is followed by a systemic oxidative burst and a micro-hypersensitive response. Changes in the levels of ROS in the local and systemic tissues are required for the expression of SAR. ROS appears to mediate SAR by regulating de novo synthesis of SA in systemic leaves. Accumulation of BR (Brassinosteroids), which functions as an immobile hormone induces a form of stress tolerance, independent of SA (Xia et al. 2011). BR-induced systemic resistance appears to be different from classic SAR response. Inhibition of ROS accumulation in BRs treated leaves abolished systemic responses.

2.3 ISR

Pathogenic attack to the plants causes great yield losses. Various disease management strategies like use of pesticides are being employed to minimize the losses. Agrochemicals are detrimental for the environment as well as for the consumers, thus use of bioresources such as plant growth promoting rhizobacteria (PGPR) is an appropriate alternative for disease management in economically important crops. PGPR compete with pathogens like fungi, bacteria, and nematodes for the substrate as well as for niche. Other mechanisms of suppression may involve secretion of allelochemicals which provide enhanced protection in plants against broad spectrum of pathogens by inducing ISR (Haas et al. 2002). Unlike SAR, ISR is stimulated by PGPR and does not involve accumulation of SA and Pathogenesis Related proteins (PR). Root exudates attract a variety of microorganisms in nutritious zone adjacent to root surface which is comprised of a complex dynamic network of synergistic and antagonistic interactions among the microorganisms. Induced systemic resistance provided by PGPR can be specific and nonspecific. ISR is a widespread mechanism of plant protection adapted by both monocotyledonous and dicotyledonous plants against a variety of pathogens. Enhanced defense capacity of the plants against pathogenic agents and development of resistance in plant tissues distantly located from site of infection are some of the hallmark features of ISR (Ryal et al. 1996). ISR is generally associated with a phenomenon known as priming, an enhanced defense response against phytopathogens. Salicylic acid or benzothiadiazole when provided to the plant cells generally induces priming and elicits a stronger response by activating defense related genes such as phenylalanine ammonia lyase. First encounter of the plant by pathogens causes accumulation of certain compounds in uninfected tissues; however, other biochemical changes are only expressed in response to further infections in plant parts where resistance is required, this phenomenon is known as priming (Conrath et al. 2001). Responses such as phytoalexin synthesis, cell wall lignification occur more strongly as compared to primary infection. Ethylene and jasmonic acid are reported to activate common responses against biotic stress in plants and their interaction pattern is determined by the type of pathogen. A group of beneficial microorganisms that colonize plant rhizosphere is able to immunize plant by providing systemic tolerance against a wide range of pathogens. Induced systemic resistance (ISR) pathway requires both ethylene and jasmonic acid (JA) (Rojo et al. 2003). Resistance induced by rhizobacteria is nonspecific and provides basal resistance to a variety of pathogens concomitantly. Jasmonic acid and ethylene during the defense response act in a sequential manner which was concluded by investigating ethylene and JA insensitive mutants (Pieterse et al. 1998). Ethylene and jasmonate converge at a point where activation of Ethylene Response Factor (ERF) gene occurs which acts downstream to regulate the response. This gene encodes a protein that is a transcriptional regulator of the genes and expressed on exposure of the plant to pathogen and thus prevents deleterious progression of disease. ERF1 gene can be activated by ethylene or jasmonate and also synergistically by both. Studies revealed that mutation in any pathway prevents the expression of ERF1 which proves that both ethylene and JA signaling pathways are required (Lorenzo et al. 2003).

When a pathogen attacks a plant it overcomes the host machinery and causes infection; this phenomenon is known as compatible response. In some cases pathogen attacks the plant but cannot establish disease successfully is known as incompatible response. During incompatible responses, pathogen elicits localized response at the site of infection. Pathogen is confined to the attacked site and prevents the infection to further escalate to other parts of the host (Heil and Bostock 2002). Generation of reactive oxygen species may sometimes lead to cell death and prevent pathogen invasion by inducing changes in the cell wall composition. Induction of de novo synthesis of antimicrobial compounds like phytoalexins and pathogenesis related proteins are some of the local responses (Kombrink and Schmelzer 2001). Following the local response, a response is also activated in distant unaffected plant parts through various signals, induced by activation and expression of genes. Phytoalexins are generally produced in local response but PR proteins are found in both local and systemic responses. PR proteins are used as ISR markers but no antimicrobial activity has been reported for them. Previously it was believed that PR proteins were absent in healthy plants and accumulate only after pathogen attack (Van Loon and Van Kammen 1970). PR proteins are reported in more than 40 plant species (Van Loon and Van Strien 1999) and are of two types. Acidic PR proteins are dominantly present in cytosol and basic PR proteins are localized in vacuoles. Some PR proteins have chitinase activity that contributes in plant protection from several fungal pathogens. Different studies suggest that plants overexpressing chitinase show decreased susceptibility to infection by fungal pathogens (Heil and Bostock 2002). However resistant phenotypes in the plants are not necessarily contributed by these proteins. Beneficial bacteria do not cause damage to host plants, thus elicitors produced by pathogens are absent in such kind of responses. Thus bacterial determinants act as elicitors in such kind of responses. Mechanism of elicitation shows similarity to that triggered by PAMPs. Both LPS and flagella of WCS 358 have been reported to elicit ISR in *Arabidopsis* plants when applied exogenously to the roots. Combination of siderophore, O-antigen, and flagella accounts for ISR response (Choudhary et al. 2007). Pathogen induced necrotic or chlorotic lesions were reduced in plants grown in soil treated with WCS 358 rhizobacterial strain (Meziane et al. 2005). Some rhizobacteria show similar response in all the plant species whereas others show narrow specificity that is an indicative of species specific recognition of the plants by bacteria. Both SAR and ISR show reduced disease severity by suppressing growth of the pathogens and reducing their colonization in induced tissues but they differ in signaling mechanism. Signaling in ISR is more complex than SAR. Defense signaling can be either SA-dependent or JA/ethylene dependent or both. SA is an important signaling molecule in local and systemic resistance in plants against pathogen but rhizobacteria mediated response is JA/ethylene dependent but phenotypically quite similar to SAR (Thomma et al. 2001). ISR was a more common response in case of necrotrophic pathogens while SAR was not much effective against *Botrytis cinerea* and *Alternaria brassicicola* (two common necrotrophic pathogens) (Thomma et al. 2000). This response is not always true and defense mechanism of different type can be effective against different pathogens attacking different plants. Arabidopsis plants when challenged with *Pseudomonas syringae* pv. tomato (Pst), SAR induced plants showed elevated expression of salicylic acid dependent PR-1. On the other hand, ISR induced plants showed accumulated mRNA of JA induced gene *vsp* to higher levels. This enhanced effect, also known as priming indicates that induced plants activate defense related gene expression to a greater extent than that of non-induced plants (Conrath et al. 2006). A study showed that *Pseudomonas chlororaphis* O6 upon colonizing cucumber root was able to trigger ISR against *Corynespora cassiicola* and the response was associated with faster and stronger accumulation of transcripts upon subsequent challenge inoculation (Kim et al. 2004).

2.4 SAR

Systemic acquired resistance (SAR) is a type of distinct long-distance signaling response in plants following the exposure to pathogens (Durrant and Dong 2004). It refers to a signaling pathway that is activated in plants as a response against attacking pathogen. Ross published the first systematic study on SAR in 1961 using TMV and demonstrated that resistance towards TMV infections was restricted to a prior infection. This resistance was effective against not only TMV but also against tobacco necrosis virus and certain bacterial pathogens. Ross coined the term "systemic acquired resistance" to refer to the inducible systemic resistance and "localized acquired resistance" to describe the resistance induced in inoculated leaves (Ross 1961a, b). Plants on exposure to a pathogen trigger formation of necrotic lesions which may lead to rapid cell death as a part of hypersensitive response (HR) or a symptom of disease. SAR is activated and results in the development of a broad spectrum systemic resistance (Neuenschwander et al. 1995). SAR is different from other defense responses in terms of type of genes induced as well as spectrum of protection it provides against the pathogens. This signaling cascade activates a set of genes known as SAR genes (Ryal et al. 1996). SAR appears to be distinct from pre-existing resistance mechanisms such as physical barriers or protein cross-linking and also from other inducible resistance mechanisms such as phytoalexin biosynthesis, the hypersensitive response, and ethylene-induced physiological changes. Furthermore, SAR is not related to responses induced by wounding or osmotic stress. SAR signal transduction pathway usually functions as a modulator of other disease resistance mechanisms. Tobacco has been well characterized for SAR. Other plants also show similar kind of responses. SAR is active in *Arabidopsis* against *Phytophthora parasitica*, *Pseudomonas syringae*, and Turnip crinkle virus and the genes expressed are similar to those expressed in tobacco (Uknes et al. 1992).

Activation of SAR signaling cascade can transform a compatible plant–pathogen interaction to an incompatible one and the converse is also true (Mauch-Mani and Slusarenko 1996). A protein is characterized as SAR protein when its activity correlates with the maintenance of resistance in plants (Neuenschwander et al. 1995). Many proteins belonging to SAR consist of pathogenesis related proteins. Variety of pathogens like bacteria, virus, and fungi can attack the plants and trigger SAR response. This phenomenon is accompanied by expression of SAR genes in plants. A number of biochemical, physiological, and molecular changes are associated with plant pathogen infection such as oxidative burst, cell death, deposition of callose and lignin, synthesis of phytoalexins and other proteins (Dangl et al. 1996; Low and Merida 1996). Number of reports suggests a close correlation between SA and SAR response and exogenous application of SA can trigger SAR response. Salicylate acts as an important signaling molecule in initiating downstream signaling cascade of SAR response (Vernooij et al. 1995; Ryal et al. 1996). The most common evidence that implicates SA as a signal in SAR comes from the experiments on transgenic tobacco which expresses the enzyme salicylate hydroxylase, encoded by the *nah* G gene from *Pseudomonas putida*. This enzyme catalyzes the conversion of salicylate to catechol. Transgenic plants with this gene were found incapable of showing SAR response against a variety of pathogens (Friedrich et al. 1995). An increased SA concentration in the phloem sap was reported in plants infected with pathogens. Labelling studies also suggest accumulation of SA in uninfected tissues of host plants (Shulaev et al. 1995). Phenylalanine is converted to trans-cinnamic acid by enzyme phenylammonia lyase. Conversion of trans-cinnamic acid to salicylic acid involves an intermediary step in which benzoic acid is formed. H_2O_2 has been reported to act as a secondary messenger to induce SAR. The first step in the development of SAR is the recognition of pathogen infection by a plant. Once the plant reacts to the pathogen, signals are released that trigger resistance in adjacent and distant tissues. Only the compatible interactions can lead to SAR induction, thus, it is not necessary for the pathogen to induce a gene-for-gene resistance reaction. A useful approach in the study of SAR is to identify easy to measure markers that precisely correlate with the biological processes. To find appropriate markers for SAR, many c-DNA have been isolated that are expressed in uninfected tissues during SAR maintenance. Through tobacco/TMV system, steady state m-RNA levels from at least nine families of genes were induced in uninfected leaves of inoculated plants. These families are collectively referred as "SAR genes." Abiotic agents that induce resistance, such as SA and 2,6-dichloroisonicotinic acid (Métraux et al. 1991) induce the same spectrum of SAR gene expression to levels comparable to those in SAR. Thus, expression of SAR genes tightly correlates with the onset of the resistant state. Along with reliable markers, some genes apparently have active role in developing resistance. Once the c-DNAs were isolated and encoded proteins purified, many genes were shown to have either direct antimicrobial or enzymatic activities of antimicrobial proteins. Several classes of SAR genes encode β-1,3-glucanases and chitinases (Linthorst et al. 1991). Thaumatin-like proteins were found to be active against fungi in vitro (Woloshuk et al. 1991) and their activity resides in the ability to disrupt membrane integrity, which is the basis for calling this class of proteins "permatins." A group of SAR genes, known to inhibit fungal growth is related to PR protein, known as PR-1. PR-1-related proteins from tobacco and tomato have in vitro activity against *Phytophthora infestans* (Cohen et al. 1992). Transgenic tobacco and *Brassica* seedlings that express a chitinase from bean were protected against damping-off caused by *Rhizoctonia* spp. (Broglie et al. 1991). High level expression of PR-1 in transgenic plants results in reduced infection by two Oomycete pathogens, *Peronospora tabaci* (which causes the downy mildew disease known as blue mold) and *Phytophthora parasitica* (black shank disease) (Alexander et al. 1993). Thus SAR genes encoding antimicrobial activities and expression of certain genes in transgenic plants that impart tolerance to the plants, strengthen the case that these genes play direct role in maintaining SAR. Each taxonomic group of plants has evolved its own set of SAR genes in response to evolutionary pressure from a specific spectrum of pathogens. One powerful method for the dissection of complex signal transduction process is the application of mutant analysis. SAR mutants have been isolated by screening ethylmethanesulfonate-mutagenized Arabidopsis plants for SAR gene expression (Lawton et al. 1993). Mutants with high constitutive PR-1 gene expression were called *cim* (constitutive immunity). Several *cim* mutants developed necrotic areas on the leaves even if they were grown under sterile conditions, a phenotype referred to as *lsd* (lesions stimulating disease). Defense response against pathogen attack involves various metabolic cues such as hormones, amino acids, and proteins. In local and systemic tissues, accumulation of SA, its glucoside derivative, and elevated expression of PR genes has been observed. Signals initiated in challenged tissues are translocated to distant parts via vascular bundles especially by phloem (Dempsey and Klessig 2012). Biologically active phloem-mobile chemical signals include Methyl salicylate (MeSA), glycerol-3-phosphate (G3P), azelaic acid (AzA), Jasmonic acid (JA), and pipecolic acid (Pip). These chemicals can induce systemic response when applied locally. Primary infection leads to accumulation of G3P and Aza which are not directly involved in SA biosynthesis and accumulation but precondition the plant for SA accumulation in response to secondary infection by the pathogen (Dempsey and Klessig 2012; Kachroo and Robin 2013). Dehydroabietinal (DA) and Pip are other two mobile SAR signals. Pip, a potential SAR signal is reported to induce its own biosynthesis and activate SAR by regulating SA accumulation in plants infected with *Pseudomonas syringae* pv. *Maculicola*. Similarly DA is reported to elevate level of SA, PR-1, SA biosynthesis enzymes and induce resistance to a variety of pathogens in tomato, *Arabidopsis*, and tobacco (Dempsey and Klessig 2012).

It is proposed that H_2O_2 acts as a second messenger in SAR signaling. A SA binding protein, identified as catalase inhibited catalase activity of this protein leading to elevated levels of H_2O_2 which further caused induction of PR-1 gene expression and was postulated to induce SAR (Chen et al. 1993, 1995). For H_2O_2 to function as a signaling agent of SA, H_2O_2 levels should be high in uninfected leaves of tobacco plants during SAR activation. This was tested by inoculating tobacco leaves with TMV and monitoring the accumulation of H_2O_2, PR-1 m-RNA, and the establishment of SAR. In the uninfected leaves of inoculated plants, SAR gene expression and establishment of SAR did not correlate with an increase in H_2O_2

levels. Apparently, induction of PR-1 expression by H_2O_2 was directly tested by infiltration of tobacco with H_2O_2. Substantial PR-1 m-RNA accumulation resulted after infiltration of 1M H_2O_2, a concentration that also caused severe tissue damage. However, high concentrations of H_2O_2 were found to induce SA synthesis in tobacco (Leon et al. 1995; Neuenschwander et al. 1995) and *Arabidopsis* (Summermatter et al. 1995), suggesting that H_2O_2 is not a second messenger of SA in the signal cascade leading to establishment of SAR.

2.5 Hormones and Plant Defense

Plant hormones play important role in regulating developmental processes and signaling network involved in plant responses to a wide range of biotic and abiotic stresses. Significant progress has been made in identifying the key components and understanding the role of salicylic acid (SA), jasmonates (JA), and ethylene (ET) in plant responses to biotic stresses. Recent studies indicate that hormones such as abscisic acid (ABA), auxin, gibberellic acid (GA), cytokinin (CK), brassino steroids (BR), and peptide hormones are also implicated in plant defense signaling pathways but their role in plant defense is not well studied. Salicylic acid (SA) is involved in the defense response to biotrophic and hemibiotrophic pathogens, while jasmonate (JA) and ethylene (ET) are the primary hormones involved in the response to necrotrophic pathogens (Neu et al. 2019). Confrontation of the plants to disease causing pathogens triggers plants to respond via activating complex cascades of signaling pathways that provide the plant with local and in some cases systemic immunity. Plant hormones are mighty chemical compounds that are an inevitable part of growth and development processes in plants. These molecules can show dramatic effect at low concentrations (Kucera et al. 2005). A set of plant factors including plant hormones (auxin, abscisic acid (ABA), ethylene, gibberellins, cytokinins, salicylic acid (SA), strigolactones, brassinosteroids (BRs), and nitrous/nitric oxide) are essential to make the plant tolerate the stress (Miransari 2016). In a study conducted by Hamayun et al. (2010), levels of ABA, jasmonate, and salicylate were high under drought stress. ABA, majorly involved in salinity and drought stress is known to play an important role in plant growth and development. It maintains dormancy, prevents precocious seed germination, regulates stomatal activity, leaf senescence, leaf hydraulic activity by regulating aquaporin activity (Tuteja 2007; Li et al. 2014). Three plant-specific hormones (salicylic acid (SA), jasmonic acid (JA), and ethylene (E)) are major endogenous signals involved in communicating the presence of an infection and triggering the defense responses in plants. Abscisic acid (ABA) and auxins may play a role in defense against pathogens. ABA negatively regulates SA-dependent resistance (Audenaert et al. 2002). However, involvement of ABA and auxins in pathogen defense is much less documented. Range of pathogen infections signaled through SA and JA or E are partially exclusive. SA has been found linked with the response towards the infection by viruses like tobacco mosaic virus (TMV) and turnip crinkle virus, as well as by biotrophic bacteria and fungi such as *Pseudomonas*, *Peronospora*, *Erysiphe*. Biotrophic pathogens require nutrients from live plant tissues. Hypersensitive responses in plants as a result of pathogen–plant interaction may lead to cell death and deprive pathogen from live host tissues (Dewdney et al. 2000; Kachroo et al. 2000). Contrary to this, HR-induced cell death proves to be beneficial for infection by necrotrophic pathogens (Govrin and Levine 2000). Other defense mechanisms against necrotrophic pathogens, evolved in plants are activated in many cases by JA and E signaling pathways. It is shown that JA and Ethylene signaling is required to develop resistance against the pathogens (mainly necrotrophic or saprophytic) such as *Alternaria*, *Botrytis*, *Septoria*, *Pythium*, *Erwinia*, *Plectosphaerella* (Berrocal-Lobo et al. 2002; Dıaz et al. 2002). However, generalizing the concept that defense against biotrophic pathogens is always attributed to SA signaling and against necrotrophic pathogens to JA and E signaling would be an oversimplification. JA and E signaling pathways have been reported in defense mechanisms against the biotrophic pathogens like *Erysiphe cichoracearum*, *Erysiphe orontii*, *Oidium lycopersicum*, *and Pseudomonas syringae* (Ellis and Turner 2001; Ellis et al. 2002). Similarly, reports exist in which SA is involved in resistance against the necrotrophic fungi *Botrytis cinerea* and *Plectosphaerella cucumerina* (Berrocal-Lobo et al. 2002; Audenaert et al. 2002; Dıaz et al. 2002). Moreover, not only these hormones have been shown to participate in activating parallel defense mechanisms against the same pathogen, but also many events of cross-talk among the SA, E, and JA signaling pathways have been reported and were shown to be significant in determining the resistance to pathogens. Recent studies demonstrated the involvement of brassinosteroids (BRs) in SAR and SAA in cucumber (Xia et al. 2011; Li et al. 2013). Although BRs are not directly involved in long-distance signaling but they affect other signals such as auxins and polyamines (Li et al. 2013). Local application of 24-epibrassinosteroid (EBR) can induce SAR and SAA to oxidative stress accompanied by local and systemic expression of known defense/acclimatory genes such as APX and catalase (Li et al. 2013).

2.6 Salicylic Acid

Salicylic acid (SA), a phenolic plant growth regulator is found to play an important role in flowering, leaf senescence, seed germination, and other physiological and developmental processes besides regulating activation of defense responses in plants under a variety of stress conditions to which plant is exposed to pathogens in particular. On exposure to pathogen, SA is synthesized in the chloroplasts via isochorismate pathway where ICS1 (isochorismate synthase 1) is the prime enzyme required for its biosynthesis. Salicylic acid turns on the transcriptional reprogramming by activating TGA transcription factors via NPR1 (non-expressor of pathogenesis-related genes 1) co-activators. Two models have been proposed for perception of SA. In one model, NPR3 and NPR4 perceive SA and then regulate NPR1 accumulation. In the second model NPR1 has itself been proposed to perceive SA and undergoes a conformational change which leads to increased transcription.

Besides the direct SA binding, NPR1 is also regulated by SA-mediated redox changes and phosphorylation. Emerging evidences show that pathogen virulence effectors target SA signaling and strengthen the importance of SA-mediated immunity (Seyfferth and Tsuda 2014). Salicylic acid at low concentration leads to generation of reactive oxygen species at non-toxic concentrations where these molecules act as secondary messengers during plant defense whereas high doses of salicylic acid lead to accumulation of reactive oxygen species at concentrations lethal for the survival of plants (Miura and Tada 2014). Hormone is able to enhance plant's survival under salinity as well as heavy metal stress by regulating redox homeostasis (Jibran et al. 2013). In plants SA is synthesized mainly via two pathways: Isochorismate (IC) and phenylalanine ammonia-lyase (PAL). Chorismate, an end product of shikimate pathway is used as a precursor of SA biosynthesis using critical enzymes, IC synthase (ICS) and PAL (Dempsey and Klessig 2012). Homologs of *ICS* and *PAL* genes are present in a variety of plants including *Arabidopsis*, tobacco, tomato, populus, sunflower, and pepper. Complete loss of SA accumulation was observed in *Arabidopsis* mutants showing dysfunctioning of ICS1 induced by pathogen (Cochrane et al. 2004; Uppalapati et al. 2007; Catinot et al. 2008; Yuan et al. 2009; Sadeghi et al. 2013; Kim and Hwang 2014). Similarly, *Arabidopsis* mutants with reduced PAL activity show reduced SA accumulation in comparison to wild type after pathogen attack (Chao et al. 2010). However, in higher plants, SA is majorly synthesized via IC pathway in chloroplast to provide immunity. ICS enzyme converts chorismate to isochorismate that is ultimately converted to SA (Strawn et al. 2007; Garcion and Métraux 2008; Dempsey and Klessig 2012). Isopyruvate lyases (IPLs) convert isochorismate to SA. Plants do not contain any gene encoding such enzyme or homologue. However the mechanism in plants for SA biosynthesis is more complex than that of bacteria (Dempsey and Klessig 2012). Export of SA from chloroplasts is mediated by EDS5 (Enhanced Disease Susceptibility 5) which has been confirmed by mutant analysis (Ishihara et al. 2008; Serrano et al. 2013). SA is synthesized from trans-cinnamic acid, via the intermediates ortho-coumaric acid or benzoic acid. Such pathway provides a link between pathogen induction of phenyl propanoid biosynthesis and SAR signal production. SA is a key player in immunizing plant against a variety of pathogens systemically and establishing SAR (Wang et al. 2006; Fu and Dong 2013). SA biosynthesis is tightly regulated as its elevated concentrations on constitutive expression can be harmful for the plant. It is often associated with stunted plant growth, resulting in reduction of plant fitness (Ishihara et al. 2008; Chandra et al. 2014). The biosynthetic pathway of SA appears to begin with the conversion of phenylalanine to trans-cinnamic acid catalyzed by PAL. The conversion of trans-cinnamic acid into SA proceeds via chain shortening to produce benzoic acid (BA) followed by hydroxylation at the C2 position to derive SA (Yalpani et al. 1993). The final step in SA synthesis is the conversion of BA to SA which is catalyzed by cytochrome P450 mono-oxygenase. Benzoic acid 2-hydroxylase is induced by either pathogen infection or by application of exogenous BA. Activity of benzoic acid 2-hydroxylase is induced by tenfold after pathogen infection and blocked by an inhibitor of protein synthesis (Leon et al. 1993). Once synthesized, the fate of SA in the cell is not clear. Considering the important role of SA as a signaling component, its biosynthesis and catabolism need to be further investigated. BA and SA can be conjugated to glucose which regulates SA level. In healthy tobacco plants, a large pool of conjugated BA is available which decreases transiently after pathogen infection. Decrease in conjugated BA level is correlated with an increase in free BA and SA (Yalpani et al. 1993).

Plant immunity can be characterized as pattern triggered immunity (PTI). Immunity triggered by recognition of patterns in pathogens specially microbes is known as microbe associated molecular pattern (MAMPs) and effector-triggered immunity (ETI). SA plays a key role in signaling associated with pathogen attack. Elicitors triggering the pathway may involve perception of PAMPs (Pathogen Associated Molecular patterns) or effector molecules produced by pathogens. Process of induction of PAMP or effector-triggered immunity involves certain molecular and biochemical events including expression of pathogen resistance, SA biosynthesis genes, production of toxic species like reactive oxygen species and NO, elevated levels of cytosolic calcium, and activation of MAPK cascades. PAMP triggered immunity induces a basal level of response whereas effector molecules may lead to hypersensitivity. Accumulation of ROS and SA influences each other in a positive manner. A loop shows that SA increases ROS accumulation and ROS in turn increases SA accumulation (Xia et al. 2015). Recognition of MAMPs is followed by a cascade of events such as calcium ion signaling, ROS generation, and activation of mitogen-activated protein kinases (MAPKs) which ultimately lead to SA production (Tsuda and Katagiri 2010; Macho and Zipfel 2014). Some pathogens can bypass plants PTI by secreting effector molecules via secretion systems such as pathogen *Pseudomonas syringae* pv. Tomato DC3000 can suppress PTI in *Arabidopsis* and tomato (Xin and He 2013). However, a second check is posed by plant to regain resistance. Plants possess intracellular receptors that recognize the effector molecules and generate a second layer of defense termed as ETI. Activation of ETI also induces SA accumulation and MAPK activation, which are also important for resistance against pathogens during ETI (Bonardi and Dangl 2012; Jacobs et al. 2013; Tsuda et al. 2013).

Salicylic acid pathway is also interlinked to other hormones like jasmonic acid (JA), ethylene, and abscisic acid. For example, JA and ethylene signaling negatively regulates SA biosynthesis at the transcriptional level (Zheng et al. 2012; Derksen et al. 2013). Salicylic acid perceiving receptors belong to a family of proteins named NPR (non-expressor of pathogenesis related genes). In *Arabidopsis*, three proteins (NPR1, NPR3, and NPR4) are bonafide SA receptors. NPR1 is a necessary protein for SA action in plants. It is an ankyrin repeat family protein comprised of BTB/POZ domain, nucleus localization signal, and an ankyrin repeat domain. It is a transcriptional co-activator. Redox change causes reduction of NPR1 oligomer in cytoplasm to monomers in the presence of SA. Monomers are then translocated to nucleus where they interact with transcriptional factors to enhance gene expression of PR proteins. Mutations in NPR1 result in complete loss of SA-mediated signaling pathway and increased susceptibility of plants towards hemibiotrophic pathogens (Dong 2004; Fu et al. 2012; Wu et al. 2012). SA-induced molecular and biochemical changes ultimately lead to activation of a downstream protein known as NPR1 which undergoes a change from oligomeric to monomeric state. NPR1 proteins then migrate towards nucleus where they interact with various transcription factors that are responsible for regulating PR gene expression. Further oligomerization of NPR1 is done by S-nitrosylation of protein which is then sequestered in cytoplasm (Mou et al. 2003; Tada et al. 2008). NPR4 has higher affinity for SA and NPR3 has low affinity in comparison to NPR1 which functions at intermediate levels of SA. ROS generation is involved in defense responses as well as cell death pathways in plants. It is not only induced by SA but also by NADPH oxidase. NADPH oxidase is involved in extracellular ROS generation whereas SA-mediated accumulation of ROS occurs in intracellular organelle such as mitochondria, chloroplast, and peroxisomes (Torres and Dangl 2005; Khokon et al. 2011). When SA signal is unavailable, NPR1 exists as an oligomer. SA perception triggers reduction of intermolecular bridges through TRXh3 and TRXh5 (Thioredoxins) causing monomerization of NPR1 followed by its transport into nucleus where it gets accumulated and regulate gene expression (Tada et al. 2008). Phosphorylation of NPR1 is also triggered in the presence of SA that leads to recruitment of ubiquitin ligase and proteasome mediates degradation of NPR1 then interact with transcription factors belonging to leucine zipper family of transcription factors, TGA (Gatz 2013). TGA2, TGA5, TGA6 repress PR1 transcription in the absence of SA and vice versa (Zhang et al. 2003).

2.7 Jasmonic Acid

Jasmonic acid (3-oxo-2-20-cis-pentenyl-cyclopentane-1-acetic acid) is an endogenous growth-regulating compound in higher plants. JA methyl ester and isoleucine conjugate (JA-Ile) are derivatives of a class of fatty acids and are collectively known as jasmonates (JAs). These compounds are vital for the plants and involved in the regulation of many physiological processes in plant growth and development. They can induce stomatal opening, transport of organic compounds like glucose, inhibit Rubisco biosynthesis, and enhance the uptake of nitrogen and phosphorus. As a signaling molecule, JAs can mediate responses against environmental stresses by inducing a series of gene expression. JA and its derivatives and the plant signaling molecules are closely related to plant defense and provide resistance against microbial pathogens, herbivorous insects, wounding, drought, salt stress, and low temperature. JAs and salicylic acid (SA)-mediated signaling pathways are mainly related to plant responses to biotic and abiotic stresses posed by environment. Response of plant towards mechanical injury and wounding is mediated via jasmonic acid pathway. Perception of signal during mechanical stress is mediated by stretch activated calcium ion channels and receptors that may lead to activation of downstream cascade of the events including Rho-GTPase, MAPK cascade, and activation of JA biosynthesis (Wolf et al. 2012). Studies on JA biosynthesis have been conducted extensively in tomato and *Arabidopsis*. Three pathways for JA biosynthesis exist in *Arabidopsis* and all of them require three reaction sites: the chloroplast, peroxisome, and cytoplasm. In the cytoplasm, JA is metabolized into different structures by various chemical reactions. These are Methyl jasmonate, cis-jasmone, and 12-hydroxyjasmonic acid. Initially, JA was identified as a stress-related hormone. Insect and pathogen attack generates oligouronides and polypeptides that initiate a signal transduction pathway involving synthesis of Jasmonic acid from linolenic acid (the octadecanoid pathway) which mediates the induction of defensive genes in plants. Hexadecane pathway starts from hexadecatrienoic acid. Cumulative evidences support a model for the activation of defensive genes in plants in response to insect and pathogen attacks in which various elicitors generated at the attack sites activate octadecanoid pathway via different recognition events to induce the expression of defensive genes in local and distal tissues of the plants (Doares et al. 1995).

Mechanical damage in plants can start accumulation of JA and its derivatives at the site of damage which activates signaling cascade and expression of defense response genes as a part of local response. Transmission of JAs to other parts of plant occurs via vascular bundles or airborne diffusion. After induction of JA biosynthesis, JAs are systemically transported via vascular bundles to distant parts of the plant (Thorpe et al. 2007). Some reports suggest that JAs are simultaneously resynthesized during transport (Heil and Ton 2008). Accumulation of JA within 15 min of mechanical injury has been reported in tomato plants by Malone (1996). JA cannot easily penetrate cell membrane but its derivative, MeJA is a volatile compound and can penetrate membrane easily. This compound can diffuse through airborne transmission to distant leaves in the same plant as well as in adjacent plants (Farmer and Ryan 1990). Jasmonic acid signaling involves interplay of various proteins and transcription factors (Li et al. 2017). Receptor for JA in JA signaling cascade that mediates its subcellular distribution is a high affinity ABC transporter (AtJAT1/AtABCG16) which is located on nuclear and on plasma membrane. It mediates JA transport across plasma membrane and bioactive isoleucine derivative (JA-Ile) across nuclear membrane. Higher JA concentration causes reduction in its intracellular concentration that in turn desensitizes JA signal. JA signal is then activated in other cells via transport across apoplast. When plant is under stress, nuclear AtJAT1/AtABCG16 transporter causes quick transport of JA-Ile into nucleus to regulate defense response. JA-Ile after entering nucleus through nuclear JAT1 transporter interacts with various MYC transcription factors by binding to JAZ, a repressor protein that blocks transcription of JA responsive genes (Chini et al. 2007). JAZ is then targeted for degradation via 26S proteasome complex. JAZ protein has two conserved domains namely Jas and ZIM. JAZ interacts with MYC2 with the help of ZIM domain and COI1 via Jas domain (Melotto et al. 2008; Chini et al. 2009). COI1 protein encodes an F-box protein, associated with SKP1 and Cullin to form a complex SCF^{COI1} which tags the repressor with ubiquitin for degradation (Zhai et al. 2015). JAZ repressor blocks JA signaling cascade and thus JAZ degradation is the crucial step to activate JA pathway. COI1 and JAZ are proposed to be coreceptors of JA (Sheard et al. 2010). JAZ sequestered transcription factors are released after JAZ degradation and lead to activation of gene expression. Studies suggest that MYB transcription factors can also be released via JA-Ile induced JAZ degradation. Transcription factors such as NAC, ERF, and WRKY are the part of JA signaling. JA signaling also activates MAPK cascade, calcium channel, and many other physiological and biochemical processes that interact with plant growth regulators like ethylene, SA, ABA (Santner and Estelle 2009; Li et al. 2017). Enhancing UVB exposure has been reported to induce JA biosynthesis in *Nicotiana* and *Brassica* genus (Mewis et al. 2012; Svyatyna and Riemann 2012). JA pathway is also induced during low temperature adaptation in plants by activating cold responsive genes (Zheng et al. 2012). JA signaling under drought has also been reported in *A. thaliana* and *Citrus*. MeJA significantly regulates stomatal closure to reduce water loss under drought stress (Evans 2003; De Ollas et al. 2013). Under salt stress, tomato, potato, and Arabidopsis plants showed significantly high JA concentration. JA concentration in salt sensitive plant varieties was significantly high than the tolerant varieties. Exogenous JA application has been reported to enhance salt tolerance in plants (Ruan et al. 2019). Higher CO_2 concentration leads to increased JA release into environment in lima bean (Ballhorn et al. 2011). Ozone also enhances level of endogenous JAs in plants (Ruan et al. 2019). Proteomics study on *A. thaliana* treated with Me-JA expressed 186 proteins which are collectively involved in physiological processes such as photosynthesis, metabolism of carbohydrates and hormones, and defense under stress, etc. (Chen et al. 2011). In tomato, a polypeptide derived from hydrolysis of precursor protein named systemin having 18 amino acids was identified that accumulated in response to mechanical damage caused by insect herbivory, which activated JA signaling cascade (Pearce et al. 1991). A similar polypeptide, (AtPEP1) from *Arabidopsis thaliana*, also activates JA signaling pathway in response to mechanical damage as well as pathogen attack (Yamaguchi et al. 2010). Wounding response pathway also involves ATP that generates ROS via activation of NADPH oxidase along with induction of JA biosynthesis (Song et al. 2006). An interplay of calcium ions and ROS help in activation of MAPK pathway in early response. However silencing of MAPK pathway and reduced ROS generation inhibit expression of late wound related genes but JA synthesis is not affected (Xia et al. 2004).

2.8 ROS and Plant Systemic Responses

Reactive oxygen species (ROS) are unavoidable products of aerobic metabolism. ROS signaling network is highly conserved among aerobic organisms and controls a broad range of biological processes such as growth, development, and responses to biotic and/or abiotic stimuli (Mittler et al. 2011; Finka et al. 2012). Increased ROS concentration can cause oxidative damage to membranes as well as other cellular macromolecules and leads to oxidative stress in the cell. ROS-scavenging enzymatic and non-enzymatic pathways maintain redox homeostasis in the cell. Proteins like superoxide dismutase (SOD), ascorbate peroxidase (APX), catalase (CAT), glutathione peroxidase (GPX), and peroxiredoxin (PRX) and antioxidants such as ascorbic acid, glutathione (GSH) are some of the scavengers of ROS inside the cell (Mittler 2002). A balance between the process of ROS generation and its sequestration exists under normal conditions; however, this balance is disturbed under stress conditions. Tissues when exposed to stress stimuli, Ca^{2+} influx inside the cell is marked as an early event in the signaling cascade followed by production of ROS along with other secondary messengers which may lead to activation of signaling cascades like MAPK pathways and protein phosphorylation that further amplify the response and decode the downstream response pathways that ultimately lead to transcriptional activation of genes involved in plant development as well as stress tolerance. ROS are well known toxic species and immense data is available regarding their toxicity and scavenging mechanisms in the cells. ROS also act as signaling molecules at non-toxic level where a balance is maintained between ROS generation and its sequestration. Activation of ROS signaling cascade occurs within few seconds or minutes on exposure to stress stimuli. ROS production in tobacco in heat stress has been reported to be triggered within 3 min of exposure. Activation of acclamatory responses of the plants to high light takes place within 5–20 min not only in local tissues but also in distant plant parts (Königshofer et al. 2008). Wounding responses in *Arabidopsis* are shown by elevated JA levels in local and in distant tissues within 30 s to 5 min after damage (Koo et al. 2009). In response to wounding, RBOHDs initiate a systemic signal that propagates at the rate of 8.4 cm/ min (Miller et al. 2009). ROS accumulation plays a key role in hormone mediated signaling in plants when exposed to external stimuli (including stress) or during normal growth and development. Accumulation of these reactive species regulates synthesis, localization, translocation of hormone. Although early research involving ROS metabolism has focused on the potential toxicity of ROS and different ROS-scavenging mechanisms, but recent studies have focused on the role of ROS as signaling molecules. To utilize ROS as signaling molecules, non-toxic levels must be maintained in a delicate balancing manner between ROS production, involving ROS-producing enzymes and the unavoidable production of ROS during basic cellular processes, and the metabolic counter-process involving ROS-scavenging pathways (Mittler et al. 2011). In plants, NADPH oxidase, respiratory burst oxidase homologues (RBOHs), oxidases and peroxidases, processes occurring at the chloroplast, mitochondria, peroxisome via different pathways are involved in ROS generation (Vaahtera et al. 2014; Gilroy et al. 2016; Mignolet-Spruyt et al. 2016). RBOHs assorted to different tissues and their differential co-expression indicates that there is some sort of specialization. This spatial coordination of signals requires cell to cell communication. ROS involvement in such systemic plant immunity, response against mechanical damage, wound, high light acclimation, etc. has been reported by Alvarez et al. (1998) and Karpinski et al. (1999). Cell to cell signal transfer is dependent on extracellular accumulation of H_2O_2 that leads to the initiation of ROS wave that is a priming signal in systemic response when plant is exposed to abiotic stress stimulus. RBOHs are transmembrane proteins that span the membrane six times and possess a C-terminal domain having a characteristic NADPH and FAD binding domains. Plant RBOHs differ from mammalian RBOHs in the presence of cytosolic N-terminal domain with two calcium binding motifs and site for phosphorylation (Lin et al. 2009; Kimura et al. 2012). RBOH proteins generate superoxide, and superoxide dismutase enzyme converts it to membrane permeable H_2O_2 which then acts as signaling molecule (Xia et al.

2009; Wi et al. 2012). Cytosolic concentration of calcium positively regulates the activity of RBOH proteins. Besides RBOHs and NADPH oxidase, other mechanisms of ROS generation in plants involve photosynthesis, respiration, glycolate and oxalate oxidase, fatty acid oxidation, etc. (Mittler 2002). Peroxidases like PRX33 and PRX34 are involved in ROS generation during fungal and bacterial attack in *Arabidopsis*. Callose deposition in plants and expression of defense genes also involve ROS. Evidences for such facts have been derived from mutant studies where peroxidases remain dysfunctional in plants and lead to decreased callose deposition and disturbance in defense gene expression (Daudi et al. 2012; Wrzaczek et al. 2013). Fluctuating environmental conditions cause plants to regulate various metabolic, physiological, and developmental processes in order to survive and all these come under long term responses. In case of pathogen attack, ROS accumulation in cells has been observed for hours or days following the infection which confers protection to plants (Torres and Dangl 2005; Kobayashi et al. 2007; Wi et al. 2012). Long term responses in plants towards high light intensity are regulated by PQ pool and its redox state. High light also alters thylakoid composition via STN7 kinase (Pesaresi et al. 2009; Mittler et al. 2011). ROS production consists of a primary and a secondary phase; both the phases vary in their time duration where primary phase lasts for minutes and secondary phase for hours or days (Nishimura and Dangl 2010). Wound response involves initial generation of superoxide radical and its subsequent conversion to H_2O_2 after 6 h (Soares et al. 2009). In a study where transcriptomic levels in light challenged leaves were compared to distant leaves, more than 70% upregulated genes were similar in both the cases. Level of amino acids such as glycine, glycerate, and serine is altered in both the tissues similarly. Responses of tissues towards heat and cold stress involved similar signaling and stress induced changes (Mühlenbock et al. 2008; Suzuki et al. 2013). There is an overlap in signals within local and systemic tissues, but studies demonstrate significant difference in transcripts and metabolic profile in both types of tissues. Metabolic profiling in pathogen infected and uninfected tissues showed variation in distribution of metabolites (Simon et al. 2010). Burst of ROS generated in response to stress stimuli acts as a signal propagating from local to systemic tissues, as a result of cell to cell communication that carries signal to distant parts (Miller et al. 2009). Ethylene biosynthesis is positively regulated by RBOH proteins (Jakubowicz et al. 2010). Under stress, PQ pool redox state and a signal causing production of ACC (1-amino-cyclopropane-1-carboxylate) are disturbed. ROS accumulation causes bleaching and programmed cell death in plants (Karpiński et al. 2013). ABA and SA treatments have been shown to increase H_2O_2 production that helps plant to tolerate salt, heat, and other abiotic stresses (Xia et al. 2009). Acclimation of plants systemically when exposed to heat stress was characterized by ROS as well as ABA accumulation in non-challenged tissues. Mutants lacking ABA signaling showed attenuated systemic response towards heat stress. Interaction between ROS and ABA is involved in systemic response (Suzuki et al. 2013). Signaling induced by insect attack and wounding are also interrelated, as they both rely on JA biosynthesis at the site of attack, at the site of wounding, and its translocation to systemic tissues (Suzuki and Mittler 2012). Mutants that show dysfunction of ROS regulation, its production and elimination were found to be more sensitive towards a range of abiotic stresses and were also found to be impaired in systemic signaling (Davletova et al. 2005; Suzuki et al. 2013).

2.9 SAR and ISR in Agriculture

Ever increasing threats to agriculture system are posing a variety of abiotic and biotic stress that directly affect the productivity in economically important crops all over the world. Abiotic stress to which plants are exposed include prolonged droughts, flooding, heat stress, and frost injuries. Insects and pests also cause significant yield loss and crop damage. Increased food production demand is the need of the hour, which can be compensated by using environmentally unfriendly pesticides and agrochemicals. SAR can be triggered either by exposing plants to pathogenic or non-pathogenic microorganisms or by using synthetic chemical elicitors such as salicylic acid, 2,6-dichloro-isonicotinic acid (INA) or benzo (1,2,3) thiadiazole-7-carbothioic acid S-methyl ester (BTH), SAR inducers, *N*-cyanomethyl-2-chloroisonicotinamide. These chemicals, at higher concentrations may show phytotoxicity (Nakashita et al. 2002; Vallad and Goodman 2004) and activate the expression of the same set of defense genes and induce resistance against the same spectrum of pathogens as in SA-induced SAR (Kessmann et al. 1994; Friedrich et al. 1996). A compound able to induce SAR was isolated from *Strobilanthes cusia* and chemically identified as 3-acetonyl-3-hydroxyoindole (AHO), which is a derivative of isatin. Resistance induced by this chemical depends on SA-mediated signaling when observed in inducing resistance against TMV and *Erysiphe cichoracearum* in tobacco. During resistance induction, AHO triggers PAL activity, elevates the SA level, and induces *PR-1* gene expression in tobacco plants (Li et al. 2008). Several reports have suggested that PAL is a key regulatory enzyme in the biosynthesis of SA and the establishment of SAR (Mauch-Mani and Slusarenko 1996). Chemical elicitors in general do not show direct antimicrobial activity against pathogenic microorganisms. However at higher concentrations these are reported to show antimicrobial activity (Rohilla et al. 2002). BTH (acibenzolar-S-methyl) was originally marketed for the protection of wheat and barley against powdery mildew (Görlach et al. 1996). Experimental studies revealed the effectiveness of BTH against a variety of diseases such as septoria leaf blotch and leaf rust in wheat, downy mildew of maize (Morris et al. 1998). Effectiveness of BTH against various bacterial, fungal, and viral diseases of solanaceous crops has been reported in experimental studies conducted in fields (Abbas et al. 2002; Matheron and Porchas 2002; Perez et al. 2003). In experimental studies, BTH was effective in reducing disease severity of bacterial spot and bacterial speck in tomato (Louws et al. 2001). Similarly, application of BTH was effective against bacterial spot in pepper and wildfire, blue mold, frogeye leaf spot, brown spot, and *Rhizoctonia* leaf spot diseases in tobacco (Buonaurio et al. 2002; Perez et al. 2003). Repeated application of INA reduced symptoms of white mold in *Glycine max* to some extent on susceptible cultivars but not against resistant varieties. BTH was also effective against the resistant varieties only when applied at tenfolds (Dann et al. 1998). Efficacy of BTH has also been checked for fruit crops such as *Malus domestica* and *Pyrus pyrifolia* (Ishii et al. 1999; Maxson-Stein et al. 2002). Application of BTH was able to control scab and rust in field grown Japanese pear trees. Generally, polycarbamate is used as a fungicide to control scab and rust diseases, but BTH was more effective as compared to polycarbamate in controlling scab disease but less effective in reducing severity of rust diseases as compared to polycarbamate (Ishii et al. 1999). Weekly spray of BTH on foliar tissues reduced fire blight caused by *Erwinia amylovora* more effectively as compared to streptomycin. Expression of PR proteins was induced after BTH application in apple seedlings under greenhouse conditions (Maxson-Stein et al. 2002). BTH application was studied in citrus fruits against scab, melanose, and *Alternaria* brown spot under greenhouse conditions by Agostini et al. (2003). Successful control of *Alternaria* leaf spot, bacterial blight, *Verticillium* wilt of cotton has also been reported with the use of BTH and INA (Colson-Hanks et al. 2000). In field trials on spinach, BTH was able to control white rust disease (Leskovar and Kolenda 2002). A number of PGPR are reported to elicit ISR response and systemically control various diseases when applied to plant roots using soil drench, transplant mix, root dip, or seed treatment. PGPR have been reported to control several diseases such as anthracnose caused by *Colletotrichum lagenarium*, angular leaf spot, and bacterial wilt in cucumber. Efficacy of most of the PGPR varies under field conditions; however, some strains like *Bacillus pumilus* INR-7 effectively protected cucumber plants for three years against anthracnose and angular leaf spot (Raupach and Kloepper 2000). This strain was also able to control downy mildew in pearl millet. PGPR strains when used against viral diseases, a limited success was obtained when applied under field conditions. However, greenhouse trials have provided some evidences of viral disease control. Application methods and efficacy of PGPR are interrelated. Repeated foliar application of *Bacillus pumilus* INR-7 and *Pseudomonas oryzihabitans* as well as seed treatment were equally effective against mixed infection of anthracnose and angular leaf spot on cucumber (Raupach and Kloepper 1998). Similarly, talc base formulations and fresh liquid suspension of PGPR provided similar results for the control of downy mildew in pearl millet (Niranjan Raj et al. 2003). Talc based formulations however allow easier storage and application in comparison to suspension. Most of the experiments conducted using BTH and INA led to the reduction of crop yield (Louws et al. 2001; Romero et al. 2001). In the field experiments on tomato across 22 fields, plots treated with standardized bacterial formulations and untreated control plots showed that BTH treated seeds were smaller than untreated plants (Louws et al. 2001). Similarly, reduction in yield, growth, and delay in plant maturity was observed in pepper plants treated with BTH (Romero et al. 2001). Resistance induced in monocots and dicots differs in the time duration for which resistance is maintained. Experimental studies showed that single application of BTH was sufficient to induce resistance in monocots for its lifespan whereas dicots required repeated application of BTH to provide longevity to the resistance (Louws et al. 2001; Romero et al. 2001). Control of disease using BTH was also influenced by stage of plant growth. Results of field experiments revealed that BTH application during late tillering stage controlled powdery mildew more efficiently as compared to that applied at mid tillering phase (Stadnik and Buchenauer 1999). Foliar application of jasmonic acid on radish under field conditions showed reduced tissue damage from subsequent insect herbivory. Jasmonate treated plants were more fit reproductively than damaged or untreated and undamaged plants (Agrawal 1998). Plant's growth stage and nutritional status do matter in physiological cost of induced resistance in plants (Heil 2001). Establishment of SAR requires a set of time that varies with type of elicitor as well as plant species. SAR is also associated with accumulation of salicylic acid and PR proteins throughout the plant. Plants deficient in ability to accumulate salicylic acid were found to lack PR proteins and showed attenuated SAR response (Uknes et al. 1992; Gaffney et al. 1993). Exposing plants to chemical elicitors like BTH and INA helped them to express PR proteins even in plants deficient in accumulating salicylic acid (Friedrich et al. 1996; Lawton et al. 1993). ISR is potentiated by beneficial PGPR and the response pathway does not involve accumulation of PR proteins and salicylic acid (Pieterse et al. 1996) but depends on ethylene and jasmonate signaling (Knoester et al. 1999; Yan et al. 2002). Some examples of ISR are also linked to production of siderophore and salicylic acid by PGPR strains (Vallad and Goodman 2004). SAR is effective across a broad range of plant species but ISR is specific for different PGPR strains of plant species and genotypes (Yan et al. 2002).

2.10 Conclusion

Till date, a number of studies have been conducted and have enlightened our understanding of plant responses towards biotic and abiotic stress stimuli. Transcriptomics, proteomics, mutant studies, and metabolic profiling has helped the researchers to understand the molecular mechanism of the process as well as genes, proteins, and other components involved to decode the events involved in downstream signaling cascade. Secondary messengers like ROS play an intermediate role in regulating biotic and abiotic stress individually as well as cross-linking the response pathways. Thus, the identification of new chemicals capable of inducing disease resistance would be useful, not only for elucidating the pathways, but also for developing new agents for plant protection. ISR and SAR have been recognized as an attractive tool in modern agriculture. Commercially available synthetic chemical inducers of SAR are commonly used to protect plant from pathogens. Similarly, ISR induced by plant growth promoting rhizobacteria are helpful in cultivating crops with optimal yield as well as reduced disease incidences. Thus, exploiting SAR and ISR and developing a combinatorial approach wherein elicitors of both the responses are used optimally to develop resistance and tolerance in plants towards different stress stimuli, can help to reduce the dependency on agrochemicals and prove to be a small step towards sustainable agriculture.

Acknowledgements Authors sincerely acknowledge the facilities provided by Department of Microbiology, CBSH, GBPUA&T, Pantnagar.

References

Abbas PA, Al-Dahmani J, Sahin F, Hoitink HAJ, Miller SA (2002) Effect of compost amendments on disease severity and yield of tomato in conventional and organic production systems. Plant Dis 86:156–161

Agostini JP, Bushong PM, Timmer LW (2003) Greenhouse evaluation of products that induce host resistance for control of scab, melanose, and *Alternaria* brown spot of citrus. Plant Dis 87:69–74

Agrawal AA (1998) Induced responses to herbivory and increased plant performance. Science 279:1201–1202

Alexander D, Goodman RM, Gut-Rella M, Glascock C, Weymann K, Friedrich L, Ward ESCHERICHIA (1993) Increased tolerance to two oomycete pathogens in transgenic tobacco expressing pathogenesis-related protein 1a. Proc Natl Acad Sci 90:7327–7331

Alvarez ME, Pennell RI, Meijer PJ, Ishikawa A, Dixon RA, Lamb C (1998) Reactive oxygen intermediates mediate a systemic signal network in the establishment of plant immunity. Cell 92:773–784

Audenaert K, Pattery T, Cornelis P, Höfte M (2002) Induction of systemic resistance to *Botrytis cinerea* in tomato by *Pseudomonas aeruginosa* 7NSK2: role of salicylic acid, pyochelin, and pyocyanin. Mol Plant-Microbe Interact 15:1147–1156

Ballhorn DJ, Reisdorff C, Pfanz H (2011) Quantitative effects of enhanced CO2 on jasmonic acid induced plant volatiles of lima bean (*Phaseolus lunatus* L.). J Appl Bot Food Qual 84:65–71

Baxter A, Mittler R, Suzuki N (2013) ROS as key players in plant stress signalling. J Exp Bot 65 (5):1229–1240

Berrocal-Lobo M, Molina A, Solano R (2002) Constitutive expression of ETHYLENE-RESPONSE-FACTOR1 in *Arabidopsis* confers resistance to several necrotrophic fungi. Plant J 29:23–32

Bonardi V, Dangl JL (2012) How complex are intracellular immune receptor signaling complexes? Front Plant Sci 3:237

Broglie K, Chet I, Holliday M, Cressman R, Biddle P, Knowlton C, Mauvais CJ, Broglie R (1991) Transgenic plants with enhanced resistance to the fungal pathogen *Rhizoctonia solani*. Science 254:1194–1197

Buonaurio R, Scarponi L, Ferrara M, Sidoti P, Bertona A (2002) Induction of systemic acquired resistance in pepper plants by acibenzolar-S-methyl against bacterial spot disease. Eur J Plant Pathol 108:41–49

Catinot J, Buchala A, Abou-Mansour E, Métraux JP (2008) Salicylic acid production in response to biotic and abiotic stress depends on isochorismate in *Nicotiana benthamiana*. FEBS Lett 582:473–478

Chandra D, Rickert J, Huang Y, Steinwand MA, Marr SK, Wildermuth MC (2014) Atypical E2F transcriptional repressor DEL1 acts at the intersection of plant growth and immunity by controlling the hormone salicylic acid. Cell Host Microbe 15:506–513

Chao YY, Chen CY, Huang WD, Kao CH (2010) Salicylic acid-mediated hydrogen peroxide accumulation and protection against Cd toxicity in rice leaves. Plant Soil 329:327–337

Chen Z, Silva H, Klessig DF (1993) Active oxygen species in the induction of plant systemic acquired resistance by salicylic acid. Science 262:1883–1886

Chen Z, Malamy J, Henning J, Conrath U, Sánchez-Casas P, Silva H, Ricigliano J, Klessig DK (1995) Induction, modification, and transduction of the salicylic acid signal in plant defense responses. Proc Natl Acad Sci 92:4134–4137

Chen Y, Pang Q, Dai S, Wang Y, Chen S, Yan X (2011) Proteomic identification of differentially expressed proteins in Arabidopsis in response to methyl jasmonate. J Plant Physiol 168:995–1008

Chini A, Fonseca S, Fernandez G, Adie B, Chico JM, Lorenzo O, Micol JL (2007) The JAZ family of repressors is the missing link in jasmonate signalling. Nature 448:666

Chini A, Fonseca S, Chico JM, Fernández-Calvo P, Solano R (2009) The ZIM domain mediates homo-and heteromeric interactions between *Arabidopsis* JAZ proteins. Plant J 59:77–87

Choudhary DK, Prakash A, Johri BN (2007) Induced systemic resistance (ISR) in plants: mechanism of action. Indian J Microbiol 47:289–297

Cochrane FC, Davin LB, Lewis NG (2004) The Arabidopsis phenylalanine ammonia lyase gene family: kinetic characterization of the four PAL isoforms. Phytochemistry 65:1557–1564

Cohen Y, Guegler K, Moesinger E and Niderman T (1992) Plant pathogenesis-related proteins. International patent application No. WO, 92: 20800

Colson-Hanks ES, Allen SJ, Deverall BJ (2000) Effect of 2, 6-dichloroisonicotinic acid or benzothiadiazole on Alternaria leaf spot, bacterial blight and *Verticillium* wilt in cotton under field conditions. Australas Plant Pathol 29:170–177

Conrath U, Thulke O, Katz V, Schwindling S, Kohler A (2001) Priming as a mechanism in induced systemic resistance of plants. Eur J Plant Pathol 107:113–119

Conrath U, Beckers GJ, Flors V, García-Agustín P, Jakab G, Mauch F, Pugin A (2006) Priming: getting ready for battle. Mol Plant-Microbe Interact 19:1062–1071

Cruickshank IAM, Mandryk M (1960) The effect of stem infestation of Tobacco with *Peronospora tabacina* Adam, on foliage reaction to blue mold. J Aust Inst Agric Sci 26(4):369–372

Dangl JL, Dietrich RA, Richberg MH (1996) Death don't have no mercy: cell death programs in plant-microbe interactions. Plant Cell 8:1793–1807

Dann E, Diers B, Byrum J, Hammerschmidt R (1998) Effect of treating soybean with 2, 6-dichloroisonicotinic acid (INA) and benzothiadiazole (BTH) on seed yields and the level of disease caused by *Sclerotinia sclerotiorum* in field and greenhouse studies. Eur J Plant Pathol 104:271–278

Daudi A, O'Brien JA, Mammarella N, Khan S, Ausubel FM, Bolwell GP (2012) The apoplastic oxidative burst peroxidase in Arabidopsis is a major component of pattern-triggered immunity. Plant Cell 24:275–287

Davies WJ, Kudoyarova G, Hartung W (2005) Long-distance ABA signaling and its relation to other signaling pathways in the detection of soil drying and the mediation of the plant's response to drought. J Plant Growth Regul 24:285–295

Davletova S, Rizhsky L, Liang H, Shengqiang Z, Oliver DJ, Coutu J, Mittler R (2005) Cytosolic ascorbate peroxidase 1 is a central component of the reactive oxygen gene network of Arabidopsis. Plant Cell 17:268–281

De Ollas C, Hernando B, Arbona V, Gómez-Cadenas A (2013) Jasmonic acid transient accumulation is needed for abscisic acid increase in citrus roots under drought stress conditions. Physiol Plant 147:296–306

Dempsey DMA, Klessig DF (2012) SOS–too many signals for systemic acquired resistance? Trends Plant Sci 17:538–545

Derksen H, Rampitsch C, Daayf F (2013) Signaling cross-talk in plant disease resistance. Plant Sci 207:79–87

Dewdney J, Reuber TL, Wildermuth MC, Devoto A, Cui J, Stutius LM, Drummond EP, Ausubel FM (2000) Three unique mutants of Arabidopsis identify eds loci required for limiting growth of a biotrophic fungal pathogen. Plant J 24:205–218

Dıaz J, ten Have A, Van Kan JA (2002) The role of ethylene and wound signaling in resistance of tomato to *Botrytis cinerea*. Plant Physiol 129:1341–1351

Doares SH, Syrovets T, Weiler EW, Ryan CA (1995) Oligogalacturonides and chitosan activate plant defensive genes through the octadecanoid pathway. Proc Natl Acad Sci 92:4095–4098

Dong X (2004) NPR1, all things considered. Curr Opin Plant Biol 7:547–552

Durrant WE, Dong X (2004) Systemic acquired resistance. Annu Rev Phytopathol 42:185–209

Ellis C, Turner JG (2001) The Arabidopsis mutant cev1 has constitutively active jasmonate and ethylene signal pathways and enhanced resistance to pathogens. Plant Cell 13:1025–1033
Ellis C, Karafyllidis I, Wasternack C, Turner JG (2002) The Arabidopsis mutant cev1 links cell wall signaling to jasmonate and ethylene responses. Plant Cell 14:1557–1566
Evans NH (2003) Modulation of guard cell plasma membrane potassium currents by methyl jasmonate. Plant Physiol 131:8–11
Farmer EE, Ryan CA (1990) Interplant communication: airborne methyl jasmonate induces synthesis of proteinase inhibitors in plant leaves. Proc Natl Acad Sci 87:7713–7716
Finka A, Cuendet AFH, Maathuis FJ, Saidi Y, Goloubinoff P (2012) Plasma membrane cyclic nucleotide gated calcium channels control land plant thermal sensing and acquired thermotolerance. Plant Cell 24:3333–3348
Friedrich L, Vernooij B, Gaffney T, Morse A, Ryals J (1995) Characterization of tobacco plants expressing a bacterial salicylate hydroxylase gene. Plant Mol Biol 29:959–968
Friedrich L, Lawton K, Ruess W, Masner P, Specker N, Rella MG, Dincher S, Staub T, Métraux JP (1996) A benzothiadiazole derivative induces systemic acquired resistance in tobacco. Plant J 10:61–70
Fu ZQ, Dong X (2013) Systemic acquired resistance: turning local infection into global defense. Annu Rev Plant Biol 64:839–863
Fu ZQ, Yan S, Saleh A, Wang W, Ruble J, Oka N, Dong X (2012) NPR3 and NPR4 are receptors for the immune signal salicylic acid in plants. Nature 486:228–232
Gaffney T, Friedrich L, Vernooij B, Negrotto D, Nye G, Uknes S, Ryals J (1993) Requirement of salicylic acid for the induction of systemic acquired resistance. Science 261:754–756
Garcion C, Métraux JP (2008) Salicylic acid. Plant Hormone Signal 24:229–255
Gatz C (2013) From pioneers to team players: TGA transcription factors provide a molecular link between different stress pathways. Plant Microbe Interact 26:151–159
Gilroy S, Białasek M, Suzuki N, Górecka M, Devireddy AR, Karpiński S, Mittler R (2016) ROS, calcium and electric signals: key mediators of rapid systemic signalling in plants. Plant Physiol 171:1606–1615
Görlach J, Volrath S, Knauf-Beiter G, Hengy G, Beckhove U, Kogel KH, Ryals J (1996) Benzothiadiazole, a novel class of inducers of systemic acquired resistance, activates gene expression and disease resistance in wheat. Plant Cell 8:629–643
Govrin EM, Levine A (2000) The hypersensitive response facilitates plant infection by the necrotrophic pathogen *Botrytis cinerea*. Curr Biol 10:751–757
Haas D, Keel C, Reimmann C (2002) Signal transduction in plant-beneficial rhizobacteria with biocontrol properties. Antonie Leeuwenhoek 81:385–395
Hamayun M, Khan SA, Khan AL, Shinwari ZK, Hussain J, Sohn EY, Lee IJ (2010) Effect of salt stress on growth attributes and endogenous growth hormones of soybean cultivar Hwangkeumkong. Pak J Bot 42:3103–3112
Heil M (2001) The ecological concept of costs of induced systemic resistance (ISR). Eur J Plant Pathol 107:137–146
Heil M, Bostock RM (2002) Induced systemic resistance (ISR) against pathogens in the context of induced plant defences. Ann Bot 89:503–512
Heil M, Ton J (2008) Long-distance signalling in plant defence. Trends Plant Sci 13:264–272
Ishihara T, Sekine KT, Hase S, Kanayama Y, Seo S, Ohashi Y, Takahashi H (2008) Overexpression of the *Arabidopsis thaliana* EDS5 gene enhances resistance to viruses. Plant Biol 10:451–461
Ishii H, Tomita Y, Horio T, Narusaka Y, Nakazawa Y, Nishimura K, Iwamoto S (1999) Induced resistance of acibenzolar-S-methyl (CGA 245704) to cucumber and Japanese pear diseases. Eur J Plant Pathol 105:77–85
Jacobs JM, Milling A, Mitra RM, Hogan CS, Ailloud F, Prior P, Allen C (2013) *Ralstonia solanacearum* requires PopS, an ancient AvrE-family effector, for virulence and to overcome salicylic acid-mediated defenses during tomato pathogenesis. MBio 4:e00875
Jakubowicz M, Gałgańska H, Nowak W, Sadowski J (2010) Exogenously induced expression of ethylene biosynthesis, ethylene perception, phospholipase D, and Rboh-oxidase genes in broccoli seedlings. J Exp Bot 61:3475–3491

Jibran R, Hunter DA, Dijkwel PP (2013) Hormonal regulation of leaf senescence through integration of developmental and stress signals. Plant Mol Biol 82:547–561

Kachroo A, Robin GPS (2013) Systemic signaling during plant defense. Curr Opin Plant Biol 16:527–533

Kachroo P, Yoshioka K, Shah J, Dooner HK, Klessig DF (2000) Resistance to turnip crinkle virus in Arabidopsis is regulated by two host genes and is salicylic acid dependent but NPR1, ethylene, and jasmonate independent. Plant Cell 12:677–690

Karpinski S, Reynolds H, Karpinska B, Wingsle G, Creissen G, Mullineaux P (1999) Systemic signaling and acclimation in response to excess excitation energy in Arabidopsis. Science 284:654–657

Karpiński S, Szechyńska-Hebda MA, Wituszyńska W, Burdiak P (2013) Light acclimation, retrograde signalling, cell death and immune defences in plants. Plant Cell Environ 36:736–744

Kessmann H, Staub T, Hofmann C, Maetzke T, Herzog J, Ward E, Ryals J (1994) Induction of systemic acquired disease resistance in plants by chemicals. Annu Rev Phytopathol 32:439–459

Khokon MAR, Okuma EIJI, Hossain MA, Munemasa S, Uraji M, Nakamura Y, Murata Y (2011) Involvement of extracellular oxidative burst in salicylic acid-induced stomatal closure in Arabidopsis. Plant Cell Environ 34:434–443

Kim DS, Hwang BK (2014) An important role of the pepper phenylalanine ammonia-lyase gene (PAL1) in salicylic acid-dependent signalling of the defence response to microbial pathogens. J Exp Bot 65:2295–2306

Kim MS, Kim YC, Cho BH (2004) Gene expression analysis in cucumber leaves primed by root colonization with *Pseudomonas chlororaphis* O6 upon challenge-inoculation with *Corynespora cassiicola*. Plant Biol 6:105–108

Kimura S, Kaya H, Kawarazaki T, Hiraoka G, Senzaki E, Michikawa M, Kuchitsu K (2012) Protein phosphorylation is a prerequisite for the Ca2+-dependent activation of Arabidopsis NADPH oxidases and may function as a trigger for the positive feedback regulation of Ca2+ and reactive oxygen species. Biochim Biophys Acta 1823:398–405

Knoester M, Pieterse CM, Bol JF, Van Loon LC (1999) Systemic resistance in Arabidopsis induced by rhizobacteria requires ethylene-dependent signaling at the site of application. Mol Plant-Microbe Interact 12:720–727

Kobayashi M, Ohura I, Kawakita K, Yokota N, Fujiwara M, Shimamoto K, Yoshioka H (2007) Calcium-dependent protein kinases regulate the production of reactive oxygen species by potato NADPH oxidase. Plant Cell 19:1065–1080

Kombrink E, Schmelzer E (2001) The hypersensitive response and its role in local and systemic disease resistance. Eur J Plant Pathol 107:69–78

Königshofer H, Tromballa HW, Löppert HG (2008) Early events in signalling high-temperature stress in tobacco BY2 cells involve alterations in membrane fluidity and enhanced hydrogen peroxide production. Plant Cell Environ 31:1771–1780

Koo AJ, Gao X, Daniel Jones A, Howe GA (2009) A rapid wound signal activates the systemic synthesis of bioactive jasmonates in Arabidopsis. Plant J 59:974–986

Kucera B, Cohn MA, Leubner-Metzger G (2005) Plant hormone interactions during seed dormancy release and germination. Seed Sci Res 15:281–307

Kunkel BN, Brooks DM (2002) Cross talk between signalling pathways in pathogen defense. Curr Opin Plant Biol 5:325–331

Lawton K, Uknes S, Friedrich L, Gaffney T, Alexander D, Goodman R, Ward E (1993) The molecular biology of systemic acquired resistance. In: Mechanisms of plant defense responses. Springer, Dordrecht, pp 422–432

Leon J, Yalpani N, Raskin I, Lawton MA (1993) Induction of benzoic acid 2-hydroxylase in virus-inoculated tobacco. Plant Physiol 103:323–328

Leon J, Lawton MA, Raskin I (1995) Hydrogen peroxide stimulates salicylic acid biosynthesis in tobacco. Plant Physiol 108:1673–1678

Leskovar DI, Kolenda K (2002) Strobilurin acibenzolar-S-methyl controls white rust without inducing leaf chlorosis in spinach. Ann Appl Biol 140:171–175

Li Y, Zhang Z, Jia Y, Shen Y, He H, Fang R, Hao X (2008) 3-Acetonyl-3-hydroxyoxindole: a new inducer of systemic acquired resistance in plants. Plant Biotechnol J 6:301–308

Li P, Chen L, Zhou Y, Xia X, Shi K, Chen Z, Yu J (2013) Brassinosteroids-induced systemic stress tolerance was associated with increased transcripts of several defence-related genes in the phloem in *Cucumis sativus*. PLoS One 8:e66582

Li C, Tan DX, Liang D, Chang C, Jia D, Ma F (2014) Melatonin mediates the regulation of ABA metabolism, free-radical scavenging, and stomatal behaviour in two Malus species under drought stress. J Exp Bot 66:669–680

Li Q, Zheng J, Li S, Huang G, Skilling SJ, Wang L, Liu P (2017) Transporter-mediated nuclear entry of jasmonoyl-isoleucine is essential for jasmonate signaling. Mol Plant 10:695–708

Lin F, Ding H, Wang J, Zhang H, Zhang A, Zhang Y, Jiang M (2009) Positive feedback regulation of maize NADPH oxidase by mitogen-activated protein kinase cascade in abscisic acid signalling. J Exp Bot 60:3221–3238

Linthorst HJ, Danhash N, Brederode FT, Van Kan JA, De Wit PJ (1991) Tobacco and tomato PR proteins homologous. Mol Plant-Microbe Interact 4:586–592

Lorenzo O, Piqueras R, Sánchez-Serrano JJ, Solano R (2003) Ethylene response factor1 integrates signals from ethylene and jasmonate pathways in plant defense. Plant Cell 15:165–178

Louws FJ, Wilson M, Campbell HL, Cuppels DA, Jones JB, Shoemaker PB, Miller SA (2001) Field control of bacterial spot and bacterial speck of tomato using a plant activator. Plant Dis 85:481–488

Low PS, Merida JR (1996) The oxidative burst in plant defence: function and signal transduction. Physiol Plant 96:533–542

Macho AP, Zipfel C (2014) Plant PRRs and the activation of innate immune signaling. Mol Cell 54:263–272

Malone M (1996) Rapid, long-distance signal transmission in higher plants. Adv Bot Res 22:163–228

Matheron ME, Porchas M (2002) Suppression of Phytophthora root and crown rot on pepper plants treated with acibenzolar-S-methyl. Plant Dis 86:292–297

Mauch-Mani B, Slusarenko AJ (1996) Production of salicylic acid precursors is a major function of phenylalanine ammonia-lyase in the resistance of Arabidopsis to *Peronospora parasitica*. Plant Cell 8:203–212

Maxson-Stein K, He SY, Hammerschmidt R, Jones AL (2002) Effect of treating apple trees with acibenzolar-S-methyl on fire blight and expression of pathogenesis-related protein genes. Plant Dis 86:785–790

Melotto M, Mecey C, Niu Y, Chung HS, Katsir L, Yao J, Zeng W, Thines B, Staswick P, Browse J, Howe GA (2008) A critical role of two positively charged amino acids in the Jas motif of Arabidopsis JAZ proteins in mediating coronatine-and jasmonoyl isoleucine-dependent interactions with the COI1 F-box protein. Plant J 55:979–988

Métraux JP, Ahlgoy P, Staub TH, Speich J, Steinemann A, Ryals J, Ward E (1991) Induced systemic resistance in cucumber in response to 2,6-dichloro-isonicotinic acid and pathogens. In: Advances in molecular genetics of plant-microbe interactions, vol 1. Springer, Dordrecht, pp 432–439

Mewis I, Schreiner M, Nguyen CN, Krumbein A, Ulrichs C, Lohse M, Zrenner R (2012) UV-B irradiation changes specifically the secondary metabolite profile in broccoli sprouts: induced signalling overlaps with defense response to biotic stressors. Plant Cell Physiol 53:1546–1560

Meziane H, Van Der Sluis I, Van Loon LC, Höfte M, Bakker PA (2005) Determinants of *Pseudomonas putida* WCS358 involved in inducing systemic resistance in plants. Mol Plant Pathol 6(2):177–185

Mignolet-Spruyt L, Xu E, Idänheimo N, Hoeberichts FA, Mühlenbock P, Brosche M, Kangasjärvi J (2016) Spreading the news: subcellular and organellar reactive oxygen species production and signalling. J Exp Bot 67:3831–3844

Miller G, Schlauch K, Tam R, Cortes D, Torres MA, Shulaev V, Mittler R (2009) The plant NADPH oxidase RBOHD mediates rapid systemic signaling in response to diverse stimuli. Sci Signal 284(2):ra45

Miransari M (2016) Soybeans and plant hormones. In: Environmental stresses in soybean production. Academic, Cambridge, pp 131–156
Mittler R (2002) Oxidative stress, antioxidants and stress tolerance. Trends Plant Sci 7:405–410
Mittler R, Vanderauwera S, Suzuki N, Miller GAD, Tognetti VB, Vandepoele K, Van Breusegem F (2011) ROS signaling: the new wave? Trends Plant Sci 16:300–309
Miura K, Tada Y (2014) Regulation of water, salinity, and cold stress responses by salicylic acid. Front Plant Sci 5:4
Morris SW, Vernooij B, Titatarn S, Starrett M, Thomas S, Wiltse CC, Uknes S (1998) Induced resistance responses in maize. Mol Plant-Microbe Interact 11:643–658
Mou Z, Fan W, Dong X (2003) Inducers of plant systemic acquired resistance regulate NPR1 function through redox changes. Cell 113:935–944
Mühlenbock P, Szechyńska-Hebda M, Płaszczyca M, Baudo M, Mateo A, Mullineaux PM, Karpiński S (2008) Chloroplast signaling and lesion simulating disease1 regulate crosstalk between light acclimation and immunity in Arabidopsis. Plant Cell 20:2339–2356
Mullineaux PM, Baker NR (2010) Oxidative stress: antagonistic signaling for acclimation or cell death. Plant Physiol 154:521–525
Mur LA, Kenton P, Atzorn R, Miersch O, Wasternack C (2006) The outcomes of concentration-specific interactions between salicylate and jasmonate signaling include synergy, antagonism, and oxidative stress leading to cell death. Plant Physiol 140:249–262
Murphy JF, Zehnder GW, Schuster DJ, Sikora EJ, Polston JE, Kloepper JW (2000) Plant growth-promoting rhizobacterial mediated protection in tomato against Tomato mottle virus. Plant Dis 84:779–784
Nakashita H, Yasuda M, Nishioka M, Hasegawa S, Arai Y, Uramoto M, Yoshida S, Yamaguchi I (2002) Chloroisonicotinamide derivative induces a broad range of disease resistance in rice and tobacco. Plant Cell Physiol 43:823–831
Nandakumar R, Babu S, Viswanathan R, Raguchander T, Samiyappan R (2001) Induction of systemic resistance in rice against sheath blight disease by *Pseudomonas fluorescens*. Soil Biol Biochem 33:603–612
Neu E, Domes HS, Menz I, Kaufmann H, Linde M, Debener T (2019) Interaction of roses with a biotrophic and a hemibiotrophic leaf pathogen leads to differences in defense transcriptome activation. Plant Mol Biol 99:299–316
Neuenschwander U, Vernooij B, Friedrich L, Uknes S, Kessmann H, Ryals J (1995) Is hydrogen peroxide a second messenger of salicylic acid in systemic acquired resistance? Plant J 8:227–233
Niranjan Raj S, Chaluvaraju G, Amruthesh KN, Shetty HS, Reddy MS, Kloepper JW (2003) Induction of growth promotion and resistance against downy mildew on pearl millet (*Pennisetum glaucum*) by rhizobacteria. Plant Dis 87:380–384
Nishimura MT, Dangl JL (2010) Arabidopsis and the plant immune system. Plant J 61:1053–1066
Pearce G, Strydom D, Johnson S, Ryan CA (1991) A polypeptide from tomato leaves induces wound-inducible proteinase inhibitor proteins. Science 253(5022):895–897
Perez L, Rodriguez ME, Rodriguez F, Roson C (2003) Efficacy of acibenzolar-S-methyl, an inducer of systemic acquired resistance against tobacco blue mould caused by *Peronospora hyoscyami* f. sp. *tabacina*. Crop Prot 22:405–413
Pesaresi P, Hertle A, Pribil M, Kleine T, Wagner R, Strissel H, Ihnatowicz A, Bonardi V, Scharfenberg M, Schneider A, Pfannschmidt T (2009) Arabidopsis STN7 kinase provides a link between short-and long-term photosynthetic acclimation. Plant Cell 21:2402–2423
Pieterse CM, Van Wees SC, Hoffland E, Van Pelt JA, Van Loon LC (1996) Systemic resistance in Arabidopsis induced by biocontrol bacteria is independent of salicylic acid accumulation and pathogenesis-related gene expression. Plant Cell 8:1225–1237
Pieterse CM, Van Wees SC, Van Pelt JA, Knoester M, Laan R, Gerrits H, Van Loon LC (1998) A novel signaling pathway controlling induced systemic resistance in Arabidopsis. Plant Cell 10:1571–1580

Raupach GS, Kloepper JW (1998) Mixtures of plant growth-promoting rhizobacteria enhance biological control of multiple cucumber pathogens. Phytopathology 88:1158–1164

Raupach GS, Kloepper JW (2000) Biocontrol of cucumber diseases in the field by plant growth-promoting rhizobacteria with and without methyl bromide fumigation. Plant Dis 84:1073–1075

Rohilla R, Singh US, Singh RL (2002) Mode of action of acibenzolar-S-methyl against sheath blight of rice, caused by *Rhizoctonia solani* Kühn. Pest Manag Sci 58:63–69

Rojo E, Solano R, Sánchez-Serrano JJ (2003) Interactions between signaling compounds involved in plant defense. J Plant Growth Regul 22:82–98

Romero AM, Kousik CS, Ritchie DF (2001) Resistance to bacterial spot in bell pepper induced by acibenzolar-S-methyl. Plant Dis 85:189–194

Ross AF (1961a) Localized acquired resistance to plant virus infection in hypersensitive hosts. Virology 14:329–339

Ross AF (1961b) Systemic acquired resistance induced by localized virus infections in plants. Virology 14:340–358

Rossel JB, Wilson PB, Hussain D, Woo NS, Gordon MJ, Mewett OP, Pogson BJ (2007) Systemic and intracellular responses to photooxidative stress in Arabidopsis. Plant Cell 19:4091–4110

Ruan J, Zhou Y, Zhou M, Yan J, Khurshid M, Weng W, Zhang K (2019) Jasmonic acid signaling pathway in plants. Int J Mol Sci 20:2479

Ryal JA, Neuenschwander UH, Willits MG, Molina A, Steiner HY, Hunt MD (1996) Systemic acquired resistance. Plant Cell 8:1809

Sadeghi M, Dehghan S, Fischer R, Wenzel U, Vilcinskas A, Kavousi HR, Rahnamaeian M (2013) Isolation and characterization of isochorismate synthase and cinnamate 4-hydroxylase during salinity stress, wounding, and salicylic acid treatment in *Carthamus tinctorius*. Plant Signal Behav 8:e27335

Santner A, Estelle M (2009) Recent advances and emerging trends in plant hormone signalling. Nature 459:1071–1078

Schenk ST, Stein E, Kogel KH, Schikora A (2012) Arabidopsis growth and defense are modulated by bacterial quorum sensing molecules. Plant Signal Behav 7:178–181

Serrano M, Wang B, Aryal B, Garcion C, Abou-Mansour E, Heck S, Métraux JP (2013) Export of salicylic acid from the chloroplast requires the multidrug and toxin extrusion-like transporter EDS5. Plant Physiol 162:1815–1821

Seyfferth C, Tsuda K (2014) Salicylic acid signal transduction: the initiation of biosynthesis, perception and transcriptional reprogramming. Front Plant Sci 5:697

Sheard LB, Tan X, Mao H, Withers J, Ben-Nissan G, Hinds TR, Kobayashi Y, Hsu FF, Sharon M, Browse J, He SY (2010) Jasmonate perception by inositol-phosphate-potentiated COI1–JAZ co-receptor. Nature 468:400–405

Shulaev V, León J, Raskin I (1995) Is salicylic acid a translocated signal of systemic acquired resistance in tobacco? Plant Cell 7:1691–1701

Simon C, Langlois-Meurinne M, Bellvert F, Garmier M, Didierlaurent L, Massoud K, Noctor G (2010) The differential spatial distribution of secondary metabolites in Arabidopsis leaves reacting hypersensitively to *Pseudomonas syringae* pv. *tomato* is dependent on the oxidative burst. J Exp Bot 61:3355–3370

Soares NC, Francisco R, Vielba JM, Ricardo CP, Jackson PA (2009) Associating wound-related changes in the apoplast proteome of Medicago with early steps in the ROS signal-transduction pathway. J Proteome Res 8:2298–2309

Song CJ, Steinebrunner I, Wang X, Stout SC, Roux SJ (2006) Extracellular ATP induces the accumulation of superoxide via NADPH oxidases in Arabidopsis. Plant Physiol 140:1222–1232

Stadnik MJ, Buchenauer H (1999) Control of wheat diseases by a benzothiadiazole-derivative and modern fungicides/Zur Bekämpfung pilzlicher Krankheiten an Weizen durch ein Benzothiadiazol-Derivat und moderne Fungizide. J Plant Dis Prot 1999:466–475

Straus MR, Rietz S, van Themaat E, Parker JE (2010) Salicylic acid antagonism of EDS1-driven cell death is important for immune and oxidative stress responses in Arabidopsis. Plant J 62:628–640

Strawn MA, Marr SK, Inoue K, Inada N, Zubieta C, Wildermuth MC (2007) Arabidopsis isochorismate synthase functional in pathogen-induced salicylate biosynthesis exhibits properties consistent with a role in diverse stress responses. J Biol Chem 282:5919–5933

Summermatter K, Sticher L, Métraux JP (1995) Systemic responses in *Arabidopsis thaliana* infected and challenged with *Pseudomonas syringae* pv. Syringe Plant Physiol 108:1379–1385

Suzuki N, Mittler R (2012) Reactive oxygen species-dependent wound responses in animals and plants. Free Radic Biol Med 53:2269–2276

Suzuki N, Miller G, Salazar C, Mondal HA, Shulaev E, Cortes DF, Shulaev V (2013) Temporal-spatial interaction between reactive oxygen species and abscisic acid regulates rapid systemic acclimation in plants. Plant Cell 25:3553–3569

Svyatyna K, Riemann M (2012) Light-dependent regulation of the jasmonate pathway. Protoplasma 249:137–145

Tada Y, Spoel SH, Pajerowska-Mukhtar K, Mou Z, Song J, Wang C, Zuo J, Dong X (2008) Plant immunity requires conformational charges of NPR1 via S-nitrosylation and thioredoxins. Science 321:952–956

Thomma BP, Eggermont K, Broekaert WF, Cammue BP (2000) Disease development of several fungi on Arabidopsis can be reduced by treatment with methyl jasmonate. Plant Physiol Biochem 38:421–427

Thomma BP, Tierens KF, Penninckx IA, Mauch-Mani B, Broekaert WF, Cammue BP (2001) Different micro-organisms differentially induce Arabidopsis disease response pathways. Plant Physiol Biochem 39:673–680

Thorpe MR, Ferrieri AP, Herth MM, Ferrieri RA (2007) 11 C-imaging: methyl jasmonate moves in both phloem and xylem, promotes transport of jasmonate, and of photoassimilate even after proton transport is decoupled. Planta 226:541–551

Torres MA, Dangl JL (2005) Functions of the respiratory burst oxidase in biotic interactions, abiotic stress and development. Curr Opin Plant Biol 8:397–403

Tsuda K, Katagiri F (2010) Comparing signaling mechanisms engaged in pattern-triggered and effector-triggered immunity. Curr Opin Plant Biol 13:459–465

Tsuda K, Mine A, Bethke G, Igarashi D, Botanga CJ, Tsuda Y, Glazebrook J, Sato M, Katagiri F (2013) Dual regulation of gene expression mediated by extended MAPK activation and salicylic acid contributes to robust innate immunity in *Arabidopsis thaliana*. PLoS Genet 9:e1004015

Tuteja N (2007) Abscisic acid and abiotic stress signaling. Plant Signal Behav 2:135–138

Uknes S, Mauch-Mani B, Moyer M, Potter S, Williams S, Dincher S, Ryals J (1992) Acquired resistance in Arabidopsis. Plant Cell 4:645–656

Uppalapati SR, Ishiga Y, Wangdi T, Kunkel BN, Anand A, Mysore KS, Bender CL (2007) The phytotoxin coronatine contributes to pathogen fitness and is required for suppression of salicylic acid accumulation in tomato inoculated with *Pseudomonas syringae* pv. tomato DC3000. Mol Plant-Microbe Interact 20:955–965

Vaahtera L, Brosché M, Wrzaczek M, Kangasjärvi J (2014) Specificity in ROS signaling and transcript signatures. Antioxid Redox Signal 21:1422–1441

Vallad GE, Goodman RM (2004) Systemic acquired resistance and induced systemic resistance in conventional agriculture. Crop Sci 44:1920–1934

Van Loon LC, Van Kammen A (1970) Polyacrylamide disc electrophoresis of the soluble leaf proteins from Nicotiana tabacum var.'Samsun'and 'Samsun NN': II. Changes in protein constitution after infection with tobacco mosaic virus. Virology 40:199–211

Van Loon LC, Van Strien EA (1999) The families of pathogenesis-related proteins, their activities, and comparative analysis of PR-1 type proteins. Physiol Mol Plant 55:85–97

Vernooij B, Friedrich L, Morse A, Reist R, Kolditz-Jawhar R, Ward E, Ryals J (1994) Salicylic acid is not the translocated signal responsible for inducing systemic acquired resistance but is required in signal transduction. Plant Cell 6:959–965

Vernooij B, Friedrich L, Goy PA, Staub T, Kessmann H, Ryals J (1995) 2,6-Dichloroisonicotinic acid-induced resistance to pathogens without the accumulation of salicylic acid. Mol Plant-Microbe Interact 8:228–234

Wang D, Amornsiripanitch N, Dong X (2006) A genomic approach to identify regulatory nodes in the transcriptional network of systemic acquired resistance in plants. PLoS Pathog 2:e123

Wi SJ, Ji NR, Park KY (2012) Synergistic biosynthesis of biphasic ethylene and reactive oxygen species in response to hemibiotrophic *Phytophthora parasitica* in tobacco plants. Plant Physiol 159:251–265

Wolf S, Hématy K, Höfte H (2012) Growth control and cell wall signaling in plants. Annu Rev Plant Biol 63:381–407

Woloshuk CP, Meulenhoff JS, Sela-Buurlage M, Van den Elzen PJ, Cornelissen BJ (1991) Pathogen-induced proteins with inhibitory activity toward *Phytophthora infestans*. Plant Cell 3:619–628

Wrzaczek M, Brosché M, Kangasjärvi J (2013) ROS signaling loops—production, perception, regulation. Curr Opin Plant Biol 16:575–582

Wu Y, Zhang D, Chu JY, Boyle P, Wang Y, Brindle ID, De Luca V, Després C (2012) The Arabidopsis NPR1 protein is a receptor for the plant defense hormone salicylic acid. Cell Rep 1:639–647

Xia Y, Suzuki H, Borevitz J, Blount J, Guo Z, Patel K, Dixon RA, Lamb C (2004) An extracellular aspartic protease functions in Arabidopsis disease resistance signaling. EMBO J 23:980–988

Xia XJ, Wang YJ, Zhou YH, Tao Y, Mao WH, Shi K, Asami T, Chen Z, Yu JQ (2009) Reactive oxygen species are involved in brassinosteroid-induced stress tolerance in cucumber. Plant Physiol 150:801–814

Xia XJ, Zhou YH, Ding J, Shi K, Asami T, Chen Z, Yu JQ (2011) Induction of systemic stress tolerance by brassinosteroid in *Cucumis sativus*. New Phytol 191:706–720

Xia XJ, Zhou YH, Shi K, Zhou J, Foyer CH, Yu JQ (2015) Interplay between reactive oxygen species and hormones in the control of plant development and stress tolerance. J Exp Bot 66:2839–2856

Xin XF, He SY (2013) *Pseudomonas syringae* pv. *tomato* DC3000: a model pathogen for probing disease susceptibility and hormone signaling in plants. Annu Rev Phytopathol 51:473–498

Yalpani N, León J, Lawton MA, Raskin I (1993) Pathway of salicylic acid biosynthesis in healthy and virus-inoculated tobacco. Plant Physiol 103:315–321

Yamaguchi Y, Huffaker A, Bryan AC, Tax FE, Ryan CA (2010) PEPR2 is a second receptor for the Pep1 and Pep2 peptides and contributes to defense responses in Arabidopsis. Plant Cell 22:508–522

Yan Z, Reddy MS, Ryu CM, McInroy JA, Wilson M, Kloepper JW (2002) Induced systemic protection against tomato late blight elicited by plant growth-promoting rhizobacteria. Phytopathology 92:1329–1333

Yuan Y, Chung JD, Fu X, Johnson VE, Ranjan P, Booth SL, Harding SA, Tsai CJ (2009) Alternative splicing and gene duplication differentially shaped the regulation of isochorismate synthase in Populus and Arabidopsis. Proc Natl Acad Sci 106:22020–22025

Zehnder GW, Murphy JF, Sikora EJ, Kloepper JW (2001) Application of rhizobacteria for induced resistance. Eur J Plant Pathol 107:39–50

Zhai Q, Zhang X, Wu F, Feng H, Deng L, Xu L, Li C (2015) Transcriptional mechanism of jasmonate receptor COI1-mediated delay of flowering time in Arabidopsis. Plant Cell 27:2814–2828

Zhang Y, Tessaro MJ, Lassner M, Li X (2003) Knockout analysis of Arabidopsis transcription factors TGA2, TGA5, and TGA6 reveals their redundant and essential roles in systemic acquired resistance. Plant Cell 15:2647–2653

Zheng XY, Spivey NW, Zeng W, Liu PP, Fu ZQ, Klessig DF, He SY, Dong X (2012) Coronatine promotes *Pseudomonas syringae* virulence in plants by activating a signaling cascade that inhibits salicylic acid accumulation. Cell Host Microbe 11:587–596

Chapter 3
Plant–Microbe Interaction: A Sustainable Strategy to Elevate Salinity Tolerance in Plants

Ajay Veer Singh, Amir Khan, and Manisha Joshi

Abstract A large portion (approx 20%) of world's agricultural land is facing salinity stress and it is continuously increasing due to the natural and anthropogenic activities Salinity is one of the most dangerous abiotic stresses and characterized as enhanced concentration of sodium, calcium, magnesium, etc., in the soil. Saline soil loses its inherent capacity to support plant growth due to imbalanced nutrient content, high pH, and reduced water potential. Therefore, most of the crops except halophytes are highly susceptible towards saline conditions. High saline stress results in physiological and biochemical alteration of cellular metabolism in plants that ends up with reduced crop yield. To meet the future demands of food for burgeoning human population and bioremediation of saline soil, researchers are trying to develop cost effective, efficient, and environmental sustainable techniques. Halotolerant microorganisms have also gained much attention in the last few decades due to their ability to confer salinity tolerance in plants via mechanisms like production of compatible solutes, antioxidants and ACC deaminase enzymes, plant hormones, VOCs, etc., along with plant growth promotion through nutrient mobilization and plant protection. Therefore, application of such potent halotolerant plant growth promoting microorganisms is a valuable, proficient, and eco-friendly approach under saline stress. Current manuscript presents a comprehensive understanding about the soil salinization and the mechanisms adopted by the plant associated microorganisms in order to improve salt tolerance and productivity of plants in a sustainable manner.

3.1 Introduction

In the present era, agriculture system is experiencing various challenges such as global warming, drought, salinity, and pathogen attack, etc., and these are major causes responsible for reduced plant growth and yield. Among all, soil salinity is of

A. V. Singh (✉) · A. Khan · M. Joshi
Department of Microbiology, College of Basic Sciences and Humanities, GBPUA&T, Pantnagar, Uttarakhand, India

A. Sharma (ed.), *Microbes and Signaling Biomolecules Against Plant Stress*, Rhizosphere Biology, https://doi.org/10.1007/978-981-15-7094-0_3

major concern due to its drastic impact on food production. It refers to the prevalence of high magnitude of soluble salts in soil or water by means of natural and anthropogenic processes. It enhances osmotic pressure around the root and causes hindrance in water uptake and creates nutrient imbalance in soil and lowers down water and nutrient availability in plants. Deficiency of water and nutrients interrupts plant metabolic machinery which is ultimately reflected as reduced plant growth and productivity. Around the world, about 832 Mha land is suffering from salt stress. Out of this about 398Mha land is experiencing saline stress, whereas 434 Mha is affected by sodicity (Dagar 2014). According to FAO and ITPS (2015), about 50% soil of Pakistan, China, India, and United States are salt affected. Moreover, every year about 1–2% of agricultural land is assaulted by soil salinization in arid and semiarid regions (FAO 2002). Plants, except halophytes are unable to grow or poorly grow in such environment because it adversely affects the biochemical pathways of plants, seed germination rate, plant vigour, and yield in terms of quantity and quality. Generally saline soil is characterized with an electrical conductivity (EC) from 4–$<$16 dS/m to $<$15 ESR, but majority of crops are susceptible for low salinity, even EC $>$ 1 dS/m can decline the growth and yield of rice plants. Food from these crops can cause undernourishment and malnutrition in people. Hence, in order to enhance global food production and achieving sustainability in agriculture, maintenance of nutrient stress is extremely vital. Developing strategies effective in mitigating negative effects of salt stress is underway. Currently researchers are trying to develop such effective methods through plant breeding practices and by applying chemical fertilizers. Hybridized salt tolerant crops are reported to have very little success rate and they also come along with many problems such as huge labour demand and intense money expenditure. On the other hand, application of chemical fertilizers on a large scale proves to be dangerous to plants in long run since excessive fertilizers become inaccessible to plants. Recently, an eco-friendly and sustainable method, i.e. the application of beneficial microorganisms in crops, has gained researchers interest to improve the plant growth and productivity. Plant growth promoting microorganisms mineralize inaccessible nutrients and make them accessible to plants. Besides, they also protect plants from pathogens; improve soil structure and composition, and biotic activities of microbes make the soil ecosystem dynamic. These microorganisms are also involved in the improvement of salinity tolerance in plants by means of various direct and indirect mechanisms. Present manuscript illustrates a comprehensive role of salt tolerant plant growth promoting microorganisms in the improvement of salinity tolerance in plants.

3.2 Causes of Soil Salinization

The term "salinity" is marked as the surplus concentration of salts such as sulphates of sodium, magnesium, calcium, etc., in soil or water. Saline or sodic soils are called salt affected. Soil salinity is caused by natural or anthropogenic processes and the outcome of this is a massing of dissolved salts in total available water present in the soil which ends up in hindering plant growth and development (Isayenkov and Maathuis 2019). Soil salinization can be of three types in accordance to its cause. These are primary salinity (termed as natural salinity), secondary salinity (referred as dry land or anthropogenic salinity), and tertiary salinity (termed as irrigation salinity). Natural salinity results from the accumulation of salts for prolonged time period through natural processes in soil or ground water. The natural processes include weathering of rocks and accumulation of oceanic salt carried by wind or rain water. Weathering of parental rocks causes emancipation of various soluble salts like sodium, chlorides, calcium, magnesium, carbonates, sulphates, etc. Wind contributes to soil salinity as it carries "the cyclic salts" which are deposited in agricultural land by rainfall. The concentration of salt deposition depends upon distance from the coast and average annual rainfall in a particular area. An increment in average annual rainfall can result in rapid charging of water table and brings ground water in close vicinity to soil surface and causes accumulation of salts near the upper surface of the soil (Shrivastava and Kumar 2015).

Secondary salinity, also called as anthropogenic or dry land salinity, is the consequence of human actions that affect hydrologic equilibrium of soil. Dry land salinity is a massing of salts in soil and ground water. It drastically affects soil health, water quality, vegetation, crops, and biodiversity. Industrialization and rigorous farming practices have caused eradication of perennial vegetation such as deep-rooted native trees and shrubs which results in more leakage of water in water table. Rise in water table brings salts close to root zone and soil surface. As the ground water comes in close vicinity to the soil surface, salt enters plant root zone and results in water logging. Moreover, surplus concentration of soil amendments such as chemical fertilizers also leads to soil salinization. Third is irrigation salinity, which happens due to the leakage of water from rainfall and irrigation. This causes ground water recharge that brings soil salts close to the soil surface. Recharge rates in irrigation salinity are much higher as compared to dry land salinity and cause soil salinization at much higher rates. Inefficient irrigation and drainage systems are the major cause of excess leakage which may increase the risk of salinity and water logging in agricultural lands. Moreover, irrigation of agriculture lands with saline water or sea water is also a major cause of soil salinization. Factors influencing irrigation salinity include soil type, climate, abundance of deep-rooted perennial vegetation, etc. Annual average rainfall and temperature of particular area also determines the salinity rates. Usage of shallow rooted vegetation in an area causes more infiltration of water into water table and leads to high soil salinization.

3.3 Effect of Salinity on Plant Growth

Salinity is a major abiotic stress which adversely affects plant physiology and causes reduced germination rate, reproductive development, yield, and quality. Generally, many cereal crops such as rice, maize, potato, tomato, etc., are highly susceptible for high salt concentrations. On the other hand, some plants like mangrove, marshes, ice

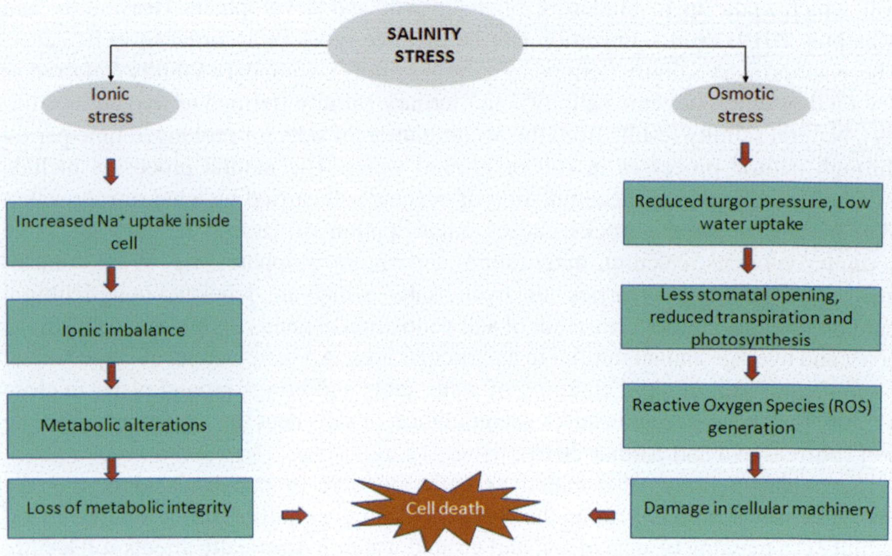

Fig. 3.1 Effect of salt stress on plant physiology and metabolism

plant, Sarcocornia, etc., able to grow under saline environment are termed as halophytes or salt tolerant plants. During saline stress, the osmotic pressure of soil solution is high in comparison to plant cells and causes problems in water uptake through root zone and hence influences plants at various growth stages (Warrence et al. 2002). Moreover, plant cells experiencing decreased turgor pressure, reduced stomata size discourage stomata opening in order to conserve water. Transpiration rate and photosynthesis rate of the plants get reduced and induce the production of reactive oxygen species (ROS) such as singlet oxygen, superoxide, etc. Reactive oxygen species (ROS) exert their effects by causing damage in nucleic acid, lipids, proteins, etc., and result in disruption of cellular machinery and ultimately cause cell death (Xie et al. 2019). An increment in soil salinity disturbs the osmotic balance of plant cells and creates difficulties for plants to imbibe nutrients from the soil. Presence of salts readily affects the hormonal balance and causes premature senescence in plants (Fig. 3.1). High salt toxicity can negatively affect plant growth by causing necrotic patches, defoliation, leaf burn, and leaf bronzing. Overabundance of some salts can disturb the ideal ionic ratio in the cell and may compete for essential salts required for plant growth and development (Numan et al. 2018). For example, the uptake of additional amount of sodium and calcium inside the cell can hinder the uptake of potassium and iron, which may lead to the disease occurrence, poor quality seed, and low yield.

3.4 Effect of Salinity on Soil Health

Soil health is referred as soil status to sustain the nutrients, biodiversity, and plant growth in an environment. Presence of higher concentration of sodium, calcium, magnesium ions in the soil may cause salinity and it is reflected in the form of white patches on the top layer of the soil. Moreover, electrical conductivity of the saturated extract of a soil determines the degree of salinity stress (Table 3.1). An elevation in soil pH (up to 82) is often observed as a result of salinity, which is linked with the inaccessibility of some nutrients such as P, Fe, Zn, etc., that are essentially required for the growth and development of plants. Therefore, saline soil loses its inherent capacity to support plant growth up to some extent and leads to the eradication of native vegetation. Salinity may result in the loss of ground cover and makes soil more prone to erosion. Increased concentration of sodium ion causes separation of soil aggregates or clay dispersion by accumulating in between them, which results in reduced permeability, porosity, hydraulic conductivity, infiltration rate, and stability of the soil (Liu et al. 2014). Soil salinization also demolishes microbial community of that soil and decreases soil fertility. Because of variations in the ability to tolerate salinity, a shift of microbial community as compared to non-saline soil is observed. Fungi tend to be more sensitive to salt stress as compared to bacteria, thus an increase in bacteria/fungi ratio occurs in saline soil. Microorganisms residing in the soil play a major role in re-cycling of soil elements which are crucial for plant growth and metabolism. Microbes are involved in the processes like phosphate and zinc mineralization, metal chelation, sulphur oxidation, and nitrogen fixation (Khan et al. 2019a). These processes result in decomposition of litter which in turn restore organic matter and essential inorganic elements in the soil. Therefore, shifting of microbial community due to the salinity stress can cause alteration in total organic matter, carbonic content, and concentration of available minerals in the soil.

3.5 Plant Adaptations to Overcome Salinity Stress

Salinity is a major biotic stress and limits the plant growth and productivity significantly. Some plants such as halophytes have adapted different mechanisms at morphological, physiological, biochemical, and molecular level to survive under saline conditions. Salinity stress has severe effect on electron transport chain (ETC) and shows disruption of normal functioning of ETC in mitochondria and chloroplast.

Table 3.1 Electrical conductivity of saline soil

Salinity class	Electrical conductivity (dS/m)
Normal soil	0–2
Poorly saline soil	2–4
Moderately saline soil	4–8
Extremely saline soil	8–16
Very extremely saline soil	>16

Disturbed ETC is the main cause of accumulation of reactive oxygen species (ROS), i.e. hydroxyl radical, hydrogen peroxide, singlet oxygen, and superoxide and cellular damages. Under saline conditions, plants stimulate the production of different antioxidant enzymes (peroxidases (POD) (e.g. glutathione peroxidases and ascorbate peroxidase), catalase (CAT), superoxide dismutase (SOD), glutathione reductase (GR) which scavenges over produced reactive oxygen species (ROS). Moreover, brassinosteroids (BRs) possess hormonal properties and stimulate the accumulation of antioxidant enzymes thereby aiding in plant survival under saline stress. Halophytic plants are also unable to bear high salts concentration inside the cell but they survive under such conditions through compartmentalization processes. In these processes, overloaded salt is either captured in older tissues or shifted into the vacuoles through Na^+/H^+ antiporters through vacuolar V-ATPases. Wang et al. (2001), reported upregulated expression of V-ATPase in a halophyte (*Suaeda salsa*) Moreover, in another halophytic plant (*Mesembryanthemum crystallinum*), a HKT (histidine kinase transporter) family transporter was identified which regulates the efflux of Na^+ and influx of K^+ to maintain the ion homeostasis in plants (Yen et al. 2000). Elevated level of nitric oxide is also a marker for salinity stress tolerance in plants because under saline stress it enhances the synthesis of H^+ ATPase which triggers efflux of excess sodium ions through Na^+/H^+ exchange. Plants also synthesize various osmolytes such as proline, glycine betaine, sugars, polyols (referred as compatible solutes) against the osmotic stress caused by salinity. Compatible solutes are soluble, uncharged, and polar in nature. The major function of osmolytes is the maintenance of turgor pressure of the cell through steady water influx. Proline is a widely accumulating osmolyte, therefore considered as one of the parameters for measurement of salinity tolerance in plants. Besides the osmotic adjustment, proline also serves as a nitrogen source and antioxidant enzyme. Up regulation of several stress responsive transcription factors such as WRKY, AP2, NAC, bZIP, and DREB is also reported under saline stress, which enhances transcription rate of the proteins involved in tolerance of salinity stress (Abbas et al. 2019).

3.6 Strategies to Elevate Salinity Stress Tolerance

Salinity drastically affects agriculture and reduces crop productivity. To conquer the effect of saline stress, researchers are practicing various techniques in order to improve salinity tolerance in plants. Some of the techniques are summarized below:

3.6.1 Breeding Techniques

In the present era, scientists are practicing several breeding and genetic engineering techniques to develop salt tolerant crop varieties to avoid drastic consequences of saline stress. Conventional breeding programme is the strategy to develop salt tolerant crops by cross-pollination of desired species. It involves two basic steps. The first step is to generate a breeding population that is highly impervious for salt tolerance and the second involves selection among the segregating progeny for the individuals that combine with most of the useful traits of the parent, i.e. high degree of salt tolerance. There are two important parameters (threshold and slope) to measure the salinity tolerance in plants. Threshold indicates the highest permissible salt level that does not affect crop production. Slope indicates percent reduction in yield per unit rise in salt level above the threshold (Hoang et al. 2016). Among the molecular breeding techniques, quantitative trait locus (QTL) is the major contributing factor to salt tolerance in crops. A major QTL designed "Saltol" was mapped on chromosome 1 of rice by using a population, generated from a cross between a sensitive variety IR29 and a salt tolerant rice, i.e. Pokkali (Waziri et al. 2016). Salt tolerance is a complex trait and the QTLs related with salt tolerance have a significant role in understanding the stress-response and to generate stress-tolerant plants. With a better understanding of the mechanisms and genetics of salt tolerance, more precise breeding approaches have been used for developing higher salt tolerance plant varieties. In order to develop salt resistance GMOs, in depth knowledge of stress responsive genes is required which may also lead to the transformation of responsible operons from salt tolerant varieties to salt sensitive plant varieties. Number of salt tolerant varieties has been released but their progress is slow. These varieties are not much successful as only a few responsible traits are considered in breeding and genetic engineering.

3.6.2 Microbial Mediated Amelioration of Salinity Tolerance

Surplus saline condition of a soil reduces its fertility and plant growth by altering the availability of the nutrients and osmotic balance. Reduced productivity and quality of the crops limit the food quantity and cause inadequacy of food. They are major reasons and found responsible for decline in the market and economic values of crops. To conquer the problem, it is mandatory to recover the immobilized nutrients in saline soil which foster the salinity tolerance in plants. Role of plant growth promoting microorganisms such as bacteria, fungi, and actinomycetes has been well established in improving the growth, development, protection, and yield of plants by improving nutrient utilization ability, phytohormone production, reducing disease incidences, decomposing organic matter, membrane transporter alteration, and also by conferring abiotic stress tolerance (Fig. 3.2). Plant associated microorganisms are very diverse and dominated by plant growth promoting rhizobacteria (PGPR) and endophytes. Their composition varies according to the nature of plant exudates, native soil, and environmental conditions. It is well known that along with plants many microorganisms are also not able to grow in high salt concentration. Elevated level of salt ions in soil not only decreases soil microbial activity but also alters microbial community composition in native soil (Van Horn et al. 2014). The microbes able to survive under saline conditions are called halotolerant, whereas

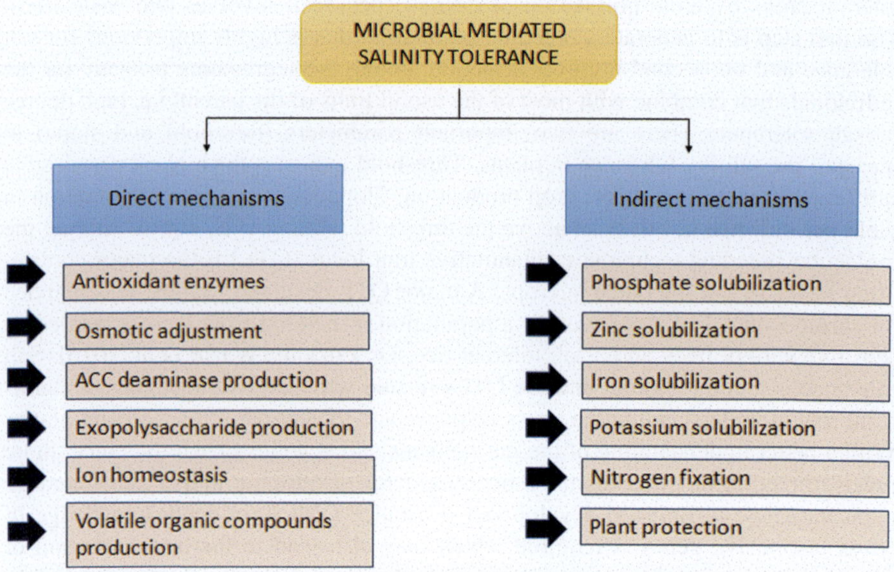

Fig. 3.2 Mechanisms of halotolerant PGPRs to confer salinity tolerance in plants

other microorganisms that require high salt concentration for their growth are called halophiles. According to preceding information, various symbiotic and free living microorganisms associated with diverse genera such as *Achromobacter, Azospirillum, Bacillus, Burkholderia, Enterobacter, Glomus, Klebsiella, Microbacterium, Rhizobium, Pseudomonas, Pantoea, Paenibacillus, and Variovorax,* etc., confer the tolerance to host plants during abiotic stresses (Abbas et al. 2019). These halophilic and halotolerant microorganisms reside around and inside the root and can foster salinity tolerance in plants to improve growth and development by lowering the negative upshot of saline stress via several direct and indirect mechanisms. Indirect mechanisms include phosphate solubilization, siderophore production, nitrogen fixation, zinc solubilization, etc., whereas direct mechanisms include production of exopolysaccharides, volatile organic compounds, antioxidants, ACC deaminase activity, ion homeostasis, osmotic balance, and induced systemic resistance (Fig. 3.2).

3.6.3 Direct Mechanisms of Salinity Stress Alleviation

Salinity tolerance prevails when plants maintain their metabolism even under high salt concentration, i.e. balance of Na^+ and K^+ ion, less reactive oxygen species (ROS) production, and maintenance of cell turgidity, etc. Many saline stress adaptive microorganisms are found around the root and release various substances that serve as signals to maintain plant metabolism. They maintain ion balance, rate of

ROS production, and ultimately improve root, shoot growth, and productivity under saline conditions.

3.6.4 Production of Antioxidant Enzymes

A plant continuously produces reactive oxygen species (ROS) during normal cellular metabolic processes. Main reactive oxygen species are hydroxyl radicals, hydrogen peroxide, and superoxide anionic radicals. Production rate of ROS becomes very high under stress of salinity, drought, and heat and which may lead to DNA damage, alteration in fatty acids structure, denaturation of membrane bound proteins, modulation in redox status, oxidative damage of pigments, etc. However, other biomolecules reduce membrane fluidity, change protein confirmation, damage the enzymatic activity, cell homeostasis and cause cell death (Sharma et al. 2016). Hydrogen peroxide (H_2O_2) stimulates ethylene synthesis in plants which leads to early senescence in plants. Many researchers have reported that seed bacterization with plant growth promoting microbes can efficiently enhance production of antioxidant enzymes (enzymatic antioxidants such as mono dehydroascorbate reductase, glutathione reductase (GR), superoxide dismutase (SOD), and non-enzymatic antioxidants like ascorbate tocopherols and cysteine) which can minimize the magnitude of ROS and reduce oxidative injury up to some extent and thus foster salinity tolerance in plants (Table 3.2) Recently, Feng et al. (2019) reported improvement

Table 3.2 Role of halotolerant microorganisms in salinity tolerance

Microorganisms	Plant	Mode of action	Reference
Bacillus aryabhattai H19-1 *B. mesonae* H20-5	Tomato	High level of proline, ABA, antioxidant enzyme and upregulation of NCED1 and AREB1 proteins	Yoo et al. (2019)
Streptomyces venezuelae	Jasmine rice	Production of ACC deaminase, antioxidant enzymes, proline, MDA	Yoolong et al. (2019)
Arthrobacter woluwensis AK1	Soybean	Production of phytohormones, siderophore, organic acids, induced expression of potassium, chloride, and Na^+/H^+ channel	Khan et al. (2019b)
Glutamicibacter halophytocola	*Limonium sinense*	Antioxidant production, compatible solute accumulation	Qin et al. (2018)
Bacillus cereus, Bacillus aerius	Safflower	Production of ACC deaminase, IAA	Hemida and Reyad (2018)
P. Yonginensis DCY84^T	*Panax ginseng*	Production of Siderophore, IAA, organic acids, antioxidant enzymes, proline accumulation	Sukweenadhi et al. (2018)
Pseudomonas frederiksbergensis OS261	Pepper	Reduced ethylene production, production of antioxidants	Chatterjee et al. (2017)
Bacillus aquimaris DY-3	Maize	Accumulation of compatible solutes, antioxidant enzymes, IAA	Li and Jiang (2017)

in saline tolerance in maize (bacterized with *Bacillus mojavensis* and *Rhodopseudomonas palustris*) with the elevation in antioxidant enzymes (Ascorbate peroxidase (APX), catalase (CAT)) under saline stress conditions. Sukweenadhi et al. (2015) observed induced expression of salinity responsive gene such as *AtRSA1* which is involved in degradation of ROS upon priming with *Paenibacillus yonginensis* in Arabidopsis seedlings. Several microorganisms belonging to genera like *Lactobacillus, Serratia, Pseudomonas, Enterococcus, Bacillus, Micrococcus, Streptomyces*, etc., having their own antioxidant enzymatic defence system are involved in eradication of ROS (Abbas et al. 2019).

3.6.5 Osmotic Adjustment

Some bacteria produce compatible solutes for maintenance of their cellular turgidity to govern regular cell functions smoothly. This is chief cellular machinery that lessens the osmotic stress in plants produced because of salinity. In saline stress, rhizobacteria accumulate compatible osmolytes such as glycine betaine and proline to assist plant growth. Proline is one of the prime osmolytes that accumulates in the plants after protein hydrolysis to lower down the osmotic stress (Krasensky and Jonak 2012). It plays multifunctional responsibilities in terms of decreasing lipid peroxidation, regulating cytosolic acidity, proteins maintenance, and scavenging of reactive oxygen species (Ilangumaran and Smith 2017). Trehalose, a compatible solute produced by some plant functions as an osmoprotectant under abiotic stress conditions, viz. osmotic stress, drought, salinity, and high temperature (Pan et al. 2019). Presence of glycine and betaine as compatible solutes in plants and microbes not only makes plants defensive against several abiotic stresses, even act as shield by stabilizing highly ordered structure of membrane and quaternary structure of proteins. Several salt tolerant microorganisms have extensive genes to endure under saline stress including Na^+/H^+ antiporter genes, etc. These genes are necessary for Na^+ detoxification and cell maintenance of cells, formation of Na^+ electrochemical gradient, adjustment of cell volume, cellular pH, and homeostasis (Abbas et al. 2019).

3.6.6 Production of ACC Deaminase

Ethylene, a gaseous plant hormone, also called a senescence hormone contributes significantly in diverse developmental processes of the plants. Under various biotic and abiotic stress conditions, production of ethylene in plants is very high which governs various activities in the plants to maintain ion homeostasis inside the cell and reduces root and shoot length ratio in plant. Elevated quantity of ethylene pessimistically affects plant physiology, reduces shoot growth, damages cell, induces early senescence, and ultimately condenses production rate. Some plants associated with microorganisms produce a unique enzyme called ACC deaminase, which catalyses the transformation of ACC (precursor for ethylene biosynthesis) into ammonia and α-ketobutyrate and suppresses the production of endogenous ethylene (Dodd and Pérez-Alfocea 2012). ACC deaminase production by microorganisms can reduce the adverse effect of ethylene and also confers the tolerance for salinity and drought stress (Bharti and Barnawal 2019). Till now, a number of ACC deaminase producing PGPR has been documented by various researchers which reduce salinity and induce shoot growth. Gupta and Pandey (2019) observed augmented root and shoot length, leaf ACC deaminase enzyme unit upon exposure of two ACC deaminase producing strains (*Aneurinibacillus* and *Paenibacillus*) in French bean under saline conditions. They correlated improved tolerance and productivity of French bean with ACC deaminase production and various metal solubilization. Similarly, enhanced salinity tolerance was also observed in tomato plants upon inoculation with ACC deaminase and IAA producing *Leclercia adecarboxylata* (Kang et al. 2019).

3.6.7 Exopolysaccharide Production

Various expolysaccharides (extrapolymeric substances) are secreted by different microorganisms around the cell. On the basis of their biological functions they can be categorized into three types, viz. capsular polysaccharide (involved in pathogenesis), extracellular polysaccharide (involved in biofilm formation and pathogenesis), and storage polysaccharide. They govern cell to cell communication, assist in adherence on a solid support, and protect cells from desiccation and phagocytosis (Etesami and Beattie 2018). EPS producing microorganisms have potential to reduce the negative effects of salinity via formation of rhizosheath, in which EPS traps Na^+ ions present around the root via hydroxyl, sulfhydryl, phosphoryl, and carboxyl groups and prevents ionic movement inside the plants and therefore reduces Na^+ ion toxicity. Paul and Lade (2014) have documented the role of plant growth promoting rhizobacteria such as *Bacillus* sp. and *Pseudomonas* sp. in salinity stress alleviation via production of EPS. Different environmental conditions (salinity, drought, C/N ratio, oxygen rate, and pH) can alter the composition of EPS and can induce the production rate of EPS. Under high salinity level bacteria generally produce uronic acid and sulphates at a higher rate. It is well known fact that under saline conditions water availability declines and leads to desertification. Production of EPS by microbes imparts gluing property in the soil and improves soil aggregation, nutrient stabilization and retains moisture in soil as it can bind with water multiple times than its weight. According to Robertson and Firestone (1992), bacterial EPS can improve moisture retaining capacity of a sandy soil in comparison to sandy soil without EPS. Presence of high water content in soil was illustrated in various researches when inoculated with *Pseudomonas* sp., which occurred due the production of EPS (Chu et al. 2019). Moreover, Jalili et al. (2009) also confirmed the efficacy of EPS producing PGPR to confer salinity tolerance in plants under different salinity stress.

3.6.8 Ion Homeostasis

Saline condition leads to distortion of ion milieu in plants due to increased uptake of Na^+ ions and slighter uptake of K^+ ions which causes various changes in physiological and biochemical pathways of plants. Na^+ toxicity can be reduced by posing hindrance in Na^+ ions uptake and also by emitting excess Na^+ ions. Similar strategy has been adapted by salt resistant halophytes, which expel Na^+ ions from the leaves and avoid toxic effect of salinity in the plants (Horie et al. 2012). Therefore, any contribution of microbes in the hindrance of Na^+ uptake and its induced expel from the plant parts can improve salinity tolerance. Application of potential PGPRs can alter Na^+ ions uptake and improves nutrient mobilization via governing ion transporter expression regulation. It hinders uptake of Na^+ ions and improves the accumulation of selective K^+ ion in the plant cell, therefore, maintains K^+/Na^+ ratio, which is beneficial for accelerating plant growth and salinity tolerance (Qin et al. 2016). Abbas et al. (2019) documented that *Glomus* (arbuscular mycorrhizal fungi) induces selective upliftment of K^+, Ca^{2+}, and Mg^{2+} and evades Na^+ uptake to enhance K^+/Na^+ ratio. Similarly, Zhang et al. (2008) also reported reduced Na^+ influx upon priming with *Bacillus subtilis* with *Arabidopsis thaliana* under salt stress conditions by down regulating the expression of HKT1/K^+ transporter. Therefore, it can be concluded that inoculation of PGPRs can improve ion milieu of plant cell and tolerance capacity and productivity of crop.

3.6.9 Volatile Compounds

Volatile compounds are low molecular weight organic compounds and have high vapour pressure at room temperature. Their composition and quantity may alter according to the environmental conditions. Concentration of volatile compounds is low under normal conditions and becomes induced under stress conditions. Induced expression of volatiles seems to be a protective mechanism against stress. Plant inhabiting microorganisms generally produce lipophilic VOCs which are slightly water soluble. These volatiles assist in cell to cell communication in order to promote growth. Besides, they induce choline, glycine betaine biosynthesis which takes part in osmotic adjustment under saline conditions. Accumulation of such osmoprotectants leads to rise in osmotic pressure of the cell thereby prevents water loss. Moreover, these VOCs also hinder Na^+ influx inside the plants and improve nutrient mobilization. A number of bacteria such as *Bacillus amyloliquefaciens, Pseudomonas chlororaphis, Pseudomonas simian*, and *Bacillus subtilis* have been documented to produce VOCs that involve in the salinity tolerance improvement and plant growth promotion (Timmusk et al. 2014; Liu and Zhang 2015). Bacterial VOCs do not directly interact with plant hormone or other growth stimuli but regulate auxin homeostasis and nutrient influx through roots (Zhang et al. 2008). Plant VOCs also assist in the plant protection from pathogens; improve biomass and other stress tolerance capacity of the plants. Cho et al. (2013) reported 2, 3-butanediol (VOC) mediated stress tolerance in terms of reduced water loss and high stomatal conductance in plants inoculated with *P chlororaphis* O6 under abiotic stress.

3.6.10 Indirect Mechanisms of Salinity Stress Alleviation

Under high salinity, soil pH becomes basic (pH ~82) wherein, most of the freely available macro and micro nutrients form complexes with other compounds. These compounds are not available to the plants and thereby negatively affect plant growth and yield in terms of the nutrition value and quality of grains. Different types of microorganisms carrying different functions are found in the rhizosphere which mobilizes nutrients through various processes. These processes including phosphorous solubilization, nitrogen fixation, iron chelation, and potassium mobilization take place via producing organic acids and secretion of cation chelators.

3.6.11 Phosphorous Solubilization

Phosphorous (P) is an important element for plant because it is involved in all the major pathways of a plant cell. It is abundantly present in the soil and found in both inorganic and organic forms About 005% of soil is comprised of P, but only 01% part of this fraction is available for plants as P is poorly soluble at neutral pH and forms insoluble complexes (Etesami and Beattie 2018). Besides, saline soil possesses high pH which enhances the complex formation capacity of P and thereby making P inaccessible to the plants. Plants influx only monobasic ($H_2PO_4^-$) and diabasic (HPO_4^{2-}) forms of phosphate (Parveen et al. 2018). This leads to a reduction in plant growth and phosphate deficiency induces disease incidences in plants. To overcome such problems, farmers generally apply NPK fertilizers but a very small amount of fertilizer is absorbed by plants and remaining fertilizer is converted into insoluble form. It is also one of the reasons for soil salinity. In this regard, application of halotolerant phosphate solubilizing PGPRs can be an effective approach to solubilize P complexes. These PGPRs solubilize P by acidification, by producing low molecular weight organic acids and by chelation and ion exchange. Whereas, to mineralize the organic phosphorous complexes bacteria secret various phytases, phosphatases (Singh and Shah 2013; Singh et al. 2018). Numerous bacteria isolated from saline environment and belonging to the genera *Arthrobacter, Azospirillum, Bacillus, Phyllobacterium*, and *Vibrio* have been documented to solubilize various phosphorous salts such as $Ca_3(PO_4)_2$, $AlPO_4$, and $FePO_4$ even under saline stress conditions (Abbas et al. 2019). Mayak et al. (2004) demonstrated that inoculation of *Achromobacter piechaudii* with *Solanum lycopersicum* under saline stress showed improvement in plant phosphorous content. Similarly, improved P content was also observed in a field trial of various crops when plants were inoculated with PGPR (Singh et al. 2013: Upadhyay and Singh 2015). Irfan et al. (2019) also communicated that instead of single bacterium inoculation, consortia (multiple bacteria) of phosphate solubilizing microorganisms (PSB) are more effective to solubilize the insoluble complexes of phosphorous under saline stress.

3.6.12 Siderophore Production

Iron is another one of the important nutrients which played a significant role in plant respiration and N_2 fixation by acting as an essential component of different enzyme complex. Iron is generally found in Fe^{2+} and Fe^{3+} forms but plants can utilize only Fe^{2+} form. Availability of iron in calcareous and saline soil is very low because most of the Fe^{2+} gets converted into Fe^{3+} form due to high pH of soil where it forms complexes such as hydroxides and oxyhydroxides and becomes inaccessible to plants (Upadhayay et al. 2018). This iron deficiency under saline stress drastically reduces plant growth and causes diseases in plants. Diverse microbial community inhabiting in rhizosphere and bulk soil possesses the ability to capture Fe^{3+} form via secretion of various type of siderophores. Bacteria sometimes convert iron-siderophore complexes into free Fe^{2+} form which is then sequestered subsequently into plants roots and assist in plant metabolic functioning. Rhizospheric region of the plants is highly capable to take up ferric iron complex due to these siderophores, and thereby improves iron availability in plants and enhances plant growth and yield even under saline stress conditions. Moreover, iron sequestration ability also assists in plant protection via competition of iron with pathogen. Yasin et al. (2018) documented that siderophore producing halotolerant *Bacillus* sp., inoculated with *Capsicum annuum* showed improved tolerance towards the saline stress. Similarly, improved salinity stress tolerance was also documented in wheat seedlings, inoculated with siderophore producing halotolerant PGPR strains (*Hallobacillus* sp. SL3 and *Bacillus halodenitrificans* PU62) (Ramadoss et al. 2013). Therefore, it can be concluded that siderophore secretion can improve iron availability in plants and improves plant growth and yield.

3.6.13 Nitrogen Fixation

Under saline condition, soil shows depletion of various nutrients; therefore, partial availability, competitive influx, and transporters may cause physiological imbalance in metabolic pathways of plants. Nitrogen is a crucial element for the health of plants and can limit plant productivity. It accounts for more than 80% of minerals absorbed by plants. Around 78% N_2 is present in the atmosphere but it is inaccessible for direct use of plants. It can be converted into usable forms, i.e. nitrite and nitrate by physical, chemical, and biological mechanisms. One third of nitrogen is fixed by chemical and physical processes, whereas remaining is converted by biological means called as biological nitrogen fixation. Nitrogenase is found in nitrogen fixing microbes which converts atmospheric nitrogen into nitrate that can be directly utilized by plants. Microbes fix nitrogen in free living form as well as in association with plants. Free living nitrogen fixing microorganisms belong to genera *Azotobacter, Bacillus, Pseudomonas*, whereas associative nitrogen fixing such as *Rhizobium* and *Bradyrhizobium* form symbiotic association with legume plants and form nodules to fix nitrogen. Salinity generally reduces the symbiosis and also hinders the nodule formation in legumes. But salt tolerant PGPRs associated with halophytes and nonhalophytes are very crucial source for availability of nitrogen in saline soil (Sharma et al. 2016). A research by Navarro-Noya et al. (2016) concluded that even under extreme salinity and high pH, PGPR strains *Paenibacillus fujiensis* and *P. durus* act as excellent nitrogen fixers. Metwali et al. (2015) documented improvement in plant growth and yield of faba beans under saline stress upon inoculation with plant growth promoting *Pseudomonas putida, P. fluorescens and Bacillus subtilis*.

3.7 Conclusion

Sustainable agriculture system refers to the maintenance and improvement of agriculture to meet up the future demands and environment protection in an efficient and cost effective manner. In the present global situation, salinity is causing drastic effects on the quality of agricultural soil, biodiversity, environment, plant growth, and productivity. Moreover, continuous conversion of agricultural land into saline area is also an issue of huge concern. Plants possess several defence mechanisms against mild salinity stress but they are not able to cope up under high saline conditions. Therefore, world is demanding an ecologically compatible, cost effective, and efficient technique to overcome the said problems. Inoculation of potential halotolerant plant growth promoting microorganisms is a sustainable and effective alternative technique as compared to other traditional agricultural methods owing to the surviving ability of PGPRs under saline environment. PGPRs associated with plants recover saline soil, improve plant growth and vigour along with alleviation of negative effects caused by saline stress by means of various mechanisms such as osmotolerance, antioxidant metabolism, ion transporter regulation, production of VOCs and EPS, and nutrient mobilization. Therefore, present manuscript concludes that the application of potential halotolerant and halophilic PGPRs and their consortium may be a promising technique to conquer the future food demands, bioremediation of saline soil, and maintenance of environmental sustainability under saline stress.

References

Abbas R, Rasul S, Aslam K, Baber M, Shahid M, Mubeen F, Naqqash T (2019) Halotolerant PGPR: a hope for cultivation of saline soils. J King Saud Univ Sci 31:1195–1201

Bharti N, Barnawal D (2019) Amelioration of salinity stress by PGPR: ACC deaminase and ROS scavenging enzymes activity. In: PGPR amelioration in sustainable agriculture. Elsevier, Woodhead Publishing, Sawston, pp 85–106

Chatterjee P, Samaddar S, Anandham R, Kang Y, Kim K, Selvakumar G, Sa T (2017) Beneficial soil bacterium *Pseudomonas frederiksbergensis* OS261 augments salt tolerance and promotes red pepper plant growth. Front Plant Sci 8:705

Cho SM, Kim YH, Anderson AJ, Kim YC (2013) Nitric oxide and hydrogen peroxide production are involved in systemic drought tolerance induced by 2R, 3R-butanediol in *Arabidopsis thaliana*. Plant Pathol J 29:427–434

Chu TN, Tran BTH, Hoang MTT (2019) Plant growth-promoting rhizobacterium *Pseudomonas* PS01 induces salt tolerance in *Arabidopsis thaliana*. BMC Res Notes 12:11

Dagar JC (2014) Greening salty and waterlogged lands through agroforestry systems for livelihood security and better environment. In: Dagar JC, Singh AK, Arunachalam A (eds) Agroforestry systems in India: livelihood security & ecosystem services. Springer, New Delhi, pp 273–332

Dodd IC, Pérez-Alfocea F (2012) Microbial amelioration of crop salinity stress. J Exp Bot 63:3415–3428

Etesami H, Beattie GA (2018) Mining halophytes for plant growth-promoting halotolerant bacteria to enhance the salinity tolerance of non-halophytic crops. Front Microbiol 9:148

FAO, ITPS (2015) Status of the world's soil resources (SWSR) Main report. Food and Agriculture organization of the United Nations and Intergovernmental Technical Panel on Soil, Rome

Feng K, Cai Z, Ding T, Yan H, Liu X, Zhang Z (2019) Effects of potassium-solubulizing and photosynthetic bacteria on tolerance to salt stress in maize. J Appl Microbiol 126:1530–1540

Food and Agriculture Organization of the United Nations (FAO) (2002) Crops and drops: making the best use of water for agriculture. FAO, Rome

Gupta S, Pandey S (2019) ACC deaminase producing bacteria with multifarious plant growth promoting traits alleviates salinity stress in French bean (*Phaseolus vulgaris*) plants. Front Microbiol 10:1506

Hemida KA, Reyad AM (2018) Improvement salt tolerance of safflower plants by endophytic bacteria. J Hortic Plant Res 5:38–56

Hoang T, Tran T, Nguyen T, Williams B, Wurm P, Bellairs S, Mundree S (2016) Improvement of salinity stress tolerance in rice: challenges and opportunities. Agronomy 6:54

Horie T, Karahara I, Katsuhara M (2012) Salinity tolerance mechanisms in glycophytes: an overview with the central focus on rice plants. Rice 5:11

Ilangumaran G, Smith DL (2017) Plant growth promoting rhizobacteria in amelioration of salinity stress: a systems biology perspective. Front Plant Sci 8:1768

Irfan M, Zahir ZA, Asghar HN, Khan MY, Tanvir H (2019) Effect of multi-strain bacterial inoculation with different carriers on growth and yield of maize under saline conditions. Int J Agric Biol 22:1407–1414

Isayenkov SV, Maathuis FJ (2019) Plant salinity stress: many unanswered questions remain. Front Plant Sci 10:80

Jalili F, Khavazi K, Pazira E, Nejati A, Rahmani HA, Sadaghiani HR (2009) Isolation and characterization of ACC deaminase-producing fluorescent pseudomonads, to alleviate salinity stress on canola (*Brassica napus* L) growth. J Plant Physiol 166:667–674

Kang SM, Shahzad R, Bilal S, Khan AL, Park YG, Lee KE, Asaf S, Khan MA, Lee IJ (2019) Indole-3-acetic-acid and ACC deaminase producing *Leclercia adecarboxylata* MO1 improves *Solanum lycopersicum* L growth and salinity stress tolerance by endogenous secondary metabolites regulation. BMC Microbiol 19:80

Khan A, Singh J, Upadhayay VK, Singh AV (2019a) Microbial biofortification: a green technology through plant growth promoting microorganisms. In: Shah S, Venkatramanan V, Prasad R (eds)

Sustainable green technologies for environmental management. Springer, Singapore, pp 255–269

Khan MA, Asaf S, Khan AL, Jan R, Kang S, Kim KM, Lee IJ (2019b) Rhizobacteria AK1 remediates the toxic effects of salinity stress via regulation of endogenous phytohormones and gene expression in soybean. Biochem J 476:2393–2409

Krasensky J, Jonak C (2012) Drought, salt, and temperature stress-induced metabolic rearrangements and regulatory networks. J Exp Bot 63:1593–1608

Li HQ, Jiang XW (2017) Inoculation with plant growth-promoting bacteria (PGPB) improves salt tolerance of maize seedling. Russian J Plant Physiol 64:235–241

Liu XM, Zhang H (2015) The effects of bacterial volatile emissions on plant abiotic stress tolerance. Front Plant Sci 6:774

Liu D, She D, Yu SE, Shao G, Chen D (2014) Predicted infiltration for sodic/saline soils from reclaimed coastal areas: sensitivity to model parameters. Sci World J 2014:1–12

Mayak S, Tirosh T, Glick BR (2004) Plant growth-promoting bacteria that confer resistance to water stress in tomatoes and peppers. Plant Sci 166:525–530

Metwali EM, Abdelmoneim TS, Bakheit MA, Kadasa NM (2015) Alleviation of salinity stress in faba bean (*Vicia faba* L.) plants by inoculation with plant growth promoting rhizobacteria (PGPR). Plant Omics 8:449

Navarro-Noya YE, Luna-Guido M, Dendooven L (2016) Cultivable nitrogen fixing bacteria from extremely alkaline-saline soils. Adv Microbiol 6:412

Numan M, Bashir S, Khan Y, Mumtaz R, Shinwari ZK, Khan A, Khan A, Ahmed AH (2018) Plant growth promoting bacteria as an alternative strategy for salt tolerance in plants: a review. Microbiol Res 209:21–32

Pan J, Peng F, Xue X, You Q, Zhang W, Wang T, Huang C (2019) The growth promotion of two salt-tolerant plant groups with PGPR inoculation: a meta-analysis. Sustainability 11:378

Parveen H, Singh AV, Khan A, Prasad B, Pareek N (2018) Influence of plant growth promoting rhizobacteria on seed germination and seedling vigour of green gram. Int J Chem Stud 6:611–618

Paul D, Lade H (2014) Plant-growth-promoting rhizobacteria to improve crop growth in saline soils: a review. Agron Sustain Dev 34:737–752

Qin Y, Druzhinina IS, Pan X, Yuan Z (2016) Microbially mediated plant salt tolerance and microbiome-based solutions for saline agriculture. Biotechnol Adv 34:1245–1259

Qin S, Feng WW, Zhang YJ, Wang TT, Xiong YW, Xing K (2018) Diversity of bacterial microbiota of coastal halophyte *Limonium sinense* and amelioration of salinity stress damage by symbiotic plant growth-promoting actinobacterium *Glutamicibacter halophytocola* KLBMP 5180. Appl Environ Microbiol 84:e01533-18

Ramadoss D, Lakkineni VK, Bose P, Ali S, Annapurna K (2013) Mitigation of salt stress in wheat seedlings by halotolerant bacteria isolated from saline habitats. SpringerPlus 2:6

Robertson EB, Firestone MK (1992) Relationship between desiccation and exopolysaccharide production in soil *Pseudomonas* sp. Appl Environ Microbiol 58:1284–1291

Sharma S, Kulkarni J, Jha B (2016) Halotolerant rhizobacteria promote growth and enhance salinity tolerance in peanut. Front Microbiol 7:1600

Shrivastava P, Kumar R (2015) Soil salinity: a serious environmental issue and plant growth promoting bacteria as one of the tools for its alleviation Saud. J Biol Sci 22:123–131

Singh AV, Shah S (2013) Plant growth-promoting rhizobacteria for sustainable agriculture. In: Kumar S, Prasad B (eds) Modern technologies for sustainable agriculture. New India Publishing Agency, New Delhi, pp 151–168

Singh AV, Chandra R, Reeta G (2013) Phosphate solubilization by *Chryseobacterium* sp and their combined effect with N and P fertilizers on plant growth promotion. Arch Agron Soil Sci 59:641–651

Singh AV, Prasad B, Goel R (2018) Plant growth promoting efficiency of phosphate solubilizing Chryseobacterium sp PSR 10 with different doses of N and P fertilizers on lentil (*Lens culinaris* var. PL-5) growth and yield. Int J Curr Microbiol Appl Sci 7:2280–2289

Sukweenadhi J, Kim YJ, Choi ES, Koh SC, Lee SW, Kim YJ, Yang DC (2015) *Paenibacillus yonginensis* DCY84T induces changes in *Arabidopsis thaliana* gene expression against aluminum, drought, and salt stress. Microbiol Res 172:7–15

Sukweenadhi J, Balusamy SR, Kim YJ, Lee CH, Kim YJ, Koh SC, Yang DC (2018) A growth promoting bacteria, *Paenibacillus yonginensis* DCY84T enhanced salt stress tolerance by activating defense-related systems in *Panax ginseng*. Front Plant Sci 9:813

Timmusk S, El-Daim IA, Copolovici L, Tanilas T, Kännaste A, Behers L, Nevo E, Seisenbaeva G, Stenström E, Niinemets U (2014) Drought-tolerance of wheat improved by rhizosphere bacteria from harsh environments: enhanced biomass production and reduced emissions of stress volatiles. PLoS One 9:e96086

Upadhayay VK, Singh AV, Pareek N (2018) An insight in decoding the multifarious and splendid role of microorganisms in crop biofortification. Int J Curr Microbiol Appl Sci 7:2407–2418

Upadhyay SK, Singh DP (2015) Effect of salt-tolerant plant growth-promoting rhizobacteria on wheat plants and soil health in a saline environment. Plant Biol 17:288–293

Van Horn DJ, Okie JG, Buelow HN, Gooseff MN, Barrett JE, Takacs-Vesbach CD (2014) Soil microbial responses to increased moisture and organic resources along a salinity gradient in a polar desert. Appl Environ Microbiol 80:3034–3043

Wang B, Luttge U, Ratajczak R (2001) Effects of salt treatment and osmotic stress on V-ATPase and V-PPase in leaves of the halophyte *Suaeda salsa*. J Exp Bot 52:2355–2365

Warrence NJ, Bauder JW, Pearson KE (2002) Basics of salinity and sodicity effects on soil physical properties Department of Land Resources and Environmental Sciences. Montana State University, Bozeman, MT, pp 1–29

Waziri A, Kumar P, Purty RS (2016) Saltol QTL and their role in salinity tolerance in rice. Austin J Biotechnol Bioeng 3:1067

Xie X, He Z, Chen N, Tang Z, Wang Q, Cai Y (2019) The roles of environmental factors in regulation of oxidative stress in plant. Biomed Res Int 2019:9732325. https://doi.org/10.1155/2019/9732325

Yasin NA, Akram W, Khan WU, Ahmad SR, Ahmad A, Ali A (2018) Halotolerant plant-growth promoting rhizobacteria modulate gene expression and osmolyte production to improve salinity tolerance and growth in *Capsicum annum* L. Environ Sci Pollut Res 25:23236–23250

Yen HE, Wu SM, Hung YH, Yen SK (2000) Isolation of 3 salt-induced low-abundance cDNAs from light-grown callus of *Mesembryanthemum crystallinum* by suppression subtractive hybridization. Physiol Plant 110:402–409

Yoo SJ, Weon HY, Song J, Sang MK (2019) Induced tolerance to salinity stress by halotolerant bacteria *Bacillus aryabhattai* H19-1 and *B mesonae* H20-5 in tomato plants. J Microbiol Biotechnol 29:1124–1136

Yoolong S, Kruasuwan W, Phạm HT, Jaemsaeng R, Jantasuriyarat C, Thamchaipenet A (2019) Modulation of salt tolerance in Thai jasmine rice (*Oryza sativa* L cv KDML105) by *Streptomyces venezuelae* ATCC 10712 expressing ACC deaminase. Sci Rep 9:127

Zhang H, Kim MS, Sun Y, Dowd SE, Shi H, Paré PW (2008) Soil bacteria confer plant salt tolerance by tissue-specific regulation of the sodium transporter HKT1. Mol Plant-Microbe Interact 21:737–744

Chapter 4
Concepts and Application of Plant–Microbe Interaction in Remediation of Heavy Metals

Geeta Bhandari and Pankaj Bhatt

Abstract Level of anthropogenic chemicals is increasing day by day in the environment due to various human activities. Elimination of such chemicals is essentially required to save life on the earth through cost effective and eco-friendly approaches as these chemicals are hazardous and toxic for the living beings. Phytoremediation is considered as one of the promising methods for the removal of such chemicals from the soil. Plants and their rhizospheric bacteria mainly participate in the removal of such toxic chemicals from the soil and nearby environment. Root exudates from almost all kind of plants play important role in degradation of toxic chemical compounds as these chemicals also act as attractant and source of carbon and energy for the development and establishment of rhizospheric bacteria in and around the plant roots. In this chapter we have emphasized plant–microbe interactions and their importance in bioremediation of anthropogenic chemicals.

4.1 Introduction

Increasing population, urbanization, industrialization, and intensive agricultural practices over the last century have made consequential impact on the environment which demands a crucial solution worldwide. Heavy metal pollution has become a significant danger to the environment and food security due to developmental activities. Important sources of heavy metals include: mining and smelting activities, batteries, coal and leaded petrol combustion, fertilizers, pesticides, and hospital waste (Memon et al. 2001; Thangavel and Subbhuraam 2004; Khan et al. 2007; Wuana and Okieimen 2011; Rodrigues et al. 2012). Heavy metal leached from these

G. Bhandari (✉)
Sardar Bhagwan Singh University, Dehradun, Uttarakhand, India

P. Bhatt
State Key Laboratory for Conservation and Utilization of Subtropical Agro-Bioresources, Guangdong Province Key Laboratory of Microbial Signals and Disease Control, Integrative Microbiology Research Centre, South China Agricultural University, Guangzhou, People's Republic of China

A. Sharma (ed.), *Microbes and Signaling Biomolecules Against Plant Stress*, Rhizosphere Biology, https://doi.org/10.1007/978-981-15-7094-0_4

sources then pollutes soils, air, ground and surface water and maybe transferred through the food chain by process of biomagnification. Metals having toxic effects include: mercury (Hg), zinc (Zn), copper (Cu), lead (Pb), manganese (Mn), chromium (Cr), cadmium (Cd), and aluminum (Al) (Omura et al. 1996; Adriano 2001; Duruibe et al. 2007) and some metalloids, such as antimony (Sb) and arsenic (As). United States Environmental Protection Agency (US EPA) and the Agency for Toxic Substances and Disease Registry (ATSDR) list of most hazardous materials comprise of some heavy metals and metalloids, e.g., As, Pb, Hg, and Cd (Smith et al. 1997; Park 2010; Wuana and Okieimen 2011). Being non-biodegradable, heavy metals persist in the soil and water for centuries which poses a considerable threat to human health (Sarwar et al. 2010). The vital health risks related to heavy metals comprise of cardiac diseases, chronic anemia, memory loss (Iqbal 2012), cancer, kidney damage (Wuana and Okieimen 2011), and impairment of nervous system, brain (Jarup 2003), skin, teeth, bones (Luo et al. 2012).

Several physicochemical approaches have been employed for remediation of toxic heavy metal, which consist of leaching, thermal treatment, excavation, electro-reclamation, and landfill. These technologies are fast but insufficient, expensive, and sometimes result in secondary pollution and thus have detrimental effects on the environment (Doble and Kumar 2005; Hooda 2007; Glick 2010; Ali et al. 2013). Due to adverse effects of the conventional physicochemical treatment methods, bioremediation has come into the existence. Bioremediation is regarded as a highly effectual technique for the remediation of toxic substances, because it is a natural, eco-friendly, and economical process (Doble and Kumar 2005). Bioremediation techniques comprise phytoremediation, bioventing, bioleaching, land farming, bioreactors, composting, bioaugmentation, and biostimulation. Phytoremediation uses plants for remediation of toxic substances from the environment. This method has been highly touted in the last decades (Glick 2010; Ali et al. 2013; Ma et al. 2016a; Sobariu et al. 2016). Phytoremediation works by several ways such as: (1) Phytoextraction is the translocation of metals from contaminated soils to the ground surface via the root system of the plants. (2) Phytostabilization employs specific plants to minimize the translocation and bioavailability of soil/water pollutants. (3) Phytodegradation is the transformation or breakdown of heavy metals by plant metabolic processes after uptake from the soils. (4) Phytovolatilization uses plants for metal extraction and conversion into volatile forms, which are then liberated to the atmosphere. (5) Rhizofiltration is associated with absorption, concentration, and precipitation of heavy metals from the water by aquatic or land plants. Several plant species have been identified as hyper accumulators due to their ability to absorb metal contaminants and accumulate them to unusually high levels. The hyper accumulator plants belong to *Pteris vittata, Sedum plumbizincicola, Thlaspi caerulescens, Alyssum serpyllifolium, Arabidopsis halleri, Phytolacca americana,* and *Solanum nigrum* (Ma et al. 2011a, b, 2015; Chen et al. 2014; Wei et al. 2014). However these hyper accumulators absorb high amounts of toxic metals but still are not competent for the remediation due to very low biomass production, slow growth rates, and selectivity for specific metals only (Abhilash et al. 2012;

Braud et al. 2006). Thus, it is of utmost importance to institute suitable phytoremediation techniques for remediation of multi-metal polluted field soils.

For upgrading the phytoremediation techniques, many researchers have focused on the interactions among plants–microbes and heavy metals in the rhizospheric soils (Glick 2010; Dharni et al. 2014; Ma et al. 2015). Inoculation of plants with screened bioaugmented microbes has accomplished eminence for phytoremediation of metal polluted soils (Lebeau et al. 2008; Glick 2010; Ma et al. 2011a, b). Microbial assisted phytoremediation comprises removing pollutants from contaminated soils by mutual interaction of plant roots and suitable microbiota, and is an effective eco-friendly, esthetic, and inexpensive method for treating a wide variety of contaminants.

4.2 Microbial Assisted Phytoremediation

The soil present in the vicinity of plant roots along with the roots is considered the rhizosphere, zone where enormous microbial activity has been described. Rhizosphere is a significant habitat and ecosystem for microbes (bacteria, fungi, algae, and protozoa). Rhizosphere is the hotspot of plant benefitting microbiome showing mutualistic plant–microbe interactions (Hinsinger et al. 2009). The rhizospheric microbes interact with the plants in several ways (Azevedo et al. 2000; Hao et al. 2012). Throughout all the developmental phases of plants, there is a mutualistic association among soil, plant, and microbes. This alliance develops when the rhizospheric microbes are triggered by the release of plant rhizo deposits. Root exudates ultimately stimulate specific microbial enzymes of the rhizospheric microbiome resulting in enhanced rhizodegradation. These microbial communities may work by influencing the metal availability and its uptake into the plant rhizosphere (Sarwar et al. 2017). Microbial revitalization using plant growth promoters had been achieved through various direct and indirect approaches including enhancing root growth, bio-fertilization and rhizoremediation, etc. (Gouda et al. 2018). According to Etesami (2018), PGPR increase plant growth and plant tolerance to biotic and abiotic stresses using different mechanisms including: phytostimulation (Glick et al. 2007; Spaepen et al. 2007), microbes assisted uptake of macro and micronutrients (Spaepen et al. 2007), and antagonistic effect of beneficial microbes against plant pathogens (Dawar et al. 2010). PGPR enhance plant growth by employing several mechanisms like phosphate solubilization, production of phytohormones, or nitrogen fixation that ultimately results in enhanced root growth; enzymatic activity; suppression of plant pathogens by production of antibiotics (Tabassum et al. 2017).

4.3 Phytoremediation of Heavy Metals Facilitated by Soil Bacteria (Table 4.1)

Plant growth promoting rhizobacteria aid the host plants to acclimatize to adverse condition of soil and elevate the ability of remediation by enhancing plant growth, reducing metal stress and metal phytotoxicity, improving metal bioavailability in soil and metal translocation in plants (Ma et al. 2011a, b). PGPR-aided metal remediation occurs by direct or indirect mechanisms. The direct mechanisms include: improved bioavailability, solubility, and accumulation of toxic metals. The indirect mechanisms involve promoting plant growth and protecting plant from pathogenic organisms. Some PGPR also help the host plants to adapt to suboptimal soil conditions and improve the capability phytoremediation by plant growth promotion, reducing metal toxicity, improving metal bioavailability in soil, and enhancing metal translocation within the plant.

4.4 Direct Mechanisms

Soil microbes are an integral part of various biogeochemical cycles of the essential nutrients (Khan et al. 2010). The direct mechanisms comprise biological nitrogen fixation, phosphate solubilization and phytohormone production and siderophores production. Some PGPR may also reduce toxic effects of heavy metals and thus promote plant growth and development through any of the abovementioned direct methods (Rajkumar et al. 2009; Pereira and Castro 2014).

4.4.1 Nitrogen Fixation

The atmosphere consists of 78% of N_2; however, this atmospheric N_2 cannot be utilized by the plants in this form. Several diazotrophic microbes are capable of fixing the atmospheric N_2 and can do so even in heavily metal contaminated soils. The usage of such nitrogen fixing bacteria as in case of legume-Rhizobium symbiosis in metal contaminated soils not only promotes plant growth and development but also enhances soil fertility (Brígido et al. 2013). Nonnoi et al. (2012) have reported *Rhizobium bv. trifolii,* which is capable of fixing atmospheric nitrogen in heavy metal contaminated soils. It has also been reported that metal stress enhances nitrogen fixation rate and nitrogen accumulation rate in plants growing in nitrogen-poor ecosystems (Gupta and Panwar 2013). A number of heavy metal resistant microorganisms belonging to rhizobial strains, including *Bradyrhizobium, Rhizobium, Mesorhizobium, Azorhizobium,* and *Sinorhizobium*, have been reported till date from various metal contaminated soils (Zheng et al. 2004; Fan et al. 2010; Wani and Khan 2013).

Table 4.1 Phytoremediation Of heavy metals assisted by soil microorganisms

Bacteria	Source	Metal	PGPR	Reference
Pseudomonas sp. TLC	*Zea mays and Helianthus annuus*	Zn, Pb, Cu, As	IAA, phosphate solubilization, enhanced plant growth and Cu uptake	Li and Ramakrishna (2011)
Serratia sp. LRE07	*Solanum nigrum* L.	Cd, Zn	IAA, ACC deaminase, phosphate solubilization, Siderophores, bioaccumulation of Cd and Zn	Luo et al. (2011)
Bacillus sp. MN3–4	*Alnus firma and Brassica napus*	Zn, Ni and Cu	IAA, Siderophores, ACC deaminase, phosphate solubilization, bioremoval and bioaccumulation of Pb	Shin et al. (2012)
Bacillus sp. SLS18	*Sorghum bicolor* L.	Cd, Mn	IAA, ACC deaminase, phosphate solubilization, Siderophores, enhanced plant growth and metal uptake	Luo et al. (2012)
Bacillus thuringiensis GDB-1	*Alnus firma*	Zn, Pb, Cu, As, Ni, Cd	IAA, Siderophores, phosphate solubilization, enhanced plant growth and accumulation of As, Cu, Pb, Ni, and Zn	Babu et al. (2013)
Rahnella sp. JN6	*Brassica napus*	Cu, Pb, Cd, and Zn	IAA, Siderophores, ACC deaminase, phosphate solubilization, increased uptake of Cu, Pb, Cd, and Zn in plant tissue	He et al. (2013)
Bradyrhizobium sp. YL-6	*Lolium multiflorum* lam *and Glycine max*	Cd	IAA, Siderophores, phosphate solubilization, enhanced plant growth and Cd uptake	Guo and Chi (2014)
Arthrobacter nicotinovorans SA40	*Alyssum pintodasilvae*	Ni	IAA, Siderophores, phosphate solubilization, growth promotion, and phytoextraction of Ni	Cabello-Conejo et al. (2014)
Acinetobacter sp. nbri05	*Cicer arietinum*	As (V & III)	IAA, Siderophores, ACC deaminase, phosphate solubilization	Srivastava and Singh (2014)
Acinetobacter sp. RG30 and *Pseudomonas putida* GN04	*Zea mays*	Cu	IAA, Siderophores phosphate solubilization	Rojas-Tapias et al. (2014)
Sinorhizobium meliloti CCNWSX0020	*Medicago lupulina*	Cu	IAA, Siderophores, ACC deaminase, enhanced plant growth and Cu uptake	Kong et al. (2015)
Bacillus pumilus E2S2	*Sedum plumbizincicola*	Zn, Cd	IAA, Siderophores, ACC deaminase, phosphate solubilization, enhanced plant growth and Cd uptake	Ma et al. (2015)

(continued)

Table 4.1 (continued)

Bacteria	Source	Metal	PGPR	Reference
Bacillus sp. E1S2	*Sedum plumbizincicola*	Zn, Cd, Pb	IAA, Siderophores, ACC deaminase, phosphate solubilization, enhanced plant growth and Zn uptake	Ma et al. (2015)
Rhodococcus erythropolis NSX2	*Sedum plumbizincicola*	Zn, Cd, Cu	IAA, phosphate solubilization, enhanced plant growth and Cd uptake	Liu et al. (2015)
Pseudomonas koreensis AGB-1	*Miscanthus sinensis*	Zn, Pb, Cu, As, Cd	Enhanced plant growth and metal uptake	Babu et al. (2015)

4.4.2 Phosphate Solubilization

Phosphorous (P) is one of vital micronutrients for plants due to its involvement in several metabolic reactions of glucose transport, root growth and development, growth promotion of microbes and plants as well as in various physiological processes (Ahemad 2015). Soils are rich source of phosphorous; however, it exists in insoluble form and thus not accessible to the plants. Several microbes have been reported till date to produce organic acids capable of solubilizing the insoluble form of phosphate. *Pseudomonas, Bacillus,* and *Rhizobium* are some of the predominating phosphate-solubilizing microbes (Chen et al. 2006; Rodríguez et al. 2006). In heavy metal-contaminated soils, PGPR capable of phosphate solubilization play a vital role since the organic acids secreted by them for phosphate solubilization can also solubilize insoluble and stable metals. Therefore these phosphate-solubilizing microbes can enhance the heavy metal bioavailability for uptake by phytoextracting or phytoaccumulating plants (Becerra-Castro et al. 2011; Ahemad and Kibret 2014). Li and Ramakrishna (2011) have reported an enhanced copper accumulation and plant biomass in maize and sunflower plants after inoculating a copper-resistant strain *Pseudomonas* sp., TLC 6–6.5-4. In another study, on inoculation of canola plants growing in heavy metal contaminated soils with PGPR *Rahnella* sp. JN6, a significant increase in plant biomass and increase in concentrations and uptake of Cd, Pb, and Zn in aerial and root tissues of canola plants were observed (He et al. 2013).

4.4.3 Siderophore Production

Iron is one of the most abundantly found elements on earth; however, it is not accessible to plants and microbes for direct assimilation, since in natural conditions it is present mostly in Fe^{3+} form. Fe^{3+} exists in insoluble forms of hydroxides and oxyhydroxides (Rajkumar et al. 2010). Since iron is a vital micronutrient for growth and development of living organisms, several PGPR obtain iron by producing low-molecular weight iron-chelating molecules called siderophores. The siderophores show high binding affinity (Ka > 1030) for Fe and thus solubilize iron from minerals or organic insoluble compounds by functioning as iron solubilizing agents (Hider and Kong 2010). Besides iron, siderophores are also capable of forming stable complexes with several other metal ions, such as Al, Cd, Cu, Ga, In, Pb, and Zn as well as with radionuclides including U and Np (Kiss and Farkas 1998; Neubauer et al. 2000). Diverse group of bacteria have been reported to produce siderophores. Siderophores production by PGPR has also been reported from multiple environmental conditions such as heavy metal contaminated soils or nutrient deficient soils (Rajkumar et al. 2010; Ullah et al. 2015). Siderophores reported from *Streptomyces tendae* F4 have remarkably enhanced plant growth and metal uptake (Fe and Cd) by sunflower (*Helianthus annuus*) (Dimpka et al. 2009). A siderophore-producing bacterium *Kluyvera ascorbata* SUD165 has been reported to reduce heavy metal (Ni, Pb, and Zn) toxicity in canola, Indian mustard, and tomato. Beside this, a siderophore overproducing mutant, SUD165/26, of this bacterium showed greater metal toxicity protection, increased plant biomass, and chlorophyll content in plants grown in Ni-contaminated soil (Burd et al. 1998).

4.4.4 Production of Organic Acids

Low-molecular weight organic acids (Citric, oxalic, and gluconic acid) production has been reported by numerous PGPR (Ullah et al. 2015). These organic acids show the capability to enhance solubility and mobility of toxic heavy metals. In addition, these organic acids also increase heavy metal uptake by plants by regulating the complexion reactions (Rajkumar et al. 2010; Ullah et al. 2015). In this context, 5-ketoglutonic acid, a derivative of gluconic acid, produced by *Gluconacetobacter diazotrophicus* was found to increase the solubilization of Zn compounds.

4.4.5 Production of Biosurfactants

PGPR are also reported to synthesize biosurfactants, which are capable of enhancing mobility and subsequent phytoremediation of toxic heavy metals. The biosurfactants form complexes with several toxic heavy metals at surface of soil resulting in desorption of metals from soil matrix. This results in increase in the heavy metal availability to plants (Gadd 2010; Rajkumar et al. 2012). Biosurfactants produced by *Pseudomonas aeruginosa* BS2 have been reported to enhance the mobility and solubility of Cd and Pb (Ullah et al. 2015).

4.4.6 Bioleaching

Oxidation-reduction reactions possess important capability to mobilize metalloids, metals, and influence compounds of organometallic nature (Bolan et al. 2014). A remarkable enhancement in the solubility of Mn(IV) and Fe(III) was reported by Ullah et al. (2015) when these metals are reduced to Mn(II) and Fe (II). Several PGPR are able to catalyze the enzymatic reduction of Hg^{2+} to metallic mercury(Hg^0) which is highly volatile in nature, thus enhancing the tolerance and detoxification process in soil (Gadd 2010). Similarly, PGPR can also solubilize Cu in Cu rich soils thus enhancing the bioavailability of Cu in the plants (Shi et al. 2012). Furthermore, bioleaching is also regarded as vital method of mobilization of minerals from their ores. *Pseudomonas fluorescence* and *Chromobacterium violaceum* capable of producing hydrogen cyanide are also able to carry out the mobilization of Cu and Ni by formation of different cyanide compounds and complexes by using various solid materials like electronic scarp and Cu-containing ores (Gadd 2010; Ullah et al. 2015).

Microorganisms can also fix CO_2 by carrying out reduction of sulfur containing compounds or oxidization of ferrous iron which results in acidification of the environment. In consequence the produced H_2SO_4 and Fe (II) improve metal solubilization (Gadd 2004). *Leptospirillum ferrooxidans* has been reported to oxidize iron, similarly *Thiobacillus ferrooxidans* has been reported to oxidize sulfur and iron and *Thiobacillus thiooxidans* to carry out oxidation of sulfur (Ullah et al. 2015). The metal oxides formed as a result of oxidation of sulfur and iron is solubilized in their acidified environment which in turn causes solubilization of various metal compounds. Metals can also be mobilized through biomethylation by microorganisms resulting in volatilization. A number of microbes possess the ability to mediate methylation of Sn, Tn, As, Se, Hg, and Pb by carrying out the transfer of methyl group to metals. The methylated metal compounds formed however differ greatly in toxicity, solubility, and volatility (Bolan et al. 2014; Ullah et al. 2015).

4.4.7 Phytochelatins

Plants and fungi are capable of carrying out biosynthesis of phytochelatins (PCs) from glutathione under heavy metal toxicity (Gadd 2010). Phytochelatins rich in cysteine are important metal binding peptides. However some recombinant bacteria are also capable of producing phytochelatin synthase that gets activated in presence of various heavy metals such as Hg, Pb, Cd, and Cu. Phytochelatin synthase is responsible for increasing mobility and bioavailability of toxic metals by synthesizing thiol complexes able to bind metals (Kang et al. 2007). A recombinant *E. coli* strain was produced by Kang et al. (2007) by adding SpPCS of *Schizosaccharomyces pombe* that caused increased PCs synthesis. This recombinant strain was found to accumulate Cd 7.5-folds greater than wild type strain (Kang et al.

2007). Cysteine rich polypeptides (Metallothioneins) possess greater binding affinity for heavy metals such as Ag, Cu, Hg, Zn, and Cd (Ullah et al. 2015). Genes encoding metallothioneins are found in some prokaryotes, plants, and animals (Sriprang et al. 2002; Ullah et al. 2015). Bacteria having greater metal binding capacity through metallothioneins possess the potential to accumulate or sequester heavy metals. Metallothionein encoding genes have been expressed in diverse organisms including bacteria for increasing the accumulation of toxic heavy metals.

4.5 Indirect Methods

4.5.1 Induced Systemic Resistance

PGP bacteria exhibit substantial ability to prevent plants from infection of pathogens, increase biomass and metal tolerance that enable plants to eliminate pollutants from the environment (Hassan et al. 2016). Several PGPR are known to diminish the stress effects in plants by suppressing damage caused by phytopathogens either via biological control of pathogens or by developing induced systemic resistance (ISR) of plants against pathogens (Harish et al. 2008). PGPR may effectively limit the phytopathogen infestation by producing antibiotics, siderophores, variety of hydrolytic enzymes (such as chitinases, proteases, and glucanases), and antimicrobial volatile organic compounds (Sheoran et al. 2015). In many biocontrol systems, one or more antibiotics have been shown to play important role in disease suppression. *Bacillus megaterium* BP 17 and *Curtobacterium luteum* TC 10 effectively suppressed burrowing nematode disease (*Radopholus similis* Thorne) by the synthesis of antibiotics (Aravind et al. 2010). PGPR can also act as effective competitors of pathogens for colonization in niches, availability of nutrients and by minimizing the adverse environmental impacts with nutrient loss and thereby indirectly promote plant productivity (Alvin et al. 2014). ISR refers to the state of systemically enhanced resistance in plants to a broad spectrum of pathogens (Kloepper and Beauchamp 1992). The ISR to various diseases caused by certain chemicals and plant growth promoting bacteria is a highly beneficial approach in protection of agroecosystem. ISR induced by PGPR has been demonstrated against various fungal, bacterial, and viral pathogens in many plant taxa (Alvin et al. 2014). Plants primed with bacterial inoculation induce a plant defense system (Sturz and Matheson 1996). Once the defense genes are expressed, ISR activates multiple potential defense mechanisms that include increased activity of chitinases, β-1,3-glucanases, superoxide dismutase, guaiacol, catalase (CAT), and peroxidases (POS) (Gechev et al. 2006). The activities of these enzymes are responsible to reduce reactive oxygen species production and protection of cell organelles against oxidative stress as a result of the activities of endophytic bacteria (Wan et al. 2012).

4.5.2 *Phytohormones Production*

PGPR are also reported to regulate a number of physiochemical processes of plants including mineral uptake, osmotic modifications, stomatal regulation, and alterations in root morphology (Amaral et al. 2016; Ma et al. 2016a). Plants often face various environmental constraints including heavy metal stress that disturb physiochemical processes of the plants. A variety of PGP bacteria from rhizosphere are reported to produce phytohormones (gibberellins, cytokinins, IAA) even under abiotic stresses. Therefore, phytohormone producing bacteria promote significant alterations in endogenous plant hormones that shape the response of plants for different environmental constraints (Verbon and Liberman 2016). Ethylene is known to play essential role in regulating different physiological responses of plants but greater production of ethylene is lethal for plant growth. However, some PGP bacteria contain ACC deaminase enzyme that breaks down ACC into a ketobutyrate and ammonia. ACC is the precursor for ethylene biosynthesis (Hassan et al. 2016). Therefore, bacteria containing ACC deaminase generates ammonia that adds to nitrogen source of plants. Moreover, PGP bacteria also act as biocontrol agents because these bacteria prevent plants from phytopathogens (Glick 2012). Antifungal metabolites including tensin, 2,4-diacetylphloroglucinol, vicosinamide, pyoluterin, HCN, pyrrolnitrin, phenazines are produced by PGP bacteria (Ahemad and Kibret 2014). In nutshell, PGP bacteria enhance plant growth in soils contaminated with metals and metalloids that further enhance phytoremediation of heavy metals.

4.6 Mechanisms Responsible for Altering Plant Metal Uptake

4.6.1 *Amelioration of Metal Stress*

The metal phytotoxicity is a critical factor affecting the success of phytoremediation (Shin et al. 2012). To overcome the metal stress, a number of bacterial mediated processes are involved in bacteria–host coevolution process either by alleviating metal toxicity or by conferring plant metal tolerance (Rajkumar et al. 2009; Ma et al. 2015). Recent studies suggest that some PGPR can reduce metal phytotoxicity via several mechanisms that include extracellular precipitation (Babu et al. 2015), intracellular accumulation and sequestration (Shin et al. 2012), biotransformation of toxic metal ions to less- or nontoxic forms (Zhu et al. 2014), and adsorption/desorption of metal ions (Guo et al. 2010; Luo et al. 2011). Abiotic or biotic stress can be alleviated or eliminated by the genes encoding metal or antibiotic resistance proteins. Shin et al. (2012) recently found that endophytic bacterial strain *Bacillus* sp. MN3–4 had evolved a better defined metal resistant mechanism, e.g. active export via a P-type ATPase efflux pump, which can transport metal ions across biological membranes against the concentration gradient using energy released by

ATP hydrolysis. Wan et al. (2012) and Zhang et al. (2010) found that bacteria could modulate the activity of plant antioxidant enzymes (such as POS, CAT, SOD, glutathione peroxidase, ascorbate peroxidase) as well as lipid peroxidation (malondialdehyde formation), which confronted plant defense mechanisms, especially resisted the heavy metal-induced oxidative stress in plants. In addition, certain PGPR can also use methylation as a metal resistance or detoxification mechanism. Some mercury resistant bacteria express MerB gene encoding organomercurial lyase which cleaves organomercurials into mercuric ion (Hg^{2+}) (Brown et al. 2003), and MerA gene encoding mercuric reductase which converts highly toxic ionic Hg^{2+} into less toxic volatile Hg^0 (Cursino et al. 2000), thus alleviate metal toxicity and improve the efficiency of phytovolatilization. The genes specifying various functions needed for degradation of organomercurials and reduction of Hg^{2+} are organized in the mercury resistance (Mer) operon, which is mostly found in gram-negative bacteria (Brown et al. 2003).Currently, several studies have focused on the role of bacteria on heavy metal tolerance and accumulation in hyperaccumulator and non-accumulator plants. However, it is not yet analyzed whether plants growing in metal polluted soils alter colonization/survival potential of specific metal resistant and/or beneficial microbes. Thus, it is critical to explore the diversity, distribution, and activity of endophytic microbial communities associated with various hyperaccumulator plants in phytoremediation studies.

4.6.2 Metal Biosorption and Bioaccumulation

Various reports have exhibited that PGPR play a significant role in reduction of metal phytotoxicity by biosorbing and bioaccumulating heavy metals (Guo et al. 2010; Luo et al. 2011). A lead resistant bacteria *Bacillus* sp. MN3–4 was obtained from roots of *Alnus firma* (a metal hyperaccumulator plant) by Shin et al. (2012). This strain was found to improve phytoremediation capacity by decreasing metal phytotoxicity by extracellular sequestration and intracellular accumulation. In a similar study conducted by Luo et al. (2011), a cadmium resistant *Serratia* sp. LRE07 was reported to absorb about 65% of Cd and 35% of Zn in cells from single metal solution by the process of biosorption and bioaccumulation. This resulted in significantly reducing phytotoxic effects of the toxic heavy metals. In this aspect, the bacterial cell wall has been considered as a significant component for metal ion absorption (Bai et al. 2008). The process of metal biosorption by micro-organism occurs in the following two steps: (1) Passive biosorption of metals by both live and dead/inactive cells that occurs by the cell wall owing to various metabolism-independent processes (Vijayaraghavan and Yun 2008). In the passive process, metal ions are first adsorbed rapidly on the microbial cell surface due to interactions between metals and functional groups (hydroxyl, carbonyl, carboxyl, sulfhydryl, thioether, sulfonate, amine, amide, and phosphonate) present on the cell surface (Ma et al. 2011b). Several mechanisms such as ion exchange, microprecipitation, sorption, complexation, electrostatic interaction, coordination, and chelation are proposed to be responsible either synergistically or independently for the complexation processes (Volesky and Schiewer 1999); (2) Active biosorption (bioaccumulation) involves metal uptake (transport into cells, accumulate intracellularly, across cell membrane through cell metabolic cycle) by live cells by a slow active metabolism-dependent transport of metal into the microbial cells (Malik 2004). As soon as the metal is inside the living microbial cells, it may get bound, precipitated, accumulated, sequestered within specific intracellular organelles or may get translocated to specific structures, depending upon the type of organism and element concerned (Ma et al. 2011b). Different bacteria carry distinct and unique metal bioaccumulation abilities that can be utilized in plant–bacteria symbiotic system for facilitating heavy metal detoxification and enhanced phytoremediation efficiency (Guo et al. 2010; Shin et al. 2012).

4.6.3 Metal Bioavailability

The translocation of heavy metals from contaminated soil to plant tissues depends on heavy metal bioavailability in the soils (Glick 2010). Lebeau et al. (2008) reported that bioavailability of heavy metals in soils is determined by different elements of soil such as particle size, nutrients, pH, redox potential, organic matter content, and presence of other metal ions. A number of studies proposed that PGPR having metal resistance/sequestration pathway (e.g. ncc-nre) may help in reduction in phytotoxicity of pollutants (Luo et al. 2011) and may enhance the phytoavailability of heavy metals by liberating metal chelating agents (e.g. siderophores, biosurfactants, and organic acid), acidification of soils, redox activity, and phosphate solubilization (Ma et al. 2011a). PGPR are able to increase metal and mineral solubilization from soils by producing low-molecular weight organic acids and metal-specific ligands like siderophore, which change the soil pH and enhance chelation efficiency (Rajkumar et al. 2009). In recent time, researchers have examined the function of microbes in heavy metal translocation by plants. Reports have indicated that organic acids production by microbes generally enhances under heavy metal stress conditions (Chen et al. 2014), thus suggesting the important role of these organic acids in heavy metal mobilization by plants. Sheng et al. (2008) examined the lead solubilizing potential of *Pseudomonas fluorescens* G10 and *Microbacterium sp.* G16 under in vitro conditions with Pb-amended and Pb-free media. They examined the application of Pb resistant bacteria on Pb uptake by *Brassica napus* and reported that both strains increased Pb availability by enhancing Pb accumulation in plant shoots from 76% to 131% (*P. fluorescens*) and from 59% to 80% (*Microbacterium* sp.), compared to the control treatments. Sheng et al. (2008) proposed that Pb accumulation in plants is due to synthesis of siderophore or organic acid by these bacterial strains.

Several endophytic bacteria have been reported to also produce biosurfactants. These biosurfactants have been reported to enhance the bioavailability of poorly soluble metals and to accelerate phytoremediation rate (Bacon and Hinton 2011).

Biosurfactants are low-molecular weight amphiphilic molecules comprised of a hydrophilic and a hydrophobic regions and contain different function groups such as mycolic acid, glycolipids, lipopeptides, polysaccharide–protein complexes, phospholipids, fatty acids, etc. (Banat et al. 2010). Biosurfactants are capable of reducing the surface tension, critical micelle concentration (CMC), and interfacial tension, thus influencing the dispersal of the metals among the different phases (Banat 1995). PGPR synthesize biosurfactants and excrete them to the host niches as root exudates. The secreted biosurfactants at first form metal complexes with the insoluble metals on the interface of rhizosphere soil particles, and then desorb metals from soil matrix, resulting in altered metal translocation and bioavailability in the soil (Rajkumar et al. 2009). The interactions of PGPR, toxic heavy metals, and biosurfactants can be considered as a vital element in facilitating metal uptake from contaminated soils (Bacon and Hinton 2011). It has also been proposed that it is feasible to enhance metal phytoextraction ability of hyper accumulating plant species by treating the plant seeds or rhizosphere soil with beneficial metal resistant bacteria.

Methylobacterium oryzae CBMB20 and *Burkholderia* sp. CBMB40 obtained from rice tissues were reported to reduce metal toxicity and accumulation and translocation of Ni and Cd in shoots of tomato plants on inoculation with these microbial strains. The reason of reduced accumulation of the toxic heavy metal may be due to immobilization of metals in rhizospheric soil by the bacterial strains. Production of various extracellular polymeric substances (EPS) mainly polysaccharides, proteins, nucleic acids, and lipids by microbes seems to play an important role in metal complexation and thus reduces metal bioaccessibility and bioavailability (Rajkumar et al. 2013). Joshi and Juwarkar (2009) have examined the capability of EPS, secreted by *Azotobacter* sp. in forming metal complexes by binding up to 15.2 mg g^{-1} of Cd and 21.9 mg g^{-1} of Cr, resulting in reduced metal accumulation by *Triticum aestivum*.

4.6.4 Metal Uptake and Translocation

PGPR can alter heavy metal bioavailability and its translocation in plants, hence modify metal toxicity and plant metal uptake by secreting a variety of metabolites including siderophores, organic acids (e.g., citric, oxalic and acetic acids), etc. (Kuffner et al. 2010; Luo et al. 2012; Visioli et al. 2014). PGPR may also alter metal accumulation capacity and its translocation in plants by using their multiple plant growth promoting traits including metal resistance, detoxification, accumulation, transformation and sequestration, thus diminish metal phytotoxicity and alter phytoavailability of heavy metals in contaminated soils (Ma et al. 2016a, b). Sun et al. (2010) reported that inoculation with the copper-resistant endophytic bacteria having multiple plant growth promoting characteristics, such as ACC deaminase and arginine decarboxylase activity, enhanced plant growth and Cu transfer from root to above-ground tissues in *B. napus*, thus improving overall phytoextraction potential. Ma et al. (2011a) showed that Ni resistant endophytic *Pseudomonas* sp. A3R3 increased plant biomass (non-host *Brassica juncea*) and Ni accumulation in plants (host *A. serpyllifolium*), grown in artificially Ni-contaminated soil. This effect can be attributed to its ability to produce plant growth promoting substances (ACC deaminase, siderophores, IAA and P solubilization) and plant polymer hydrolyzing enzymes (cellulase and pectinase). Therefore, the plant associated PGPR are now being considered as a bio-tool to improve the efficiency of phytoremediation of metal polluted soils. The metal phytotoxicity is a critical factor inhibiting the success of phytoremediation (Shin et al. 2012). A lead resistant bacterium, *Bacillus* sp. MN3–4 isolated from the roots of the metal hyperaccumulator *Alnus firma* enhanced phytoremediation potential by reducing the phytotoxic effects of metals via extracellular sequestration and intracellular accumulation (Shin et al. 2012). PGPB also have great potential to enhance metal and mineral mobilization by the production of organic acids and metal-specific ligands (e.g. siderophores), and subsequently increase nutrient levels and heavy metal uptake, which may result in improved metal phytoextraction potential of host plants (Chen et al. 2014). Moreover, PGPB may also alter metal bioavailability and its translocation in plants, hence modifying metal phytotoxic effects and increasing plant metal uptake. Copper-resistant PGPB having multiple plant growth promoting traits enhanced plant growth and Cu translocation from roots to aerial parts of *B. napus*, thus improved overall phytoextraction potential of the plant (Sun et al. 2010). On the other hand, some PGPB are effective in reducing metal availability in soils and plant metal accumulation. A recent study reported that two metal resistant bacteria (*B. megaterium* H3 and *Neorhizobium huautlense* T1–17) decreased Cd bioavailability in rhizosphere soils, resulting in reduced Cd accumulation in polished rice (Li et al. 2017). Reduction in Cd uptake was dependent on the bacterial strain used and the level of metal contamination.

4.7 Reports of PGP Bacteria-Assisted Phytoremediation

Several reports on PGPR capable of enhancing phytoremediation are available in the literature (Glick 2010). A metal resistant PGPR was obtained from *Polygonum pubescens* plants growing in metal contaminated soil by Jing et al. (2014). These PGPR were then identified as *Enterobacter* sp. and *Klebsiella* sp. *Brassica napus* was observed to hyper accumulate toxic heavy metals (Zn, Pb, and Cd) on inoculation with these metal resistant PGPR strains. In another study, inoculating *Brassica napus* with *Pseudomonas tolaasii* ACC23, *Pseudomonas fluorescence* ACC9, and *Mycobacterium* sp. ACC14, an improved metal (Cd) uptake in aerial parts of plants was reported by Dell'Amico et al. (2008). Rye grass displayed an enhanced metal (Cr, Cd, Zn, Cu, and Pb) tolerance ability on inoculation with *Microbacterium arborescens*, isolated from the rhizosphere of *Prosopis juliflora* (Khan et al. 2015a, b). A notable amount of toxic metals were found in the root and shoot of the plants inoculated with *Microbacterium arborescens* strains. Enhanced plant growth and development was also observed on the inoculation with *Microbacterium arborescens* due to plant growth promoting properties (siderophore and IAA production, P solubilization, and ACC deaminase activity) of this strain. Inoculation of *Frankia* sp. decreased accumulation of Ni and increased nodulation in *Alnus glutinosa* (Wheeler et al. 2001). A *Streptomyces mirabilis* P16B-1 strain was isolated from soil of uranium mining area; it showed tolerance against multiple metals including Zn, Co, Cd, Ni, and Cu. Reduced metal bioavailability and enhanced plant growth were observed in *Sorghum bicolor* plants inoculated with *Streptomyces mirabilis* P16B-1 strain as compared to untreated plants (Schutze et al. 2013). Similarly, treating maize plants with *Streptomyces* sp. HM1 showed higher tolerance for Cd and exhibited significantly higher chlorophyll contents (El Sayed et al. 2015). In another study, *Rhodococcus erythropolis* was obtained from metal polluted soil; it was shown to decrease metal uptake and enhance plant growth in *Pisum sativum*. Improved plant growth was related to P solubilization and IAA production ability of *Rhodococcus erythropolis* (Trivedi et al. 2007). Likewise on treating green chili with *Cellulosimicrobium cellulans,* it showed reduced Cr uptake. *Cellulosimicrobium cellulans* possessing P solubilization and IAA production ability makes it an ideal candidate for plant growth promoting bacteria (Chatterjee et al. 2009). It has been proposed that heavy metals removal by phytobacterial systems is given more preference over traditional phytoremediation system alone. Accumulation of heavy metals by plants is determined by different elements such as soil pH, oxidation state of metal and bacterial species. PGPR that are also involved in phytoremediation are capable of synthesizing siderophores, IAA (Indole-3-acetic acid), ACC deaminase, and solubilize phosphorous which then results in improved plant growth in addition to increased solubility and bioavailability of heavy metals. Consequently, plants are able to hyperaccumulate heavy metals due to enhanced solubility and bioavailability of toxic heavy metals.

4.8 Phytoremediation Through Genetically Engineered Bacteria

A number of organic pollutants have been bioremediated in situ through a symbiotic relationship between plants and genetically transformed bacteria (Ullah et al. 2015). The application of such genetically engineered PGPR for phytoremediation of toxic heavy metals has been proposed as a novel phytobacterial approach for removal of toxic heavy metals. Genetically engineered PGPR carry one or more genes that are inserted so as to enhance heavy metal remediation. Genes responsible for metal chelation, transportation, metal homeostasis, degradative enzyme, and metal uptake regulators are significant for consideration for production of recombinant strains (Singh et al. 2011). Gene for phytochelatin synthase (PCS) from *Schizosaccharomyces pombe* was inserted in *Pseudomonas putida* KT2440. Phytochelatins are molecules having greater binding affinity toward several toxic heavy metals due to presence of cysteine-rich peptides. The recombinant strain

KT2440-spPCS was reported to increase tolerance to Ag, Hg, and Cd, accumulated Cd in greater quantities which subsequently enhanced the phytoremediation efficiency. In addition the recombinant strain KT2440-spPCS remarkably increased germination and plant growth in wheat. This indicates the symbiosis behavior of *Pseudomonas putida* KT2440 in improving phytoremediation of lethal metals (Yong et al. 2014). A recombinant strain *Pseudomonas putida* containing metal binding peptide (EC 20) was found to have higher Cd binding and mitigated Cd induced cellular toxicity. On inoculating sunflower roots with this recombinant strain *Pseudomonas putida* 06909 a significant decrease in phytotoxicity of Cd and 40% rise in Cd accumulation in plant roots was observed. Enhanced plant growth and phytoextraction ability for heavy metals through symbiotic association of recombinant PGP bacteria was reported by Wu et al. (2006). Sriprang et al. (2002) have constructed a recombinant strain of *Mesorhizobium huakuii* sb. Sp. *rengei* B3 by inserting human tetrameric metallothionein (MLT4) and reported improved heavy metal remediation ability through this symbiotic phytobacterial system. In another study, Sriprang et al. (2003) have constructed a recombinant of *Mesorhizobium huakuii* sb. Sp. *rengei* B3 by introducing the *Arabidopsis thaliana* gene AtPCS for enhanced production of PCS that showed significant increase in Cd uptake. Similarly, symbiotic association of recombinant bacteria with *Astragalus sinicus* was found to enhance Cd accumulation in root nodules (Ullah et al. 2015). Likewise, a recombinant strain of *Mesorhizobium huakuii* sb. Sp. *rengei* B3 with MLT4 and AtPCS resulted in enhanced accumulation of Cd in cells. Symbiotic association of these strains with *Astragalus sinicus* caused threefold increase in Cd accumulation in plant roots. Furthermore, abovementioned strains AtPCS or MLT4 strain B3 were modified by introducing iron-regulated transporter 1 gene of *Arabidopsis thaliana*. A significant rise in the buildup of As and Cu level in nodules than Zn and Cd was observed when AtPCS or MLT4 strain B3 (recombinant strains) established a symbiotic association with *Astragalus sinicus* (Ike et al. 2008).

4.9 Conclusion

Heavy metals induce serious effects on plant biosphere and increase the pollution load in the environment. Thus, there is dire need to find out some solutions to eliminate these pollutants from the environment. Different biological and physiochemical approaches are being employed to eliminate heavy metals from polluted soils. Of these approaches, microbial assisted phytoremediation is a promising sustainable, viable, inexpensive, and safe method. Phytobacterial extraction of heavy metals safely eliminates heavy metals without rendering any toxic effect on the environment. PGPR with multifunctional activities such as production of siderophore, IAA, ACC deaminase, organic acids, biomethylation, induce redox reactions, enhanced synthesis of biosurfactants, and anti-pathogen metabolites prove to be a better choice in assisting the phytoremediation process in metal contaminated soils. PGP bacteria transform heavy metals into sparingly soluble metal complexes and bioavailable forms thus accelerating phytoremediation. Moreover, recombinant bacterial strains and overexpression of genes responsible for metal degradation can also be explored for remediation of heavy metals. In this context, a combination of different bacterial strains could be more effective in enhancing reclamation of polluted soils. It is imperative to establish some novel strategies to eliminate pollutants from our environment so that we may provide safe environment to our future generations.

Acknowledgement We sincerely acknowledge all the authors whose contributions have been cited in this chapter.

References

Abhilash PC, Powell JR, Singh HB, Singh BK (2012) Plant–microbe interactions: novel applications for exploitation in multipurpose remediation technologies. Trends Biotechnol 30:416–420

Adriano DC (2001) Biogeochemistry, bioavailability and risks of metals. Trace elements in the terrestrial environment, 2nd edn. Springer-Verlag, New York

Ahemad M (2015) Phosphate-solubilizing bacteria-assisted phytoremediation of metalliferous soils: a review. 3 Biotech 5:111–121

Ahemad M, Kibret M (2014) Mechanisms and applications of plant growth promoting rhizobacteria: current perspective. J King Saud UnivSci 26:1–20

Ali H, Khan E, Sajad MA (2013) Phytoremediation of heavy metals—concepts and applications. Chemosphere 91:869–881

Alvin A, Miller KI, Neilan BA (2014) Exploring the potential of endophytes from medicinal plants as sources of antimycobacterial compounds. Microbiol Res 169:483–495

Amaral FP, Pankievicz VC, Arisi ACM, de Souza EM, Pedrosa F, Stacey G (2016) Differential growth responses of *Brachypodium distachyon* genotypes to inoculation with plant growth promoting rhizobacteria. Plant Mol Biol 90:689–697

Aravind R, Eapen SJ, Kumar A, Dinu A, Ramana KV (2010) Screening of endophytic bacteria and evaluation of selected isolates for suppression of burrowing nematode (*Radopholus similis* Thorne) using three varieties of black pepper (*Piper nigrum* L.). Crop Prot 29:318–324

Azevedo JL, Maccheroni W Jr, Pereira JO, de Araújo WL (2000) Endophytic microorganisms: a review on insect control and recent advances on tropical plants. Electron J Biotechnol 3:15–16

Babu AG, Kim JD, Oh BT (2013) Enhancement of heavy metal phytoremediation by Alnus firma with endophytic *Bacillus thuringiensis* GDB-1. J Hazard Mater 250:477–483

Babu AG, Shea PJ, Sudhakar D, Jung IB, Oh BT (2015) Potential use of Pseudomonas koreensis AGB-1 in association with *Miscanthus sinensis* to remediate heavy metal (loid)-contaminated mining site soil. J Environ Manag 151:160–166

Bacon CW, Hinton DM (2011) In planta reduction of maize seedling stalk lesions by the bacterial endophyte *Bacillus mojavensis*. Can J Microbiol 57:485–492

Bai H-J, Zhang Z-M, Yang G-E, Li B-Z (2008) Bioremediation of cadmium by growing Rhodobacter sphaeroides: kinetic characteristic and mechanism studies. Bioresour Technol 99 (16):7716–7722

Banat IM (1995) Characterization of biosurfactants and their use in pollution removal–state of the art. Acta Biotechnol 15:251–267

Banat IM, Franzetti A, Gandolfi I, Bestetti G, Martinotti MG, Fracchia L, Marchant R (2010) Microbial biosurfactants production, applications and future potential. Appl Microbiol Biotechnol 87:427–444

Becerra-Castro C, Prieto-Fernández Á, Álvarez-López V, Monterroso C, Cabello-Conejo MI, Acea MJ, Kidd PS (2011) Nickel solubilizing capacity and characterization of rhizobacteria isolated from hyperaccumulating and non-hyperaccumulating subspecies of *Alyssum serpyllifolium*. Int J Phytoremediation 13:229–244

Bolan N, Kunhikrishnan A, Thangarajan R, Kumpiene J, Park J, Makino T, Scheckel K (2014) Remediation of heavy metal (loid) s contaminated soils–to mobilize or to immobilize? J Hazard Mater 266:141–166

Braud A, Jézéquel K, Vieille E, Tritter A, Lebeau T (2006) Changes in extractability of Cr and Pb in a polycontaminated soil after bioaugmentation with microbial producers of biosurfactants, organic acids and siderophores. Water Air Soil Pollut 6:261–279

Brígido C, Nascimento FX, Duan J, Glick BR, Oliveira S (2013) Expression of an exogenous 1-aminocyclopropane-1-carboxylate deaminase gene in *Mesorhizobium* spp. reduces the negative effects of salt stress in chickpea. FEMS Microbiol Lett 349:46–53

Brown NL, Stoyanov JV, Kidd SP, Hobman JL (2003) The MerR family of transcriptional regulators. FEMS Microbiol Rev 27:145–163

Burd GI, Dixon DG, Glick BR (1998) Plant growth promoting bacteria that decrease heavy metal toxicity in plants. Can J Microbiol 46:237–245

Cabello-Conejo MI, Becerra-Castro C, Prieto-Fernández A, Monterroso C, Saavedra-Ferro A, Mench M (2014) Rhizobacterial inoculants can improve nickel phytoextraction by the hyperaccumulator *Alyssum pintodasilavae*. Plant Soil 379:35–50

Chatterjee S, Sau GB, Mukherjee SK (2009) Plant growth-promotion by hexavalent chromium reducing bacterial strain, *Cellulosimicrobium cellulans* KUCr3. World J Microbiol Biotechnol 25:1829–1836

Chen Y, Rekha P, Arun A, Shen F, Lai W (2006) Phosphate solubilizing bacteria from subtropical soil and their tricalcium phosphate solubilizing abilities. Appl Soil Ecol 34:33–41

Chen L, Luo SL, Li XJ, Wan Y, Chen JL, Liu CB (2014) Interaction of Cd hyperaccumulator *Solanum nigrum L.* and functional endophyte *Pseudomonas sp.* Lk9 on soil heavy metals uptake. Soil Biol Biochemist 68:300–308

Cursino L, Mattos SV, Azevedo V, Galarza F, Bucker DH, Chartone-Souza E, Nascimento A (2000) Capacity of mercury volatilization by mer (from *Escherichia coli*) and glutathione S-transferase (from *Schistosoma mansoni*) genes cloned in *Escherichia coli*. Sci Total Environ 261:109–113

Dawar S, Wahab S, Tariq M, Zaki MJ (2010) Application of *Bacillus* species in the control of root rot diseases of crop plants. Archives Phytopathol Plant Prot 43:412–418

Dell'Amico E, Cavalca L, Andreoni V (2008) Improvement of *Brassica napus* growth under cadmium stress by cadmium-resistant rhizobacteria. Soil Biol Biochem 40:74–84

Dharni S, Srivastava AK, Samad A, Patra DD (2014) Impact of plant growth promoting *Pseudomonas monteilii* PsF84 and *Pseudomonas plecoglossicida* PsF610 on metal uptake and production of secondary metabolite (monoterpenes) by rose-scented geranium (*Pelargonium graveolens* cv. bourbon) grown on tannery sludge amended soil. Chemosphere 117:433–439

Dimpka CO, Merten D, Svatos A, Buchel G, Kothe E (2009) Siderophores mediate reduced and increased uptake of cadmium by *Streptomyces tendae* F4 and sunflower (*Helianthus annuus*), respectively. J Appl Microbiol 107:1687–1696

Doble M, Kumar A (2005) Biotreatment of industrial effluents. Elsevier, Butterworth-Heinemann

Duruibe JO, Ogwuegbuand MOC, Egwurugwu JN (2007) Heavy metal pollution and human biotoxic effects. Int J Phys Sci 2:112–118

El Sayed HE, Othaimen HS, Aburas MMA, Jastaniah SD (2015) Efficiency of an Cd-tolerant actinomycete isolate obtained from wastewater in removal of heavy metals and enhancing plant growth of *Zea mays* L. plant. Int J Curr Microbiol Appl Sci 4:553–565

Etesami H (2018) Ecotoxicology and environmental safety bacterial mediated alleviation of heavy metal stress and decreased the accumulation of metals in plant tissues: mechanisms and future prospects. Ecotoxicol Environ Saf 147:175–191

Fan LM, Maa ZQ, Liang JQ, Li HF, Wangc ET, Wei GH (2010) Characterization of a copper-resistant symbiotic bacterium isolated from *Medicago lupulina* growing in minc tailings. Bioresour Technol 102:703–709

Gadd GM (2004) Microbial influence on metal mobility and application for bioremediation. Geoderma 122:109–119

Gadd GM (2010) Metals, minerals and microbes: geomicrobiology and bioremediation. Microbiology 156:609–643

Gechev TS, Van Breusegem F, Stone JM, Denev I, Laloi C (2006) Reactive oxygen species as signals that modulate plant stress responses and programmed cell death. Bioassays 28:1091–1101

Glick BR (2010) Using soil bacteria to facilitate phytoremediation. Biotechnol Adv 28:367–374

Glick BR (2012) Plant growth-promoting bacteria: mechanisms and applications. Scientifica, Cairo

Glick BR, Todorovic B, Czarny J, Cheng Z, Duan J, McConkey B (2007) Promotion of plant growth by bacterial ACC deaminase. Crit Rev Plant Sci 26:227–242

Gouda S, Kerry RG, Das G, Paramithiotis S, Shin HS, Patra JK (2018) Revitalization of plant growth promoting rhizobacteria for sustainable development in agriculture. Microbiol Res 206:131–140

Guo J, Chi J (2014) Effect of Cd-tolerant plant growth promoting rhizobium on plant growth and Cd uptake by *Lolium multiflorum* Lam. and Glycine max (L.) Merr.in Cd-contaminated soil. Plant Soil 375:205–214

Guo HJ, Luo SL, Chen L, Xiao X, Xi Q, Wei WZ, Zeng G, Liu C, Wan Y, Chen J, He Y (2010) Bioremediation of heavy metals by growing hyperaccumulator endophytic bacterium *Bacillus* sp. L14. Bioresour Technol 101:8599–8605

Gupta G, Panwar J, Jha PN (2013) Natural occurrence of *Pseudomonas aeruginosa*, a dominant cultivable diazotrophic endophytic bacterium colonizing *Pennisetum glaucum* (L.). R Br Appl Soil Ecol 64:252–261

Hao X, Xie P, Johnstone L, Miller SJ, Rensing C, Wei G (2012) Genome sequence and mutational analysis of plant-growth-promoting bacterium *Agrobacterium tumefaciens* CCNWGS0286 isolated from a zinc-lead mine tailing. Appl Environ Microbiol 78:5384–5394

Harish S, Kavino M, Kumar N, Saravanakumara D, Soorianathasundaramb K, Samiyappana R (2008) Biohardening with plant growth promoting rhizosphere and endophytic bacteria induces systemic resistance against banana bunchy top virus. Appl Soil Ecol 39:187–200

Hassan W, Bashir S, Ali F, Ijaz M, Hussain M, David J (2016) Role of ACC deaminase and/or nitrogen fixing rhizobacteria in growth promotion of wheat (*Triticum aestivum* L.) under cadmium pollution. Environ Earth Sci 75:114

He H, Ye Z, Yang D, Yan J, Xiao L, Zhong T, Yuan M, Cai X, Fang Z, Jing Y (2013) Characterization of endophytic *Rahnella* sp. JN6 from *Polygonum pubescens* and its potential in promoting growth and Cd, Pb, Zn uptake by *Brassica napus*. Chemosphere 90:1960–1965

Hider RC, Kong X (2010) Chemistry and biology of siderophores. Nat Prod Rep 27:637–657

Hinsinger P, Bengough AG, Vetterlein D, Young IM (2009) Rhizosphere: biophysics, biogeochemistry and ecological relevance. Plant Soil 321:117–152

Hooda V (2007) Phytoremediation of toxic metals from soil and waste water. J Environ Biol 28:367–376

Ike A, Sriprang R, Ono H, Murooka Y (2008) Promotion of metal accumulation in nodule of *Astragalus sinicus* by the expression of the iron-regulated transporter gene in *Mesorhizobium huakuii* subsp. rengei B3. J Biosci Bioeng 105:642–648

Iqbal M (2012) Lead pollution-a risk factor for cardiovascular disease in Asian developing countries. Pak J Pharm Sci 25:289

Jarup L (2003) Hazards of heavy metal contamination. Br Med Bull 68:167–182

Jing Y, Yan J, He H, Yang D, Xiao L (2014) Characterization of bacteria in the rhizosphere soils of *Polygonum pubescens* and their potential in promoting growth and Cd, Pb, Zn uptake by *Brassica napus*. Int J Phytoremediation 16:321–333

Joshi P, Juwarkar A (2009) In vivo studies to elucidate the role of extracellular polymeric substances from *Azotobacter* in immobilization of heavy metals. Environ Sci Technol 43:5884–5889

Kang S, Singh S, Kim J, Lee W (2007) Bacteria metabolically engineered for enhanced phytochelatin production and cadmium accumulation. Appl Environ Microbiol 73:6317–6320

Khan MA, Ahmad I, Rahman IU (2007) Effect of environmental pollution on heavy metals content of *Withania somnifera*. J Chin Chem Soc 54:339–343

Khan S, Hesham A, Qiao M, Rehman S (2010) Effects of Cd and Pb on soil microbial community structure and activities. Sci Pollut Res 17:288–296

Khan A, Khan S, Khan MA, Qamar Z, Waqas M (2015a) The uptake and bioaccumulation of heavy metals by food plants, their effects on plants nutrients, and associated health risk: a review. Environ Sci Pollut Res 22:13772–13799

Khan MU, Sessitsch A, Harris M, Fatima K, Imran A, Arslan M, Shabir G, Khan QM, Afzal M (2015b) Cr-resistant rhizo- and endophytic bacteria associated with *Prosopis juliflora* and their potential as phytoremediation enhancing agents in metal-degraded soils. Front Plant Sci 5:755–760

Kiss T, Farkas E (1998) Metal binding ability of desferrioxamine B. J Incl Phenom Macrocycl Chem 32:385–403

Kloepper JW, Beauchamp CJ (1992) A review of issues related to measuring colonization of plant roots by bacteria. Can J Microbiol 38:1219–1232

Kong Z, Mohamad OA, Deng Z, Liu X, Glick BR, We G (2015) Rhizobial symbiosis effect on the growth, metal uptake, and antioxidant responses of *Medicago lupulina* under copper stress. Environ Sci Pollut Res 22:12479–12489

Kuffner M, De Maria S, Puschenreiter M, Fallmann K, Wieshammer G, Gorfer M, Strauss J, Rivelli AR, Sessitsch A (2010) Culturable bacteria from Zn- and Cd-accumulating *Salix caprea* with differential effects on plant growth and heavy metal availability. J Appl Microbiol 108:1471–1484

Lebeau T, Braud A, Jezequel K (2008) Performance of bioaugmentation-assisted phytoextraction applied to metal contaminated soils: a review. Environ Pollut 153:497–522

Li K, Ramakrishna W (2011) Effect of multiple metal resistant bacteria from contaminated lake sediments on metal accumulation and plant growth. J Hazard Mater 189:531–539

Li Y, Pang HD, He LY, Wang Q, Sheng XF (2017) Cd immobilization and reduced tissue Cd accumulation in rice (*Oryza sativa* wuyun-23) in the presence of heavy metal- resistant bacteria. Ecotoxicol Environ Saf 138:56–63

Liu W, Wang Q, Wang B, Hou J, Luo Y, Tang C (2015) Plant growth promoting rhizobacteria enhance the growth and Cd uptake of *Sedum plumbizincicola* in a Cd-contaminated soil. J Soils Sediments 15:1191–1199

Luo SL, Wan Y, Xiao X, Guo H, Chen L, Xi Q, Zeng G, Liu C, Chen J (2011) Isolation and characterization of endophytic bacterium LRE07 from cadmium hyperaccumulator *Solanum nigrum* L. and its potential for remediation. Appl Microbiol Biotechnol 89:1637–1644

Luo S, Xu T, Chen L, Chen J, Rao C, Xiao X (2012) Endophyte-assisted promotion of biomass production and metal-uptake of energy crop sweet sorghum by plant-growth-promoting endophyte *Bacillus sp.* SLS18. Appl Microbiol Biotechnol 93:1745–1753

Ma Y, Rajkumar M, Luo YM, Freitas H (2011a) Inoculation of endophytic bacteria on host and non-host plants effects on plant growth and Ni uptake. J Hazard Mater 195:230–237

Ma Y, Prasad MNV, Rajkumar M, Freitas H (2011b) Plant growth promoting rhizobacteria and endophytes accelerate phytoremediation of metalliferous soils. Biotechnol Adv 29:248–258

Ma Y, Oliveira RS, Nai FJ, Rajkumar M, Luo YM, Rocha I, Freitas H (2015) The hyperaccumulator *Sedum plumbizincicola* harbors metal-resistant endophytic bacteria that improve its phytoextraction capacity in multi-metal contaminated soil. J Environ Manag 156:62–69

Ma Y, Rajkumar M, Zhang C, Freitas H (2016a) Beneficial role of bacterial endophytes in heavy metal phytoremediation. J Environ Manag 174:14–25

Ma Y, Rajkumar M, Zhang C, Freitas H (2016b) Inoculation of *Brassica oxyrrhina* with plant growth promoting bacteria for the improvement of heavy metal phytoremediation under drought conditions. J Hazard Mater 320:36–44

Malik A (2004) Metal bioremediation through growing cells. Environ Int 30:261–262

Memon AR, Aktoprakligil D, Ozdemir A, Vertii A (2001) Heavy metal accumulation and detoxification mechanisms in plants. Turk J Bot 25:111–121

Neubauer U, Nowak B, Furrer G, Schulin R (2000) Heavy metal sorption on clay minerals affected by the siderophore desferrioxamine B. Environ Sci Technol 34:2749–2755

Nonnoi F, Chinnaswamy A, De La Torre V (2012) Metal tolerance of rhizobial strains isolated from nodules of herbaceous legumes (*Medicago* spp. and *Trifolium* spp.) growing in mercury-contaminated soils. Appl Soil Ecol 61:49–59

Omura Y, Shimotsuura Y, Fukuoka A, Fukuoka H, Nomoto T (1996) Significant mercury deposits in internal organs following the removal of dental amalgam, & development of pre-cancer on the gingiva and the sides of the tongue and their represented organs as a result of inadvertent exposure to strong curing light (used to solidify synthetic dental filling material) & effective treatment: a clinical case report, along with organ representation areas for each tooth. Acupunct Electrother Res 21:133–160

Park JD (2010) Heavy metal poisoning. Hanyang Med Rev 30:319–325

Pereira SIA, Castro PML (2014) Diversity and characterization of culturable bacterial endophytes from *Zea mays* and their potential as plant growth promoting agents in metal-degraded soils. Environ Sci Pollut Res 21:14110–14123

Rajkumar M, Ae N, Freitas H (2009) Endophytic bacteria and their potential to enhance heavy metal phytoextraction. Chemosphere 77:153–160

Rajkumar M, Ae N, Prasad M, Freitas H (2010) Potential of siderophore producing bacteria for improving heavy metal phytoextraction. Trends Biotechnol 28:142–149

Rajkumar M, Sandhya S, Prasad M, Freitas H (2012) Perspectives of plant associated microbes in heavy metal phytoremediation. Biotechnol Adv 30:1562–1574

Rajkumar M, Prasad MNV, Sandhya S, Freitas H (2013) Climate change driven plant-metal-microbe interactions. Environ Int 53:74–86

Rodrigues S, Henriques B, Reis A, Duarte A, Pereira E, Romkens PFAM (2012) Hg transfer from contaminated soils to plants and animals. Environ Chem Lett 10:61–67

Rodríguez H, Fraga R, Gonzalez T, Bashan Y (2006) Genetics of phosphate solubilization and its potential applications for improving plant growth-promoting bacteria. Plant Soil 287:15–21

Rojas-Tapias DF, Bonilla R, Dussán J (2014) Effect of inoculation and co-inoculation of *Acinetobacter* sp. RG30 and *Pseudomonas putida* GN04 on growth, fitness, and copper accumulation of maize (*Zea mays*). Water Air Soil Pollut 225(12):2232

Sarwar N, Saifullah Malhi SS, Zia MH, Naeem A, Bibi S, Farid G (2010) Role of plant nutrients in minimizing cadmium accumulation by plant. J Sci Food Agric 90:925–937

Sarwar N, Imran M, Shaheen MR, Ishaq W, Kamran A, Matloob A, Rehimb A, Hussain S (2017) Phytoremediation strategies for soils contaminated with heavy metals: modifications and future perspectives. Chemosphere 171:710–721

Schutze E, Weist A, Klose M, Wach T, Schumann M, Nietzsche S, Merten D, Baumert J, Majzlan J, Kothe E (2013) Taking nature into lab: biomineralization by heavy metal resistant Streptomycetes in soil. Biogeosciences 10:2345–2375

Sheng XF, Xia JJ, Jiang CY, He LY, Qian M (2008) Characterization of heavy metal resistant endophytic bacteria from rape (*Brassica napus*) roots and their potential in promoting the growth and lead accumulation of rape. Environ Pollut 156:1164–1170

Sheoran N, Valiya Nadakkakath A, Munjal V, Kundu A, Subaharan K, Venugopal V, Rajamma S, Eapen SJ, Kumar A (2015) Genetic analysis of plant endophytic *Pseudomonas putida* BP25 and chemo-profiling of its antimicrobial volatile organic compounds. Microbiol Res 173:66–78

Shi R, Weber G, Koster J, Reza-Hajirezaei M (2012) Senescence-induced iron mobilization in source leaves of barley (*Hordeum vulgare*) plants. New Phytol 195:372–383

Shin M, Shim J, You Y, Myung H, Bang KS, Cho M, Kamala-Kannan S, Oh BT (2012) Characterization of lead resistant endophytic *Bacillus* sp. MN3-4 and its potential for promoting lead accumulation in metal hyperaccumulator *Alnus firma*. J Hazard Mater 199:314–320

Singh J, Abhilash P, Singh H, Singh R, Singh D (2011) Genetically engineered bacteria: an emerging tool for environmental remediation and future research perspectives. Gene 480:1–9

Smith SR, Jaffe DM, Skinner MA (1997) Case report of metallic mercury injury. Pediatr Emerg Care 13:114–116

Sobariu DL, Fertu DIT, Diaconu M, Pavel LV, Hlihor RM, Dragoi EN (2016) Rhizobacteria and plant symbiosis in heavy metal uptake and its implications for soil bioremediation. New Biotechnol 39:125–134

Spaepen S, Vanderleyden J, Remans R (2007) Indole-3-acetic acid in microbial and microorganism-plant signaling. FEMS Microbiol Rev 31:425–448

Sriprang R, Hayashi M, Yamashita M, Ono H (2002) A novel bioremediation system for heavy metals using the symbiosis between leguminous plant and genetically engineered rhizobia. J Biotechnol 99:279–293

Sriprang R, Hayashi M, Ono H, Takagi M (2003) Enhanced accumulation of Cd^{2+} by a *Mesorhizobium* sp. transformed with a gene from *Arabidopsis thaliana* coding for phytochelatin synthase. Appl Environ Microbiol 69:1791–1796

Srivastava S, Singh N (2014) Mitigation approach of arsenic toxicity in chickpea grown in arsenic amended soil with arsenic tolerant plant growth promoting *Acinetobacter* sp. Ecol Eng 70:146–153.

Sturz AV, Matheson BG (1996) Populations of endophytic bacteria which influence host-resistance to Erwinia-induced bacterial soft rot in potato tubers. Plant Soil 184:265–271

Sun LN, Zhang YF, He LY, Chen ZJ, Wang QY, Qian M, Sheng XF (2010) Genetic diversity and characterization of heavy metal-resistant-endophytic bacteria from two copper-tolerant plant species on copper mine wasteland. Bioresour Technol 101:501–509

Tabassum B, Khanb A, Tariqa M, Ramzanc M, Khana MSI, Shahida N, Aaliya K (2017) Review bottlenecks in commercialisation and future prospects of PGPR. Appl Soil Ecol 121:102–117

Thangavel P, Subbhuraam C (2004) Phytoextraction: role of hyperaccumulators in metal contaminated soils. Proc NatlAcad SciIndia B 70:109–130

Trivedi P, Pandey A, Sa T (2007) Chromate reducing and plant growth-promoting activities of psychrotrophic *Rhodococcus erythropolis* MTCC 7905. J Basic Microbiol 47:513–517

Ullah A, Heng S, Munis MFH, Fahad S, Yang X (2015) Phytoremediation of heavy metals assisted by plant growth promoting (PGP) bacteria: a review. Environ Exp Bot 117:28–40

Verbon EH, Liberman LM (2016) Beneficial microbes affect endogenous mechanisms controlling root development. Trends Plant Sci 21:218–229

Vijayaraghavan K, Yun YS (2008) Bacterial biosorbents and biosorption. Biotechnol Adv 26:266–291

Visioli G, D'Egidio S, Vamerali T, Mattarozzi M, Sanangelantoni AM (2014) Culturable endophytic bacteria enhance Ni translocation in the hyperaccumulator *Noccaea caerulescens*. Chemosphere 117:538–544

Volesky B, Schiewer S (1999) Biosorption of metals. In: Flickinger M, Drew SW (eds) Encyclopedia of bioprocess technology. Wiley, New York, pp 433–453

Wan Y, Luo S, Chen J, Xiao X, Chen L, Zeng G, Liu C, He Y (2012) Effect of endophyte-infection on growth parameters and Cd-induced phytotoxicity of Cd-hyperaccumulator *Solanum nigrum* L. Chemosphere 89:743–750

Wani PA, Khan MS (2013) Nickel detoxification and plant growth promotion by multi metal resistant plant growth promoting *Rhizobium* species RL9. Bull Environ Contam Toxicol 91:117–124

Wei Y, Hou H, Xian Y, Guan S, Li JN, Li FS (2014) Genetic diversity of endophytic bacteria of the manganese-hyperaccumulating plant *Phytolacca Americana* growing at manganese mine. Eur J Soil Biol 62:15–21

Wheeler CT, Hughes LT, Oldroyd J, Pulford ID (2001) Effect of nickel on Frankia and its symbiosis with *Alnus glutinosa* (L.) gaertn. Plant Soil 231:81–90

Wu C, Wood T, Mulchandani A (2006) Engineering plant-microbe symbiosis for rhizoremediation of heavy metals. Appl Environ Microbiol 72:1129–1134

Wuana RA, Okieimen FE (2011) Heavy metals in contaminated soils: a review of sources, chemistry, risks and best available strategies for remediation. ISRN Ecol 2011:402647

Yong X, Chen Y, Liu W, Xu L, Zhou J (2014) Enhanced cadmium resistance and accumulation in Pseudomonas putida KT2440 expressing the phytochelatin synthase gene of *Schizosaccharomyces pombe*. Lett Appl Microbiol 58:255–261

Zhang XX, Li CJ, Nan ZB (2010) Effects of cadmium stress on growth and antioxidative systems in *Achnatherum inebrians* symbiotic with *Neotyphodium gansuense*. J Hazard Mater 175:703–709

Zheng H, Ouyang Z, Wang X, Miao H, Zhao T, Peng T (2004) Effects of forest restoration types on soil quality in red soil eroded region, Southern China. Acta Ecologica Sinica 24:1994–2002

Zhu LJ, Guan DX, Luo J, Rathinasabapathi B, Ma LQ (2014) Characterization of arsenic-resistant endophytic bacteria from hyperaccumulators Pteris vittata and Pteris multifida. Chemosphere 113:9–16

Chapter 5
Abiotic Stress: Its Outcome and Tolerance in Plants

P. Rawat, D. Shankhdhar, and S. C. Shankhdhar

Abstract The onset of nineteenth century along with anthropogenic pressure paved the way for global climatic variation which is a major factor for global undernourishment, malnutrition and endangered food security. The major upshot of climate change is abiotic stress like salinity, flood and drought that declines the agricultural productivity. Abiotic stress hampers the survival of the plants and restricts their growth and development. Each abiotic stress confers negative impact on plants by altering its physiology, morphology and metabolism. Production of reactive oxygen species during stress condition alters the structure and metabolic function in plants and restricts its growth. Drought is one of the serious threats to crop productivity among the abiotic stress that imposes multidimensional effects on plants. Drought alters physiology and anatomy of the plants and is the main reason for economic loss in terms of livestock and grain yield in both developed and developing countries. Plants adapt several resistance mechanisms to cope up with the drastic impact of stress. Main tolerance mechanisms are alteration in membrane structure, escaping the unfavourable conditions, activation of antioxidant defense system, production of compatible solutes for maintaining osmotic balance of the cell. Present manuscript focuses on the outcome of major abiotic stress in plants and their tolerance strategies against the variable environmental conditions.

5.1 Introduction

Contemporary civilization and industrialization have led to global climate change whose consequences like flood, drought, high temperature and global warming are quite perceptible and hazardous for soil heath, microbial diversity and sustainable crop production. Climate fluctuation also threatens the vulnerable and endangered organisms in the ecosystem that are incompetent to endure such environmental changes in their habitat. Climate variability has immense impact on food security.

P. Rawat (✉) · D. Shankhdhar · S. C. Shankhdhar
Department of Plant Physiology, College of Basic Sciences and Humanities, G. B. Pant University of Agriculture and Technology, Pantnagar, Uttarakhand, India

A. Sharma (ed.), *Microbes and Signaling Biomolecules Against Plant Stress*, Rhizosphere Biology, https://doi.org/10.1007/978-981-15-7094-0_5

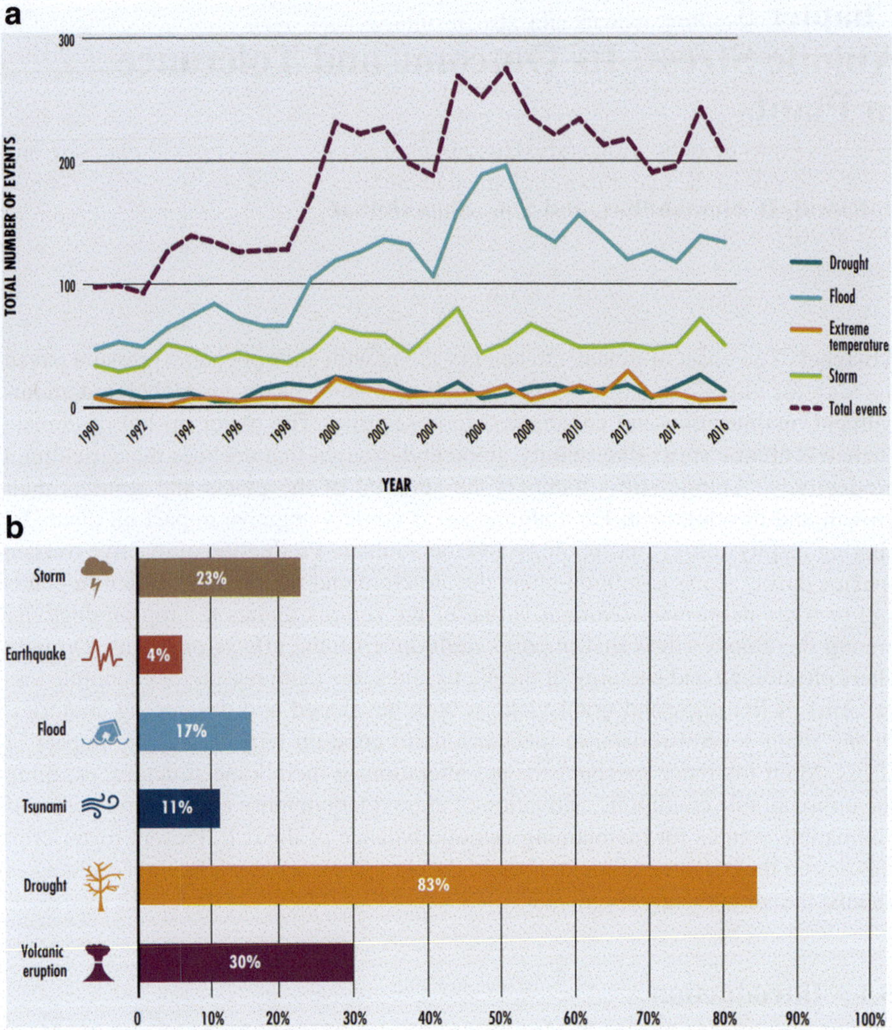

Fig. 5.1 (**a**) Rise in the calamitous events (1990–2016). Adopted from FAO (2018). (**b**) Damage and loss in agriculture as share of total damage and loss across all sectors by type of hazards. Adopted from FAO (2018)

It hits the underdeveloped regions of the world causing yield and economic losses. It has been estimated that 26% of total loss was in agriculture sector due to calamitous events from 2006 to 2016 (FAO 2018). Figure 5.1a, b illustrates the severity of disastrous events in the past years along with its impact on agricultural sector. There has been a dire inflation in the number of undernourished people across the world from 806 million in 2016 to 821 million in 2017 as an outcome of climate shock (FAO 2018). In view of the above facts, the foremost challenge is to provide food security to the projected global population of about 9.7 billion in 2050 under climate fluctuation and dwindling crop productivity.

Abiotic stress is an aftermath of climatologically variation that ceases plant growth, development and productivity below optimum levels. A surfeit or scant in energetic, chemical or physical condition in the nature paved the origin of abiotic stress in plants. The utmost abiotic stress comprises of drought, salinity, flooding and oxidative stress that abolishes metabolic equilibrium of plants thereby threatening food security (Fedoroff et al. 2010). Economic forfeiture in the developing countries in terms of crop and livestock production was discerned up to USD 96 billion from 2005 to 2015 with a 2.6% decrement in national agricultural value added growth due to occurrence of each climatic hazards (FAO 2018). An alarming upshot of climate shock is forecasted to decline agriculture productivity and influence five billion people by 2050 (UNESCO 2018; Watts 2018).

5.2 Abiotic Stress in Plants

5.2.1 Salt Stress

One of the most detrimental climate shocks is salinity stress which is prevalent in arid and semiarid regions in the world as depicted in Fig. 5.2. Approximately 33% of the world's irrigated land and 20% of overall cultivated land are devastated by salinity stress (Shrivastava and Kumar 2015). It has been forecasted that by 2050 around 50% of arable land would be influenced by salinity (Jamil et al. 2011). In India seven million hectares of land is sheathed by saline soil (Patel and Dave 2011). An electrical conductivity of precisely 4 dSm^{-1} of saturation extract from root zone at 25 °C marks the saline soil (Jamil et al. 2011). Salinity damages the agricultural productivity as the major food crops are susceptible to it. A remarkable yield loss of

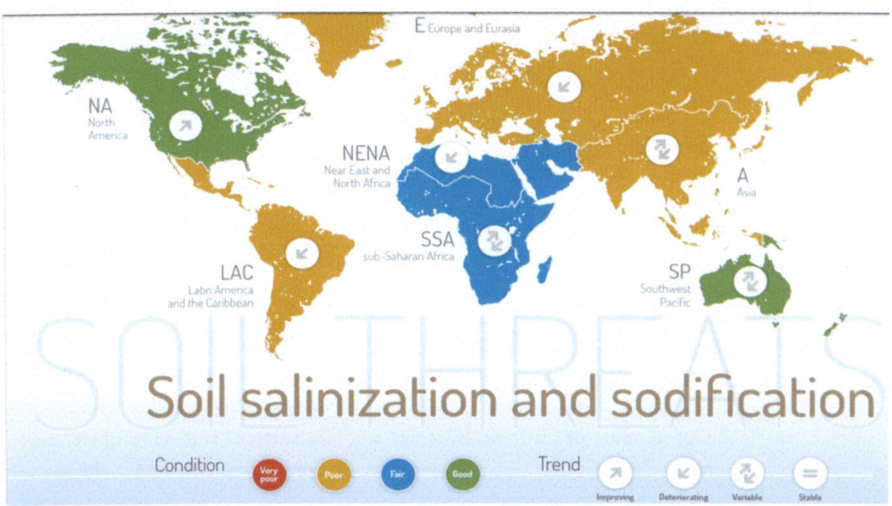

Fig. 5.2 Salt affected regions of the world. Adopted from FAO (2016)

about 65% in wheat was evident as an outcome of salinity stress (Ahmad et al. 2013). Yield reduction of about 55% in corn and cotton was observed in case of salinity in the range of about 8–10 dSm^{-1} and 18 dSm^{-1}, respectively (Satir and Berberoglu 2016; Zörb et al. 2019).

5.2.2 *Impact of Salinity Stress in Plants*

Salinity stress devastates the plants by two ways: (a) Osmotic effect that diminishes plant potential for water uptake due to excess salt in soil and (b) ion excessive effect that occurs due to accumulation of excess salt in the leaf tissues (Greenway and Munns 1980). Salinity amends the major physiological processes like photosynthesis, transpiration, nutrient regulation and osmotic balance as an outcome and affects plant growth and yield. Influence of salinity stress in plants has been illustrated in Table 5.1.

Table 5.1 Outcome of salinity stress in plants

Parameters	Alterations	References
Germination	Hinders water imbibition by seeds	Khan and Gul (2006)
	Alters the enzyme activity requisite for metabolic functions	Gomes-Filho et al. (2008)
	Deteriorates seed coat, diminishes the seed vigour index and surges seed dormancy	Panuccio et al. (2014)
Water relation	Osmotic imbalance leading to soaring salt concentration near roots	Munns (2005)
	Decrement in relative water content and turgor pressure	Ghoulam et al. (2002)
Plant growth	Minimizes water retaining capacity of plant	Munns (2002b)
	Enhances ion toxicity that results in early leaves senescence	Munns (2002a, 2005)
Photosynthesis	Accumulation of Na^+ and Cl^- ions in chloroplasts	Sudhir and Murthy (2004)
	Alteration in electron transport chain by impeding photon phosphorylation and carbon metabolism	Farahbakhsh et al. (2017).
	Decrement in chlorophyll content	Saha et al. (2010)
	Modification in cytoplasmic structure, retardation in carboxylation reaction and generation of ROS	Maxwell and Johnson (2000)
Nutrient imbalance	High osmotic potential lessens nitrogen content in plants	Lea-Cox and Syvertsen (1993); Rozeff (1995); Bar et al. (1997)
	Low phosphorous content	Qadir and Schubert (2002)
	Reduced influx of Ca^{2+} and K^+ ions	Suhayda et al. (1990); Hu and Schmidhalter (1997); Asch et al. (2000)

5.2.3 Plant Tolerance Strategies against Salinity Stress

Resistance strategies of plants to salinity rely on environmental conditions and plant species. Plants mainly endure salt stress by three ways:

(a) Endurance to osmotic stress
(b) Efflux of sodium from leaf blades
(c) Tissue tolerance

5.2.3.1 Osmotic Stress Endurance

Osmotic stress is marked by generation of osmolytes like glycine betaine, proline, polyols, etc. to maintain the osmotic balance. Osmotic tolerance also aids in balancing stomatal conductance and leaf expansion during salinity (Rajendran et al. 2009). Osmolytes stabilize the membrane integrity, proteins and pigments in the cell and also alleviate reactive oxygen species (Krasensky and Jonak 2012; Gupta and Huang 2014). It was evident that proline accumulation improved the salinity tolerance in maize (Tang et al. 2019). Likewise, trehalose accumulation also assisted in boosting salinity endurance in wheat (Sadak 2019).

5.2.3.2 Sodium Ion Efflux from Leaf Blades

Outcome of salinity stress in plants comprises ion toxicity due to sodium ion (Na^+) accumulation inside the cell. This can be overcome by down regulating the expression of ionic channels and Na^+ transporters (Munns and Tester 2008; Rajendran et al. 2009). An antiporter (AtNHX1) confined in tonoplast of *Arabidopsis* maintains Na^+ balance in the cell under salinity condition by directing the influx of Na^+ ions from cytosol to vacuole (Apse et al. 1999). Likewise NAX1gene in durum wheat impedes Na^+ translocation in the shoots and leaf sheath from roots, whereas NAX2 gene confers high potassium ion (K^+)translocation in leaf (Flagella et al. 2006). Sodium exclusion theory for salt resistance has been validated in rice (Ren et al. 2005) and durum wheat (James et al. 2006). It was studied that potassium transporter (PpHKT1) gene from rootstock of almond also enhanced salt tolerance in transgenic *Arabidopsis* by balancing ion homeostasis (Kaundal et al. 2019).

5.2.3.3 Tissue Tolerance

For hindering the senescence of leaves and enhancing its survival rate, tissue tolerance is a salient attribute during stress. Generation of compatible solutes that balances the osmotic homeostasis in the cell and antioxidant defense system that scavenges the reactive oxygen species under stress condition is an attribute of tolerance of plants to stress condition (Hasegawa et al. 2000; Zhu 2001; Sakamoto and Murata 2002). Tissue resistance also incites the ion compartmentalization in the cell to diminish toxicity of ion (Munns and Tester 2008). A notable rise in osmolyte production like proline up to 14.87 μmol g^{-1} DW and an increment in antioxidants like ascorbate peroxidase up to 77%, catalase up to 15% and glutathione reductase up to 138% were recorded in chickpea as a tolerance response to salinity stress (Arefian et al. 2018).

5.3 Flooding Stress

Prolonged submergence and soil waterlogging result in hypoxic condition that constitutes flood stress in plants. One of the key limitations to crop productivity in high rainfall areas of the world is flood stress. Climate change promotes submergence to be more recurring and drastic (Pachauri et al. 2014). Approximately 10% of irrigated land in India, China, Russia and 16% of cultivable land in the USA are damaged by waterlogging (FAO 2015). Figure 5.3 explains the ubiquity of flood hazard in the world. According to statistical approximates of food and agriculture organization of United Nations, flood stress upshots 17% of total loss and damage in agriculture (FAO 2018).

It was discerned that annually 20–50% production loss in wheat occurred under waterlogging condition (Hossain and Uddin 2011). Overall 15% of maize production in South Asia is damaged by floods. India accounts for 25–30% maize production loss every year on account of flood stress (Bailey-Serres et al. 2012).

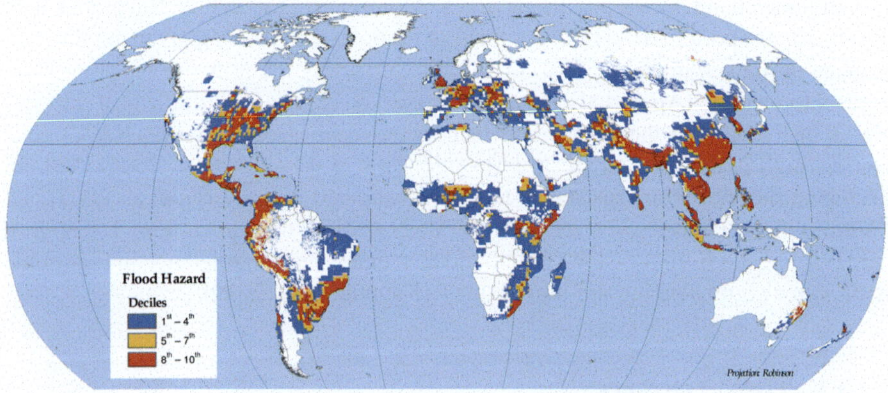

Fig. 5.3 Global flood hazard prevalence. Adopted from SEDAC (n.d.)

5.3.1 *Outcome of Flood Stress in Plants*

The prime consequence of flood stress in plant comprises anoxia or oxygen limitation. Necrosis, senescence, low nitrogen fixation and ultimately plant death are the consequences of flood stress (Hasanuzzaman et al. 2012). Overall effects of flood stress in plants are outlined in Table 5.2.

5.3.2 *Adaptation of Plants to Flood Stress*

5.3.2.1 Morphological Alterations

For escaping anoxic conditions under flood stress, plants showcase morphological adaptations. Submerged leaves have low stomatal counts and are often dissected and filamentous (Sculthorpe 1967). Vigorous adventitious root growth is one of the prominent adaptation as evident in *Sesbania javanica* (Jackson 2006) and *Chrysanthemum* (Yin et al. 2009). Decrement in root hairs and diameter, rapid shoot elongation are some of the evident variations (Jackson 2008). Under submergence, formation of leaf gas film improves carbohydrate supply, generation of adventitious roots and aeration of roots. In rice, resistance to radial oxygen loss is enhanced by

Table 5.2 Consequences of flood stress in plants

Parameters	Alterations	References
Anatomical variation	Over growth of aerenchyma, lenticels and adventitious roots	Ashraf (2009)
	Decrement in stele and seminal root diameter, reduced cortex thickness	Grzesiak et al. (1999)
	Increment in width of phloem tissue and number of xylem vessels	de Souza et al. (2013)
	Formation of suberized exodermis on root cortex periphery	Kulichikhin et al. (2014)
Physiology and metabolism	Impaired stomatal conductance, CO_2 assimilation	Ashraf (2009)
	Decrement in net photosynthetic rate, transpiration rate, PSII efficiency and rubisco activity	Zheng et al. (2009)
	Low relative water content and membrane stability	Kumar et al. (2013)
	Accumulation of hexose sugars in roots and organic acids in leaves	Vandoorne et al. (2014)
Nutrient availability	Enhanced nitrogen content and reduced phosphorous and potassium content in plant tissues	Reddy and mittra (1985)
	Reduced uptake of copper and zinc	Tarekegne et al. (2000)
Growth and yield	Reduction in leaf number, diameter, height and leaf area of plant, leaf epinasty	Parent et al. (2008); Gonzalez et al. (2009)
	Decline in dry matter accumulation, chlorosis, and flower abortion	Mensah et al. (2006)

development of effective barriers. This aids in shielding the plants from toxins generated from anoxic environment and boosts root tip aeration under flood stress (Kirk et al. 2014; Yamauchi et al. 2018).

5.3.2.2 Anatomical Alterations

To hinder the loss of radial oxygen from submerged roots, plants show formation of lateral diffusion barrier (Sauter 2013). Lenticels development is also evident in flood affected plants as this aids in oxygen diffusion in downward direction (Parelle et al. 2006). Aerenchyma formation is one of the unique characters of plants under flood or submerged condition as these gas spaces allow oxygen transport from roots to shoots (Shiono et al. 2008). Exodermis thickness was improved by 23.70% compared to control in *Garcinia brasiliensis* under flood state in addition to increment in root xylem vessels and phloem width (de Souza et al. 2013). In flood tolerant barley cultivar (TX9425 and Yerong), more well developed intercellular spaces in leaves and mesophyll cells were observed that enhanced photosynthesis due to quick gaseous exchange Zhang et al. 2016). In rice seedlings, rapid coleoptile elongation during germination is a marker for submergence tolerance that facilitates aeration of developing embryo (Guru and Dwivedi 2018).

5.3.2.3 Biochemical Alterations

Under flood stress, switch in the metabolism of plants from aerobic to anaerobic condition with high activity of alcohol dehydrogenase, a key enzyme in anaerobic fermentation is observed (Sairam et al. 2008). Proline content in *Casuarina* was high in tolerant cultivar as it acts as a signal molecule, an osmolyte and showcases antioxidant defense response under stress condition (Bajpai and Chandra 2015). Enhanced activity of alpha aminobutyric acid has been evident in soybean nodules in flooded conditions as this metabolite balances intercellular pH, acts as nitrogen reservoir under flood condition and also serves as antioxidant (Souza et al. 2016). Elevated activity of enzymes like phosphoenolpyruvate carboxylase, glutamate dehydrogenase and alcohol dehydrogenase was observed in flooded nodules of *Sesbania herbacea* (Krishnan et al. 2019).

5.4 Heavy Metal Stress

One of the major aftermaths of industrial revolution in the twenty-first century is heavy metal pollution in the environment. Heavy metals are potentially noxious elements and present in trace amount in soil. They are comprised of metalloids, lanthanoides and transition metals with atomic density ⍰6 g cm^{-3}except selenium, boron and arsenic (Park et al. 2011). The root cause of heavy metals in the environment encompasses usage of pesticides and fertilizers, combustion of fuels, mining, sewage sludge, oil spills, etc. Heavy metals are diverse in their functioning. Metals like iron, zinc, copper, nickel, manganese, chromium impart proper functioning of plant's metabolic processes contrarily lead, arsenic, mercury have negative effect on plant growth and productivity (Pierart et al. 2015). These noxious elements enter the food chain and attack human, animal and soil health (Augustsson et al. 2015).

Uptake of heavy metal by plants interrupts the chief metabolic processes like photosynthesis, nitrogen fixation and nutrients uptake and results in biomass and yield reduction and also death of the plants (Buendía-González et al. 2010). In China, about 20,000,000 acres of farmland is polluted by heavy metal with an annual crop loss of 1,000,000 million tons per year (Wu et al. 2010). About 12% of world's agricultural land is estimated to be affected by heavy metal pollution (Dziubanek et al. 2015). One of the most detrimental heavy metal for human health is lead. Lead exposure resulted in loss in economic productivity of about $977 billion annually in low to middle-income count (Attina and Trasande 2013). Statistical estimates of Institute for Health Metrics and Evaluation disclosed that lead exposure caused nearly 0.5 million deaths and 9.3 million life years lost among adults of 15 years and older (Landrigan et al. 2018).

5.4.1 *Outcome of heavy metal stress in plants*

Heavy metal stress deteriorates soil and plant health. Metal toxicity alters the morphology and physiology of plants at different growth stage as a consequence there is decline in crop productivity. Table 5.3 outlines the impact of heavy metal stress in plants.

5.4.2 *Plant tolerance to heavy metal stress*

5.4.2.1 Antioxidant enzymes

To escape heavy metal toxicity plants evolve many alterations. Exclusion of metal ions from the cell or binding with ligand prevents entry of metal ions to the roots is the first line of defense mechanism (Zeng et al. 2011). Metal ions also make complexes with carboxyl group of pectin in the cell wall thereby restricting its entry in the cell (Jiang and Liu 2010). Secondary response for metal detoxification is generation of antioxidant enzymes like superoxide dismutase, catalase and peroxidase. Compounds like glutathione, proline, cysteine, ascorbic acid also serve as scavengers of reactive oxygen species (Shahid et al. 2012; Pourrut et al. 2013).

Glutathione is a tripeptide antioxidant with thiol group and forms mercaptide bond with metals due to its nucleophilic thiol group and its reduced form also

Table 5.3 Out-turn of heavy metal stress in plants

Parameters	Modifications	References
Germination	Inhibition of water uptake by copper, cadmium	Kranner and Colville (2011)
	Reduction in germination percentage	Chigbo and Batty (2013)
	Formation of abnormal embryos and decrement in seed viability	Stvolinskaya (2000)
	Slow activity of alpha amylases, phosphatases. Decrement in remobilization of storage reserves	Kalai et al. (2014)
	Lipid peroxidation and proline accumulation in the radical	Kalai et al. (2014)
Plant growth	Chlorosis, decline in growth and yield, reduced nutrient uptake and nitrogen fixing potential	Guala et al. (2010)
	Generation of reactive oxygen species	Romero-Puertas et al. (2004)
Photosynthesis	Disruption in energy transfer in light harvesting complex, decline in chlorophyll and carotenoid content	Kuzminov et al. (2013)
	Electron transport inhibition between photosystem I and II	Rama Devi and Prasad (1999)
	Deformation of chloroplast ultrastructure	Kalaji and Loboda (2007)
Nutrient uptake	Low concentration of zinc, iron, calcium and manganese	Chatterjee et al. (2004)
	Decline in shoot nitrate content and nitrate reductase activity	Xiong et al. (2006)

scavenges peroxide radicals. Glutathione conjugates with compounds of heavy metals that are translocated to vacuole to protect cell from its harmful effects (Klein et al. 2006; Yazaki 2006). It was also studied that in response to cadmium toxicity there is a modulation in the activity of antioxidants like guaiacol peroxidase, ascorbate peroxidase and catalase in tolerant plants (Alves et al. 2017; Borges et al. 2018). In *Brassica juncea* L. such kind of high efficiency antioxidative defense system was evident during caesium toxicity (Lai and Luo 2019).

5.4.2.2 Metallothioneins and Phytochelatins

Vacuolar sequestration of heavy metals is an important response of plants to metal stress. Complex formation of metal ion with ligand reduces the toxicity of heavy metals. Metallothioneins and phytochelatins rich in cysteine are natural metal chelators in plants that reduce metal toxicity by forming mercaptide bonds with metals and sequester them to vacuoles (Gupta et al. 2010; Jiang and Liu 2010). In *Brassica napus,* cadmium toxicity results in strong expression of MT4 metallothionein in germinating seeds as a defense mechanism (Mierek-Adamska et al. 2018). Likewise in *Oryza sativa,* phytochelatin synthase genes (OsPCS5 and OsPCS15) were profoundly induced under cadmium stress as tolerance mechanism.

These genes encode phytochelatins that bind heavy metals in cytosol and detoxify them in the vacuoles (Park et al. 2019).

5.5 Low Temperature Stress

Chilling or low temperature stress amends the plant morphology, physiology and metabolism. Occurrence of chilling stress at temperature 0 °C to 15 °C whereas freezing stress at ambient temperature for ice crystal formation causes cellular injury and osmotic dehydration in the cell (Bhatla 2018).

5.5.1 Impact of low temperature stress in plants

Chilling in plants is advantageous for breaking seed dormancy, for vernalization and cold acclimation. In contrast, chilling in susceptible crops has many aftermaths as listed in Table 5.4.

5.5.2 Adaptations to low temperature stress

5.5.2.1 Cold acclimation

Increment in phospholipid proportion and decrement in ceramide and sugar containing lipids in plasma membrane results in reduction in expansion induced lysis. Expression of sugar producing enzymes like acid invertase, galactinol synthase and sucrose synthase to improve membrane stability is induced in response to cold acclimation (Turhan and Ergin 2012). Accumulation of compatible solutes like

Table 5.4 Consequences of chilling stress in susceptible crops

Parameters	Modifications	References
Lipid membrane	Membrane becomes leaky and dysfunctional	Bhatla (2018)
	Membrane transport and receptors disabled	
	Cellular structure distorted	
	Osmotic dehydration of cell	
Photosynthetic apparatus	Impaired electron transport chain	Bhatla (2018); Wise et al. (1987)
	Generation of reactive oxygen species	
	Cell plasmolysis	
	Chloroplast swelling and dilation of thylakoids, disintegration of chloroplast	
	Depletion of starch and phosphorylated metabolites in cell	

hydroxyl proline and arabinose in the cell maintains osmotic balance during cold acclimation (Bhatla 2018). Expression of hydrophilic and LEA (late embryogenesis abundance protein) proteins declines the denaturation of proteins under cold and drought stress. In *Brassica napus,* dehydrin proteins, a class of LEA proteins was highly expressed under cold stress in tolerant cultivar. It is used as biomarker for selection of cold tolerant lines (Maryan et al. 2019).

5.5.2.2 Modification in gene expression

Expression of cold regulated genes (COR) confers cold tolerance in plants by stabilizing the plasma membrane during low temperature stress. Hydrophilic proteins encoded by COR genes expressed during low temperature stress are involved in the production of anthocyanin and play indirect role in cold acclimatization. Extracellular production of antifreeze proteins (AFP) prevents ice crystal formation during low temperature stress. Molecular chaperons like HSP90 restrict denaturation of proteins during chilling stress. Other proteins like MAP (mitogen activated protein) kinase and calmodulin related proteins are encoded by COR genes and impart role in cold temperature signalling in plants (Bhatla 2018). A chromatin remodelling factor PICKLE (PKL) was found responsible for cold stress tolerance in *Arabidopsis*. It also modulates expression of other cold responsive genes like RD29A and COR15A (Zhao et al. 2019). In *Brassica rapa* L., genes such as temperature-induced lipocalin-1, zinc finger protein ZAT12, dehydrin ERD10 and adenosylhomocysteinase 2 were analysed and found to be involved in cold stress tolerance (Ma et al. 2019).

5.6 Drought Stress

One of the most disastrous outcomes of climate change is drought that restricts plant growth and development in both developed and developing countries. The utmost driver of undernourishment and hunger is geographical and global drought that declines agricultural productivity (FAO 2017).US, a technologically advanced country encounters an annual loss of $ 6 billion in agricultural productivity due to drought and such losses are more severe in underdeveloped and developing countries (CIA 2017). Around 20% of world's cultivable land is hit by drought and more than 80% damage in livestock and agricultural production is because of drought stress (FAO 2018). Major drought exposed regions in the world have been delineated in Fig. 5.4.

Drought is a climatic variation with perpetuate periods of rainfall scarcity that results in hunger, malnutrition, deficit food productivity and ultimately food insecurity. Plants encounter severe drought when water loss via transpiration is high enough or when roots are deficient in water supply (Anjum et al. 2011). A severe drought can threaten the national food availability and results in pervasiveness of undernourishment. Approximate global yield loss of 21% in wheat and 40% in

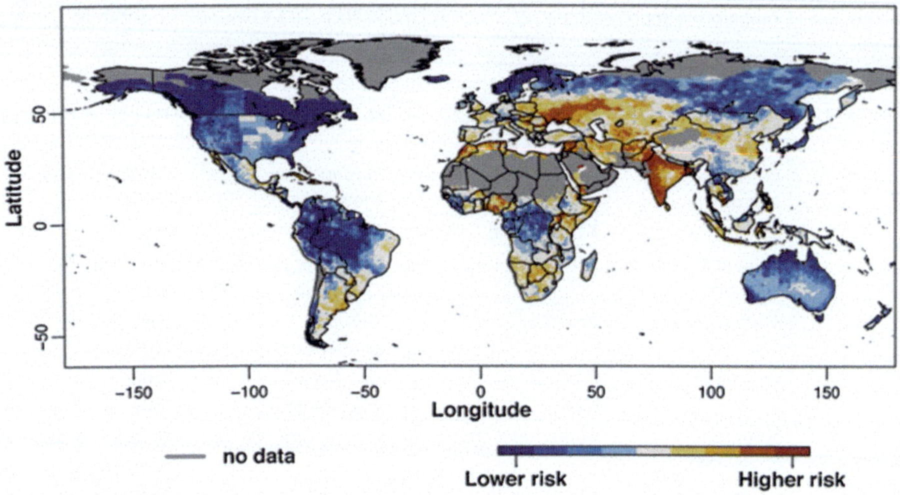

Fig. 5.4 Global map of drought affected regions. Adapted from Carrao et al. (2016)

maize was observed from 1980 to 2015 (Daryanto et al. 2016). Table 5.5 illustrates susceptible and resistant crops under major abiotic stress.

Drought alters photosynthesis, water balance, nutrient acquisition, oxidative metabolism and assimilates partitioning in plants. This modification in plants depends on species, growth stage of the plants and environmental conditions. Apparent symptoms of drought in plants are depicted in Fig. 5.5. Reduction in grain filling rate, harvest index and solar use efficiency are some key drivers responsible for diminished yield under drought stress. Table 5.6 describes deleterious impact of drought stress in plants ranging from morphological, biochemical to molecular level and is discernible at every phenological stage of crop growth at any period of time (Farooq et al. 2009; Deepak et al. 2019).

5.6.1 *Outcome of Drought Stress on Morphological Attributes*

The foremost impact of drought stress on morphology of plant is restricted seed germination and seedling growth (Farooq et al. 2009). Drought negatively alters seed vigour, seedling growth and also results in poor seed quality as reported in *Medicago sativa* (Zhang et al. 2019a, b), *Oryza sativa* (Bhavyasree et al. 2019), *Glycine max* (Reddy et al. 2019), *Zea mays* and *Sorghum* (Queiroz et al. 2019).

Drought impedes mitosis and cell enlargement because of restricted water flow from xylem to neighbouring cells (Hussain et al. 2008). Reduction in plant size and biomass is one of the major morphological alterations in plants during drought stress. Fifty percent reduction in biomass in wheat was observed as compared to control

Table 5.5 Comparative representation of crops for resistance and tolerance to different abiotic stress

Stress	Sensitive crop	Reference	Resistant crop	Reference
Salinity	*Brassica napus* L.	Dolatabadi et al. (2019)	*Hordeum vulgare*	Han et al. (2018)
	Phaseolus vulgaris	Lahaye and Epstein (1971)	*Echinochloa frumentaceae*	Williams et al. (2019)
	Oryza sativa L.	Khatun et al. (1995)	*Beta vulgaris* L.	Wu et al. (2018)
	Triticum aestivum	Ahmad et al. (2013)	*Vigna unguiculata*	Pan et al. (2019)
	Lupinus albus	Jeschke (1984)	*Solanum lycopersicum*	Amjad et al. (2019)
Flood	*Zea mays*	Panozzo et al. (2019)	*Saccharum officinarum*	Singh et al. (2019)
	Solanum lycopersicum	Ezin et al. (2010)	*Oryza sativa* L.	Dwivedi et al. (2018)
	Glycine max	Tewari et al. (2016)	*Solanum dulcamara* L.	Dawood et al. (2014)
	Triticum aestivum	Zhou (2010)	*Zea nicaraguensis*	Yamauchi et al. (2018)
	Cicer arietinum	Solaiman et al. (2007)	*Populus euphratica*	Yu et al. (2015)
Heavy metal	*Triticum aestivum* (Lead)	Alamri et al. (2018)	*Lupinus* (Cadmium, nickel)	Rathaur et al. (2012)
	Oryza sativa L. (Cadmium, arsenic)	Makino et al. (2019)	*Brassica napus* (Nickel, cadmium, copper)	Marchiol et al. (2004)
	Lolium perenne L. (Aluminium)	Parra-Almuna et al. (2018)	*Raphanus sativus* (Nickel, cadmium, copper)	Marchiol et al. (2004)
	Brassica juncea (Copper)	Yadav et al. (2018)	*Setaria italica L.* (Cadmium)	Chiang et al. (2011)
	Solanum lycopersicum (Cadmium)	Carvalho et al. (2018)	*Miscanthus sinensis* (Aluminium)	Ezaki et al. (2008)
Chilling	*Solanum lycopersicum*	Ronga et al. (2018)	*Arabidopsis thaliana*	Schulz et al. (2016)
	Zea mays	Bilska-Kos et al. (2017)	*Brassica juncea* L.	Sinha et al. (2015)
	Cicer arietinum	Selpair (2018)	*Mentha arvensis*	Dhawan et al. (2018)
	Brassica oleracea	Zhang et al. (2019a, b)	*Pisum sativa* L.	Zhang et al. (2016)
	Vigna radiata	Batra et al. (2018)	*Capsicum annuum*	Shawon et al. (2017)
Drought	*Oryza sativa* L.	Lawas et al. (2018)	*Setaria italica*	Han et al. (2019)

(continued)

Table 5.5 (continued)

Stress	Sensitive crop	Reference	Resistant crop	Reference
	Triticum aestivum	Bakhshandeh et al. (2019)	*Sorghum*	Ohnishi et al. (2019)
	Zea mays	Daryanto et al. (2016)	*Arachis hypogaea* L.	Banavath et al. (2018)
	Cicer arietinum	Kaloki et al. (2019)	*Macrotyloma uniflorum*	Sharma et al. (2018)
	Glycine max	Popović et al. (2015)	*Brassica napus* L.	Mahmood et al. (2019)
	Helianthus annuus	Pekcan et al. (2015)	*Hordeum vulgare*	Carter et al. (2019)

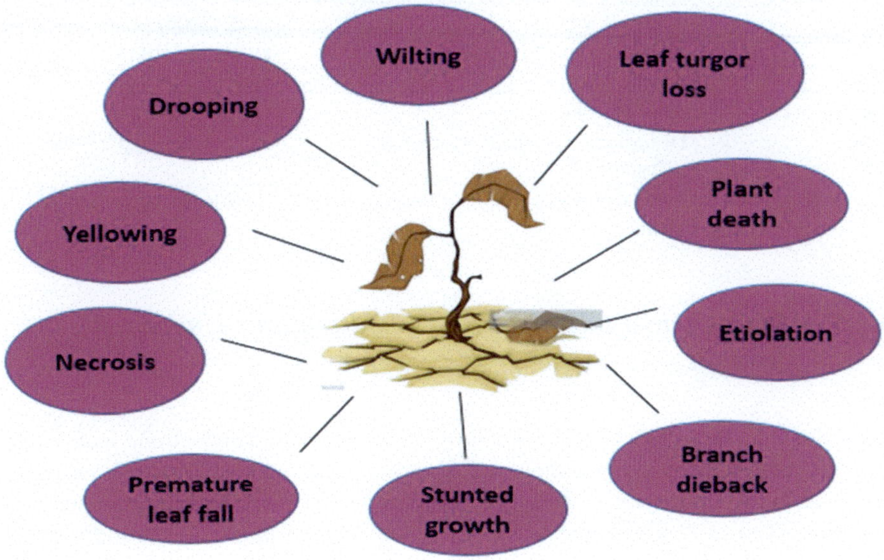

Fig. 5.5 Drought stress prodrome in plants

under drought stress (Paul et al. 2019). Dry weight and fresh weight of root and shoot as well as root shoot ratio tend to cut down in drought induced *Brassica napus* L. (Khan et al. 2019).

Decrement in stomatal aperture, leaf rolling, cutinization of leaf surfaces are some observable drought stress induced morphological parameters (Hussain et al. 2008). Drought hampers the leaf size, grain yield in crops, for instance, yield parameters like panicle length, seed setting rate and grain weight declined in *Oryza sativa* L. under drought (Wei et al. 2017). Likewise in wheat, 62% of grain loss was observed in drought condition compared to well-watered conditions (Paul et al. 2019).

Table 5.6 Salient modifications in drought stricken plants

Parameters	Modification	Reference
Morphology	Inhibition of seed germination	Farooq et al. (2009)
	Reduce shoot and root dry weight	Manickavelu et al. (2006)
	Diminished leaf area and number of stomata	Jaleel et al. (2009)
	Impaired mitosis and cell elongation	Hussain et al. (2008)
	Loss of turgor pressure and cell growth	Taiz and Zeiger (2006)
Water and nutrient relation	Low water use efficiency and transpiration rate	Turner et al. (2001)
	Low uptake of phosphorous, calcium and magnesium	Barber (1995)
	Enhanced nitrogen uptake	Barber (1995)
Photosynthesis	Reduction in RUBP generation and photo-system II activity	Wise et al. (2004); Camejo et al. (2006)
	Low chlorophyll content	Din et al. (2011)
Assimilate partitioning	Decline in translocation of assimilates from source to sink	Kim et al. (2000)
	Disturbance in phloem loading and unloading	Zinselmeier et al. (1999)
	Low potential of sink to utilize the assimilates	Zinselmeier et al. (1999)

5.6.2 Physiological and Biochemical Alterations Under Drought

Physiological attributes like transpiration rate, stomatal conduction, leaf water potential and relative water content are adversely affected by drought. Low water potential, transpiration rate, relative water content and enhanced leaf temperature were observed in *Oryza sativa* under drought (Fahad et al. 2017). Drought stress induces the formation of reactive oxygen species (ROS) along with enhanced generation of osmolytes like proline, glycine betaine. Production of ROS leads to oxidative stress that ultimately results in cell death in plants (Silva et al. 2019).

Another important trait influenced by drought is photosynthesis. Drought stress induces reduction in leaf expansion, damages the photosynthetic apparatus and activity of photosynthetic enzymes. Reduction in phosphorylation and disruption in ATP synthesis are key drivers for impaired photosynthesis under drought (Fahad et al. 2017). Reduction in thylakoid membrane proteins and PSII activity under PEG induced drought stress have been evident in many studies (Wang et al. 2019).

Activity of rubisco, a motor enzyme of photosynthesis is declined drastically under drought. Reduction in amount of substrate, i.e. ribulose 1, 5 bisphosphate for carboxylation, modification in ultrastructure of rubisco and chloroplast and acidification of chloroplast stroma are the driving factors for decrement in rubisco activity. Other enzymes like fructose-1, 6-bisphosphatase, NADP-glyceraldehyde phosphate dehydrogenase, phosphoribulokinase, phosphoenol pyruvate carboxylase and pyruvate orthophosphate dikinase also decline under drought stress (Farooq et al. 2009).

Drought restricts nutrient movement from soil to plants. Under drought, root growth and proliferation are hindered that limit nutrient translocation in plants. Limited flow of oxygen, carbon and nitrogen assimilation in root nodules impedes nitrogen fixing ability of legumes under drought (Ladrera et al. 2007). Nitrate reduction in leaves was limited by declined photosynthesis, which supplies reducing power, generated during photosynthetic via electron transport for the process of nitrate reduction (Chen et al. 2018).

5.6.3 Adaptation of Plants to Drought Stress

5.6.3.1 Escape

Drought escape is characterized by short life cycle that enables plants to flower early before onset of drought. Growing season is dependent on environmental variation and plant genotype. Drought escape is enhanced when development of plant synergizes with duration when soil moisture is available. Although drought escape helps in plants survival during stress yet there is a decrement in the yield (Farooq et al. 2009). High grain yield in *Lens culinaris* was recorded under early drought as a result of drought escape. Maximum leaf area, stomatal conductance, high stomatal density, early flowering with short life cycle were also observed as an outcome of drought escape strategy in lentil (Sanchez-Gomez et al. 2019).

5.6.3.2 Avoidance

Avoidance is marked by maintaining the water potential of plants during drought. It is also characterized by reduced water deprivation through transpiration and balancing the water uptake from soil. For efficient water uptake, a deep and extensive root system is required along with thick cuticle whose formation requires high input of energy that ultimately results in low yield. Therefore, plants with avoidance strategies are generally of compact size (Lisar et al. 2012). In *Sorghum,* aquaporin genes PIP2;3 were strongly expressed in roots under drought compared to roots under watered conditions. Aquaporins are the membrane proteins that allow water and solute passage through the membrane into cell and their expression under drought is an adaptive strategy (Schulze et al. 2019).

5.6.3.3 Tolerance

Tolerance to drought stress is perceptible by limited area and number of leaves during water deficit condition. Plants show traits like formation of trichomes and hairy leaves which reduce leaf temperature during drought. Root architecture like root density, root proliferation, root growth rate are some key alterations in plants during drought. Reduction in stomata number, increment in root shoot ratio, accumulation of osmolytes and induction of antioxidant defense system are other drought tolerance traits in plants (Sapeta et al. 2013). In sorghum, leaf rolling, leaf waxiness, stay green trait, root architecture, abscisic acid metabolism, ion transporter, transpiration efficiency and osmotic adjustments are drought tolerant traits that enable it to survive under drought stress (Badigannavar et al. 2018).

5.7 Conclusion

Major constraint for sustainable agricultural productivity and global food security is abiotic stress which is an outcome of global climate change. Extreme weather conditions attack morphological, physiological and biochemical attributes of the plants and threat their survival in the environment. Multiple abiotic stresses like salinity, flood, heavy metal and drought impact the crop yield. Drought is the most devastating abiotic stress that declines the global crop productivity. Drought hampers seed germination, reduces the stomatal frequency and diminishes leaf area and water potential. At physiological level, drought restricts photosynthesis, thylakoid structure and enzyme production. All these modifications alter the metabolism and limit growth and development of plants. Despite such abiotic constraints, plants have inherent tolerance mechanisms that enable them to cope up and survive under extreme conditions. Activation of antioxidant defense system, changes in the membrane lipid composition and enzyme production as well as morphological alterations in plants are some of the tolerance traits in plants. In spite of such adaptive strategies, improvements in the tolerance of plants against the extreme calamities are required. Strategies like alteration in expression of stress responsive genes and transcription factors are potential candidates to develop stress tolerant crops. More emphases should be there on molecular research for exploiting the stress tolerance traits and minimizing the aftermath of stress in plants.

References

Ahmad M, Shahzad A, Iqbal M, Asif M, Hirani AH (2013) Morphological and molecular genetic variation in wheat for salinity tolerance at germination and early seedling stage. Aust J Crop Sci 7:66–74

Alamri SA, Siddiqui MH, Al-Khaishany MY, Nasir Khan M, Ali HM, Alaraidh IA, Alsahli AA, Al-Rabiah H, Mateen M (2018) Ascorbic acid improves the tolerance of wheat plants to lead toxicity. J Plant Interact 13:409–419

Alves LR, Monteiro CC, Carvalho RF, Ribeiro PC, Tezotto T, Azevedo RA, Gratão PL (2017) Cadmium stress related to root-to-shoot communication depends on ethylene and auxin in tomato plants. Environ Exp Bot 134:102–115

Amjad M, Ameen N, Murtaza B, Imran M, Shahid M, Abbas G, Naeem MA, Jacobsen SE (2019) Comparative physiological and biochemical evaluation of salt and nickel tolerance mechanisms in two contrasting tomato genotypes. Physiol Plant 168:27–37

Anjum SA, Wang LC, Farooq M, Hussain M, Xue LL, Zou CM (2011) Brassinolide application improves the drought tolerance in maize through modulation of enzymatic antioxidants and leaf gas exchange. J Agron Crop Sci 197:177–185

Apse MP, Aharon GS, Snedden WA, Blumwald E (1999) Salt tolerance conferred by overexpression of a vacuolar Na+/H+ antiport in Arabidopsis. Science 285:1256–1258

Arefian M, Vessal S, Shafaroudi SM, Bagheri A (2018) Comparative analysis of the reaction to salinity of different chickpea (Cicer aretinum L.) genotypes: a biochemical, enzymatic and transcriptional study. J. Plant Growth Regul 37:391–402

Asch F, Dingkuhn M, Dörffling K, Miezan K (2000) Leaf K/Na ratio predicts salinity induced yield loss in irrigated rice. Euphytica 113:109–118

Ashraf M (2009) Biotechnological approach of improving plant salt tolerance using antioxidants as markers. Biotechnol Adv 27:84–93

Attina TM, Trasande L (2013) Economic costs of childhood lead exposure in low-and middle-income countries. Environ Health Perspect 121:1097–1102

Augustsson AL, Uddh-Söderberg TE, Hogmalm KJ, Filipsson ME (2015) Metal uptake by homegrown vegetables–the relative importance in human health risk assessments at contaminated sites. Environ Res 138:181–190

Badigannavar A, Teme N, de Oliveira AC, Li G, Vaksmann M, Viana VE, Ganapathi TR, Sarsu F (2018) Physiological, genetic and molecular basis of drought resilience in sorghum [Sorghum bicolor (L.) Moench]. Indian J Plant Physiol 23(4):670–688

Bailey-Serres J, Lee SC, Brinton E (2012) Waterproofing crops: effective flooding survival strategies. Plant Physiol 160:1698–1709

Bajpai S, Chandra R (2015) Effect of waterlogging stress on growth characteristics and sod gene expression in sugarcane. Int J Sci Res Publ 5:1–8

Bakhshandeh S, Corneo PE, Yin L, Dijkstra FA (2019) Drought and heat stress reduce yield and alter carbon rhizodeposition of different wheat genotypes. J Agron Crop Sci 205:157–167

Banavath JN, Chakradhar T, Pandit V, Konduru S, Guduru KK, Akila CS, Podha S, Puli CO (2018) Stress inducible overexpression of AtHDG11 leads to improved drought and salt stress tolerance in peanut (*Arachis hypogaea* L.). Front Chem 6:1–24

Bar Y, Apelbaum A, Kafkafi U, Goren R (1997) Relationship between chloride and nitrate and its effect on growth and mineral composition of avocado and citrus plants. J Plant Nutr 20:715–731

Barber SA (1995) Soil nutrient bioavailability: a mechanistic approach, 2nd edn. New York, Wiley

Batra NG, Kumari N, Sharma V (2018) Chlorophyll fluorescence based assessment of low temperature stress in different varieties of *Vigna radiata* (L.). Int J Plant Res 31:146–155

Bhatla SC (2018) Abiotic stress. In: Plant physiology, development and metabolism. Springer, Singapore, pp 969–1028

Bhavyasree RK, Joel AJ, Jeyaprakash P, Raveendran M (2019) The effect of induced drought stress on seedling vigour and antioxidant enzymes in wild and cultivated Oryza species. Int J Chem Stud 7:1078–1081

Bilska-Kos A, Solecka D, Dziewulska A, Ochodzki P, Jończyk M, Bilski H, Sowiński P (2017) Low temperature caused modifications in the arrangement of cell wall pectins due to changes of osmotic potential of cells of maize leaves (*Zea mays* L.). Protoplasma 254:713–724

Borges KLR, Salvato F, Alcântara BK, Nalin RS, Piotto FA, Azevedo RA (2018) Temporal dynamic responses of roots in contrasting tomato genotypes to cadmium tolerance. Ecotoxicology 27:245–258

Buendía-González L, Orozco-Villafuerte J, Cruz-Sosa F, Barrera-Díaz CE, Vernon-Carter EJ (2010) Prosopis laevigata a potential chromium (VI) and cadmium (II) hyperaccumulator desert plant. Bioresour Technol 101:5862–5867

Camejo D, Jiménez A, Alarcón JJ, Torres W, Gómez JM, Sevilla F (2006) Changes in photosynthetic parameters and antioxidant activities following heat-shock treatment in tomato plants. Funct Plant Biol 33:177–187

Carrao H, Naumann G, Barbosa P (2016) Mapping global patterns of drought risk: an empirical framework based on sub-national estimates of hazard, exposure and vulnerability. Glob Environ Chang 39:108–124

Carter AY, Ottman MJ, Curlango-Rivera G, Huskey DA, D'Agostini BA, Hawes MC (2019) Drought-tolerant barley: II. Root tip characteristics in emerging roots. Agronomy 9:1–10

Carvalho MEA, Piotto FA, Franco MR, Borges KLR, Gaziola SA, Castro PRC, Azevedo RA (2018) Cadmium toxicity degree on tomato development is associated with disbalances in B and Mn status at early stages of plant exposure. Ecotoxicology 27:1293–1302

Chatterjee C, Dube BK, Sinha P, Srivastava P (2004) Detrimental effects of lead phytotoxicity on growth, yield, and metabolism of rice. Commun Soil Sci Plant Anal 35:255–265

Chen H, Zhang Q, Lu Z, Xu F (2018) Accumulation of ammonium and reactive oxygen mediated drought-induced rice growth inhibition by disturbed nitrogen metabolism and photosynthesis. Plant Soil 431:107–117

Chiang PN, Chiu CY, Wang MK, Chen BT (2011) Low-molecular-weight organic acids exuded by Millet (*Setaria italica* (L.) Beauv.) roots and their effect on the remediation of cadmium-contaminated soil. Soil Sci 176:33–38

Chigbo C, Batty L (2013) Effect of combined pollution of chromium and benzo (a) pyrene on seed growth of *Lolium perenne*. Chemosphere 90:164–169

CIA (2017) The world factbook. https://www.cia.gov/library/publications/the-world-factbook/. Accessed 23 Sept 2019

Daryanto S, Wang L, Jacinthe PA (2016) Global synthesis of drought effects on maize and wheat production. PLoS One 11:1–15

Dawood T, Rieu I, Wolters-Arts M, Derksen EB, Mariani C, Visser EJ (2014) Rapid flooding-induced adventitious root development from preformed primordia in *Solanum dulcamara*. AoB Plants 6:1–13

Deepak SB, Thakur A, Singh S, Bakshi M, Bansal S (2019) Changes in crop physiology under drought stress: a review. J Pharmacogn Phytochem 8:1251–1253

de Souza TC, dos Santos SE, Dousseau S, de Castro EM, Magalhães PC (2013) Seedlings of *Garcinia brasiliensis* (Clusiaceae) subjected to root flooding: physiological, morphoanatomical, and antioxidant responses to the stress. Aquat Bot 111:43–49

Devi SR, Prasad MNV (1999) Membrane lipid alterations in heavy metal exposed plants. In: Prasad MNV (ed) Heavy metal stress in plants. Springer, Berlin, Heidelberg, pp 99–116

Dhawan SS, Mishra A, Gupta P, Bahl JR, Bansal RP (2018) Phylogentic relationship of cold tolerant Mentha arvensis variety 'CIM Kranti' with some released varieties as assessed through physiological and molecular analysis. J Appl Res Med Aromat Plants 10:67–74

Din J, Khan SU, Ali I, Gurmani AR (2011) Physiological and agronomic response of canola varieties to drought stress. J Anim Plant Sci 21:78–82

Dolatabadi N, Toorchi M, Valizadeh M, Bandehagh A (2019) The proteome response of salt-sensitive rapeseed (Brassica napus L.) genotype to salt stress. Not Bot Horti Agrobot Cluj Napoca 47:17–23

Dwivedi SK, Kumar S, Bhakta N, Srivastava AK, Mishra JS, Kumar V, Kumara BH, Bhatt BP, Singh S (2018) Physiological mechanism and nutrient management strategies for flood tolerance in rice grown in lowland flood prone ecosystem. J Crop Sci Biotechnol 21:321–331

Dziubanek G, Piekut A, Rusin M, Baranowska R, Hajok I (2015) Contamination of food crops grown on soils with elevated heavy metals content. Ecotoxicol Environ Saf 118:183–189

Ezaki B, Nagao E, Yamamoto Y, Nakashima S, Enomoto T (2008) Wild plants, *Andropogon virginicus* L. and *Miscanthus sinensis* Anders, are tolerant to multiple stresses including aluminum, heavy metals and oxidative stresses. Plant Cell Rep 27:951–961

Ezin V, Pena RDL, Ahanchede A (2010) Flooding tolerance of tomato genotypes during vegetative and reproductive stages. Braz J Plant Physiol 22:131–142

Fahad S, Bajwa AA, Nazir U, Anjum SA, Farooq A, Zohaib A, Sadia S, Nasim W, Adkins S, Saud S, Ihsan MZ (2017) Crop production under drought and heat stress: plant responses and management options. Front Plant Sci 8:1–16

FAO (2015) Food and Agriculture Organization of the United Nations. http://www.fao.org/3/a-bc600e.pdf. Accessed 1 Oct 2016

FAO (2016) Food and Agriculture Organization of the United Nations. http://www.fao.org/3/a-i6471e.pdf. Accessed 1 Oct 2017

FAO (2017) How close we are to zero Hunger. http://www.fao.org/state-of-food-securitynutrition/en/geos/kz.html. Accessed 1 Mar 2019

FAO (2018) The state of food security and nutrition in the world. http://www.fao.org/state-of-food-security-nutrition/en/. Accessed 15 Sept 2019

Farahbakhsh H, Pasandi Pour A, Reiahi N (2017) Physiological response of henna (*Lawsonia inermise* L.) to salicylic acid and salinity. Plant Prod Sci 20:237–247

Farooq M, Wahid A, Kobayashi N, Fujita D, Basra SMA (2009) Plant drought stress: effects, mechanisms and management. In: Lichtfouse E, Navarrete M, Debaeke P, Véronique S, Alberola C (eds) Sustainable agriculture. Springer, Dordrecht, pp 153–188

Fedoroff NV, Battisti DS, Beachy RN, Cooper PJ, Fischhoff DA, Hodges CN, Knauf VC, Lobell D, Mazur BJ, Molden D, Reynolds MP (2010) Radically rethinking agriculture for the 21st century. Science 327:833–834

Flagella Z, Trono D, Pompa M, Di Fonzo N, Pastore D (2006) Seawater stress applied at germination affects mitochondrial function in durum wheat (*Triticum durum*) early seedlings. Funct Plant Biol 33:357–366

Ghoulam C, Foursy A, Fares K (2002) Effects of salt stress on growth, inorganic ions and proline accumulation in relation to osmotic adjustment in five sugar beet cultivars. Environ Exp Bot 47:39–50

Gomes-Filho E, Lima CRFM, Costa JH, da Silva ACM, Lima MDGS, de Lacerda CF, Prisco JT (2008) Cowpea ribonuclease: properties and effect of NaCl-salinity on its activation during seed germination and seedling establishment. Plant Cell Rep 27:147–157

Gonzalez JA, Gallardo M, Hilal M, Rosa M, Prado FE (2009) Physiological responses of quinoa (*Chenopodium quinoa* Willd.) to drought and waterlogging stresses: dry matter partitioning. Bot Stud 50:35–42

Greenway H, Munns R (1980) Mechanisms of salt tolerance in nonhalophytes. Annu Rev Plant Physiol 31:149–190

Grzesiak S, Hura T, Grzesiak MT, Pieńkowski S (1999) The impact of limited soil moisture and waterlogging stress conditions on morphological and anatomical root traits in maize (*Zea mays* L.) hybrids of different drought tolerance. Acta Physiol Plant 21:305–315

Guala SD, Vega FA, Covelo EF (2010) The dynamics of heavy metals in plant–soil interactions. Ecol Model 221:1148–1152

Gupta B, Huang B (2014) Mechanism of salinity tolerance in plants: physiological, biochemical, and molecular characterization. Int J Genom 2014:1–18

Gupta DK, Huang HG, Yang XE, Razafindrabe BHN, Inouhe M (2010) The detoxification of lead in Sedum alfredii H. is not related to phytochelatins but the glutathione. J Hazard Mater 177:437–444

Guru A, Dwivedi P (2018) Physiological, biochemical and molecular mechanism of submergence tolerance in rice (*Oryza sativa* L.). J Pharmacogn Phytochem 7:1116–1121

Han Y, Yin S, Huang L, Wu X, Zeng J, Liu X, Qiu L, Munns R, Chen ZH, Zhang G (2018) A sodium transporter HvHKT1; 1 confers salt tolerance in barley via regulating tissue and cell ion homeostasis. Plant Cell Physiol 59:1976–1989

Han Y, Yang H, Wu M, Yi H (2019) Enhanced drought tolerance of foxtail millet seedlings by sulfur dioxide fumigation. Ecotoxicol Environ Saf 178:9–16

Hasanuzzaman M, Hossain MA, da Silva JAT, Fujita M (2012) Plant response and tolerance to abiotic oxidative stress: antioxidant defense is a key factor. In: Tuteja N, Gill SS (eds) Crop stress and its management: perspectives and strategies. Springer, Dordrecht, pp 261–315

Hasegawa PM, Bressan RA, Zhu JK, Bohnert HJ (2000) Plant cellular and molecular responses to high salinity. Annu Rev Plant Biol 51:463–499
Hossain MA, Uddin SN (2011) Mechanisms of waterlogging tolerance in wheat: morphological and metabolic adaptations under hypoxia or anoxia. Aust J Crop Sci 5:1094–1101
Hu Y, Schmidhalter U (1997) Interactive effects of salinity and macronutrient level on wheat. II. Composition. J Plant Nutr 20:1169–1182
Hussain M, Malik MA, Farooq M, Ashraf MY, Cheema MA (2008) Improving drought tolerance by exogenous application of glycinebetaine and salicylic acid in sunflower. J Agron Crop Sci 194:193–199
Jackson MB (2006) Plant survival in wet environments: resilience and escape mediated by shoot systems. In: Bobbink R, Beltman B, Verhoeven JTA, Whigham DE (eds) Wetlands: functioning, biodiversity conservation, and restoration. Springer, Berlin, Heidelberg, pp 15–36
Jackson MB (2008) Ethylene-promoted elongation: an adaptation to submergence stress. Ann Bot 101:229–248
Jaleel CA, Manivannan P, Wahid A, Farooq M, Al-Juburi HJ, Somasundaram R, Panneerselvam R (2009) Drought stress in plants: a review on morphological characteristics and pigments composition. Int J Agric Biol 11:100–105
James RA, Davenport RJ, Munns R (2006) Physiological characterization of two genes for Na+ exclusion in durum wheat, Nax1 and Nax2. Plant Physiol 142:1537–1547
Jamil A, Riaz S, Ashraf M, Foolad MR (2011) Gene expression profiling of plants under salt stress. Crit Rev Plant Sci 30:435–458
Jeschke WD (1984) K+-Na+ exchange at cellular membranes, intracellular compartmentation of cations, and salt tolerance. In: Staples RC, Toenniessen GH (eds) Salinity tolerance in plants strategies for crop improvement. Wiley, Toronto, pp 37–66
Jiang W, Liu D (2010) Pb-induced cellular defense system in the root meristematic cells of *Allium sativum* L. BMC Plant Biol 10:40–48
Kalai T, Khamassi K, Teixeira da Silva JA, Gouia H, Bettaieb Ben-Kaab L (2014) Cadmium and copper stress affect seedling growth and enzymatic activities in germinating barley seeds. Arch Agron Soil Sci 60:765–783
Kalaji HM, Loboda T (2007) Photosystem II of barley seedlings under cadmium and lead stress. Plant Soil Environ 53:511–516
Kaloki P, Devasirvatham V, Tan DK (2019) Chickpea abiotic stresses: combating drought, heat and cold. In: Alexandre DO (ed) Abiotic and biotic stress in plants. IntechOpen, London, pp 725–729
Kaundal A, Sandhu D, Duenas M, Ferreira JF (2019) Expression of the high-affinity K+ transporter 1 (PpHKT1) gene from almond rootstock 'Nemaguard' improved salt tolerance of transgenic Arabidopsis. PLoS One 14:e0214473
Khan MA, Gul B (2006) Halophyte seed germination. In: Khan MA, Weber DJ (eds) Ecophysiology of high salinity tolerant plants. Springer, Dordrecht, pp 11–30
Khan MN, Zhan J, Luo T, Liu J, Ni F, Rizwan M, Fahad S, Hu L (2019) Morpho-physiological and biochemical responses of tolerant and sensitive rapeseed cultivars to drought stress during early seedling growth stage. Acta Physiol Plant 41:1–13
Khatun S, Rizzo CA, Flowers TJ (1995) Genotypic variation in the effect of salinity on fertility in rice. Plant Soil 173:239–250
Kim JY, Mahé A, Brangeon J, Prioul JL (2000) A maize vacuolar invertase, IVR2, is induced by water stress. Organ/tissue specificity and diurnal modulation of expression. Plant Physiol 124:71–84
Kirk GJD, Greenway H, Atwell BJ, Ismail AM, Colmer TD (2014) Adaptation of rice in flooded soil. In: Lüttge U, Beyschlag W, Cushman J (eds) Progress in botany. Springer, Berlin, pp 215–253
Klein M, Burla B, Martinoia E (2006) The multidrug resistance-associated protein (MRP/ABCC) subfamily of ATP-binding cassette transporters in plants. FEBS Lett 580:1112–1122

Kranner I, Colville L (2011) Metals and seeds: biochemical and molecular implications and their significance for seed germination. Environ Exp Bot 72:93–105

Krasensky J, Jonak C (2012) Drought, salt, and temperature stress-induced metabolic rearrangements and regulatory networks. J Exp Bot 63:1593–1608

Krishnan HB, Oehrle NW, Alaswad AA, Stevens WG, Maria John KM, Luthria DL, Natarajan SS (2019) Biochemical and anatomical investigation of *Sesbania herbacea* (Mill.) McVaugh nodules grown under flooded and non-flooded conditions. Int J Mol Sci 20:1–20

Kulichikhin K, Yamauchi T, Watanabe K, Nakazono M (2014) Biochemical and molecular characterization of rice (Oryza sativa L.) roots forming a barrier to radial oxygen loss. Plant Cell Environ 37:2406–2420

Kumar P, Pal M, Joshi R, Sairam RK (2013) Yield, growth and physiological responses of mung bean [*Vigna radiata* (L.) Wilczek] genotypes to waterlogging at vegetative stage. Physiol Mol Biol Plants 19:209–220

Kuzminov FI, Brown CM, Fadeev VV, Gorbunov MY (2013) Effects of metal toxicity on photosynthetic processes in coral symbionts, *Symbiodinium* spp. J Exp Mar Biol Ecol 446:216–227

Ladrera R, Marino D, Larrainzar E, González EM, Arrese-Igor C (2007) Reduced carbon availability to bacteroids and elevated ureides in nodules, but not in shoots, are involved in the nitrogen fixation response to early drought in soybean. Plant Physiol 145:539–546

Lahaye PA, Epstein E (1971) Calcium and salt toleration by bean plants. Physiol Plant 25:213–218

Lai JL, Luo XG (2019) High-efficiency antioxidant system, chelating system and stress-responsive genes enhance tolerance to cesium ionotoxicity in Indian mustard (*Brassica juncea* L.). Ecotoxicol Environ Saf 181:491–498

Landrigan PJ, Fuller R, Acosta NJ, Adeyi O, Arnold R, Baldé AB, Bertollini R, Bose-O'Reilly S, Boufford JI, Breysse PN, Chiles T (2018) The lancet commission on pollution and health. Lancet 391:462–512

Lawas LMF, Shi W, Yoshimoto M, Hasegawa T, Hincha DK, Zuther E, Jagadish SK (2018) Combined drought and heat stress impact during flowering and grain filling in contrasting rice cultivars grown under field conditions. Field Crops Res. 229:66–77

Lea-Cox JD, Syvertsen JP (1993) Salinity reduces water use and nitrate-N-use efficiency of citrus. Ann Bot 72:47–54

Lisar SY, Motafakkerazad R, Hossain MM, Rahman IM (2012) Water stress in plants: causes, effects and responses. In: Ismail MD, Mofizur R, Hiroshi H (eds) Water stress. IntechOpen, London, pp 1–15

Ma L, Coulter JA, Liu L, Zhao Y, Chang Y, Pu Y, Zeng X, Xu Y, Wu J, Fang Y, Bai J (2019) Transcriptome analysis reveals key cold-stress-responsive genes in winter rapeseed (*Brassica rapa* L.). Int J Mol Sci 20:1–19

Mahmood T, Mustafa HSB, Aftab M, Ali Q, Malik A (2019) Super canola: newly developed high yielding, lodging and drought tolerant double zero cultivar of rapeseed (*Brassica napus* L.). Genet Mol Res 18:gmr16039951

Makino T, Murakami M, Ishikawa S, Abe T (2019) Regulations for cadmium in Rice and soil in Japan and countermeasures to reduce the concentrations. In: Himeno S, Aoshima K (eds) Cadmium toxicity. Springer, Singapore, pp 103–114

Manickavelu A, Nadarajan N, Ganesh SK, Gnanamalar RP, Babu RC (2006) Drought tolerance in rice: morphological and molecular genetic consideration. Plant Growth Regul 50:121–138

Marchiol L, Assolari S, Sacco P, Zerbi G (2004) Phytoextraction of heavy metals by canola (*Brassica napus*) and radish (*Raphanus sativus*) grown on multicontaminated soil. Environ Pollut 132:21–27

Maryan KE, Lahiji HS, Farrokhi N, Komeleh HH (2019) Analysis of *Brassica napus* dehydrins and their co-expression regulatory networks in relation to cold stress. Gene Expr Patterns 31:7–17

Maxwell K, Johnson GN (2000) Chlorophyll fluorescence—a practical guide. J Exp Bot 51:659–668

Mensah JK, Obadoni BO, Eruotor PG, Onome-Irieguna F (2006) Simulated flooding and drought effects on germination, growth, and yield parameters of sesame (*Sesamum indicum* L.). Afr J Biotechnol 5:1249–1253
Mierek-Adamska A, Dąbrowska GB, Blindauer CA (2018) The type 4 metallothionein from *Brassica napus* seeds folds in a metal-dependent fashion and favours zinc over other metals. Metallomics 10:1430–1443
Munns R (2002a) Comparative physiology of salt and water stress. Plant Cell Environ 25:239–250
Munns R (2002b) Salinity, growth and phytohormones. In: Läuchli A, Lüttge U (eds) Salinity: environment-plants-molecules. Springer, Dordrecht, pp 271–290
Munns R (2005) Genes and salt tolerance: bringing them together. New Phytol 167:645–663
Munns R, Tester M (2008) Mechanisms of salinity tolerance. Annu Rev Plant Biol 59:651–681
Ohnishi N, Wacera F, Sakamoto W (2019) Photosynthetic responses to high temperature and strong light suggest potential post-flowering drought tolerance of Sorghum Japanese landrace Takakibi. Plant Cell Physiol 60:2086–2099
Pachauri RK, Allen MR, Barros VR, Broome J, Cramer W, Christ R, Church JA, Clarke L, Dahe Q, Dasgupta P, Dubash NK (2014) Climate change 2014: synthesis report. In: Pachauri R, Meyer L (eds) Contribution of working groups I, II and III to the fifth assessment report of the intergovernmental panel on climate change. Ipcc, Geneva, p 151
Pan L, Yu X, Shao J, Liu Z, Gao T, Zheng Y, Chen C (2019) Transcriptomic profiling and analysis of differentially expressed genes in asparagus bean (*Vigna unguiculata* ssp. *sesquipedalis*) under salt stress. PLoS One 14:e0219799
Panozzo A, Dal Cortivo C, Ferrari M, Vicelli B, Varotto S, Vamerali T (2019) Morphological changes and expressions of AOX1A, CYP81D8 and putative PFP genes in a large set of commercial maize hybrids under extreme waterlogging. Front Plant Sci 10:1–14
Panuccio MR, Jacobsen SE, Akhtar SS, Muscolo A (2014) Effect of saline water on seed germination and early seedling growth of the halophyte quinoa. AoB Plants 6:plu047
Parelle J, Roudaut JP, Ducrey M (2006) Light acclimation and photosynthetic response of beech (*Fagus sylvatica* L.) saplings under artificial shading or natural Mediterranean conditions. Ann For Sci 63:257–266
Parent C, Capelli N, Berger A, Crèvecoeur M, Dat JF (2008) An overview of plant responses to soil waterlogging. Plant Stress 2:20–27
Park JH, Lamb D, Paneerselvam P, Choppala G, Bolan N, Chung JW (2011) Role of organic amendments on enhanced bioremediation of heavy metal (loid) contaminated soils. J Hazard Mater 185:549–574
Park HC, Hwang JE, Jiang Y, Kim YJ, Kim SH, Nguyen XC, Kim CY, Chung WS (2019) Functional characterisation of two phytochelatin synthases in rice (*Oryza sativa* cv. Milyang 117) that respond to cadmium stress. Plant Biol 21:854–861
Parra-Almuna L, Diaz-Cortez A, Ferrol N, de la Luz Mora M (2018) Aluminium toxicity and phosphate deficiency activates antioxidant systems and up-regulates expression of phosphate transporters gene in ryegrass (Lolium perenne L.) plants. Plant Physiol Biochem 130:445–454
Patel BB, Dave RS (2011) Studies on infiltration of saline-alkali soils of several parts of Mehsana and Patan districts of North Gujarat. J Appl Technol Environ Sanit 1:87–92
Paul K, Pauk J, Kondic-Spika A, Grausgruber H, Allahverdiyev T, Sass L, Vass I (2019) Co-occurrence of mild salinity and drought synergistically enhances biomass and grain retardation in wheat. Front Plant Sci 10:1–15
Pekcan V, Evci G, Yilmaz MI, Nalcaiyi ASB, Erdal SÇ, Cicek N, Ekmekci Y, Kaya Y (2015) Drought effects on yield traits of some sunflower inbred lines. Poljopr Sumar 61:101–107
Pierart A, Shahid M, Sejalon-Delmas N, Dumat C (2015) Antimony bioavailability: knowledge and research perspectives for sustainable agricultures. J Hazard Mater 289:219–234
Popović V, Miladinović J, Vidić M, Vučković S, Dolijanović Ž, Ikanović J, Zivanović LJ, Kolarić L (2015) Drought–limiting factors in soybean production. The effect of irrigation on yield of soybean [Glycine max (L.) Merr.]. Proceedings. Institute of PKB Agroekonomik, Belgrade, pp 11–21

Pourrut B, Shahid M, Douay F, Dumat C, Pinelli E (2013) Molecular mechanisms involved in lead uptake, toxicity and detoxification in higher plants. In: Gupta DK (ed) Heavy metal stress in plants. Springer, Berlin, Heidelberg, pp 121–147

Qadir M, Schubert S (2002) Degradation processes and nutrient constraints in sodic soils. Land Degrad Dev 13:275–294

Queiroz MS, Oliveira CE, Steiner F, Zuffo AM, Zoz T, Vendruscolo EP, Silva MV, Mello BFFR, Cabra RC, Menis FT (2019) Drought stresses on seed germination and early growth of maize and sorghum. J Agric Sci 11:310–318

Rajendran K, Tester M, Roy SJ (2009) Quantifying the three main components of salinity tolerance in cereals. Plant Cell Environ 32:237–249

Rathaur P, Ramteke PW, Raja W, John SA (2012) Isolation and characterization of nickel and cadmium tolerant plant growth promoting rhizobacteria from rhizosphere of *Withania somnifera*. J Environ Sci 6:253–261

Reddy MD, Mittra BN (1985) Effect of complete plant submergence at different growth stages on grain yield, yield components and nutrient content of rice. Plant Soil 86:379–386

Reddy KR, Krutz L, Gao W, Bellaloui N (2019) Poor seed quality, reduced germination, and decreased seedling vigor in soybean is linked to exposure of the maternal lines to drought stress. BioRxiv:590059 (in press)

Ren ZH, Gao JP, Li LG, Cai XL, Huang W, Chao DY, Zhu MZ, Wang ZY, Luan S, Lin HX (2005) A rice quantitative trait locus for salt tolerance encodes a sodium transporter. Nat Genet 37:1141–1146

Romero-Puertas MC, Rodríguez-Serrano M, Corpas FJ, Gomez MD, Del Rio LA, Sandalio LM (2004) Cadmium-induced subcellular accumulation of O2·− and H2O2 in pea leaves. Plant Cell Environ 27:1122–1134

Ronga D, Rizza F, Badeck FW, Milc J, Laviano L, Montevecchi G, Pecchioni N, Francia E (2018) Physiological responses to chilling in cultivars of processing tomato released and cultivated over the past decades in Southern Europe. Sci Hortic 231:118–125

Rozeff N (1995) Sugarcane and salinity—a review paper. Sugarcane 5:8–19

Sadak MS (2019) Physiological role of trehalose on enhancing salinity tolerance of wheat plant. Bull Natl Res Cent 43:1–10

Saha P, Chatterjee P, Biswas AK (2010) NaCl pretreatment alleviates salt stress by enhancement of antioxidant defense system and osmolyte accumulation in mungbean (Vigna radiata L. Wilczek). Indian J. Exp Biol 48:593–600

Sairam RK, Kumutha D, Ezhilmathi K, Deshmukh PS, Srivastava GC (2008) Physiology and biochemistry of waterlogging tolerance in plants. Biol Plant 52:401

Sakamoto A, Murata N (2002) The role of glycine betaine in the protection of plants from stress: clues from transgenic plants. Plant Cell Environ 25:163–171

Sánchez-Gómez D, Cervera MT, Escolano-Tercero MA, Vélez MD, de María N, Diaz L, Sánchez-Vioque R, Aranda I, Guevara MA (2019) Drought escape can provide high grain yields under early drought in lentils. Theor Exp Plant Physiol 31(2):273–286

Sapeta H, Costa JM, Lourenco T, Maroco J, Van der Linde P, Oliveira MM (2013) Drought stress response in *Jatropha curcas*: growth and physiology. Environ Exp Bot 85:76–84

Satir O, Berberoglu S (2016) Crop yield prediction under soil salinity using satellite derived vegetation indices. Field Crops Res 192:134–143

Sauter M (2013) Root responses to flooding. Curr Opin Plant Biol 16:282–286

Schulz E, Tohge T, Zuther E, Fernie AR, Hincha DK (2016) Flavonoids are determinants of freezing tolerance and cold acclimation in *Arabidopsis thaliana*. Sci Rep 6:1–14

Schulze ED, Beck E, Buchmann N, Clemens S, Müller-Hohenstein K, Scherer-Lorenzen M (2019) Water deficiency (Drought). In: Schulze ED, Beck E, Buchmann N (eds) Plant Ecology. Springer, Heidelberg, pp 162–202

Sculthorpe CD (1967) The biology of aquatic vascular plants. Edward Arnold Press, London, p 610

SEDAC (n.d.) Global flood hazard frequency and distribution, v1 (1985 – 2003). http://sedac.ciesin.columbia.edu/data/set/ndh-flood-hazard-frequency-distribution. Accessed 15 Oct 2019

Selpair S (2018) Analyzing reproductive capabilities of chickpea in cold environment. http://210.212.34.21/handle/32116/1952. Accessed 2 Mar 2018

Shahid M, Arshad M, Kaemmerer M, Pinelli E, Probst A, Baque D, Pradere P, Dumat C (2012) Long-term field metal extraction by Pelargonium: phytoextraction efficiency in relation to plant maturity. Int J Phytoremediation 14:493–505

Sharma N, Bisht SS, Gupta S, Panda AK, Rana M (2018) Analysis of proteomic diversity and calcium binding protein(s) in seeds of horse gram (*Macrotyloma uniflorum*) cultivars from Uttarakhand. Int J Pharm Sci Res 9:3274–3280

Shawon RA, Kang DH, Ryu CS, Kim DE, Lee SY, Bae JH, Kim YO, Ku YG (2017) Physiological responses and antioxidative enzyme activities in pepper (*Capsicum annuum*) seedlings under low temperature stress. J Agric Life Sci 51:67–77

Shiono K, Takahashi H, Colmer TD, Nakazono M (2008) Role of ethylene in acclimations to promote oxygen transport in roots of plants in waterlogged soils. Plant Sci 175:52–58

Shrivastava P, Kumar R (2015) Soil salinity: a serious environmental issue and plant growth promoting bacteria as one of the tools for its alleviation. Saudi J Biol Sci 22:123–131

Silva EN, Silveira JA, Aragão RM, Vieira CF, Carvalho FE (2019) Photosynthesis impairment and oxidative stress in *Jatropha curcas* exposed to drought are partially dependent on decreased catalase activity. Acta Physiol Plant 41:1–12

Singh S, Singh SP, Pathak AD, Pandey N (2019) Assessment of waterlogging induced physiobiochemical changes in sugarcane varieties and its association with waterlogging tolerance. J Environ Biol 40:384–392

Sinha S, Raxwal VK, Joshi B, Jagannath A, Katiyar-Agarwal S, Goel S, Kumar A, Agarwal M (2015) De novo transcriptome profiling of cold-stressed siliques during pod filling stages in Indian mustard (*Brassica juncea* L.). Front Plant Sci 6:1–17

Solaiman Z, Colmer TD, Loss SP, Thomson BD, Siddique KHM (2007) Growth responses of cool-season grain legumes to transient waterlogging. Aust J Agric Res 58:406–412

Souza SC, Mazzafera P, Sodek L (2016) Flooding of the root system in soybean: biochemical and molecular aspects of N metabolism in the nodule during stress and recovery. Amino Acids 48:1285–1295

Stvolinskaya NS (2000) Viability of *Taraxacum officinale* Wigg. in populations of the city of Moscow in relation to motor transport pollution. Russ J Ecol 31:129–131

Sudhir P, Murthy SDS (2004) Effects of salt stress on basic processes of photosynthesis. Photosynthetica 42:481–486

Suhayda CG, Giannini JL, Briskin DP, Shannon MC (1990) Electrostatic changes in *Lycopersicon esculentum* root plasma membrane resulting from salt stress. Plant Physiol 93:471–478

Taiz L, Zeiger E (2006) Plant physiology, 4th edn. Sinauer Associates, Sunderland, MA, pp 345–376

Tang H, Niu L, Wei J, Chen X, Chen Y (2019) Phosphorus limitation improved salt tolerance in maize through tissue mass density increase, osmolytes accumulation, and Na+ uptake inhibition. Front Plant Sci 10:1–10

Tarekegne A, Bennie ATP, Labuschagne MT (2000) Effects of soil waterlogging on the concentration and uptake of selected nutrients in wheat genotypes differing in tolerance. In: The eleventh regional wheat workshop for eastern, central and southern Africa, Addis Abeba, Ethiopia. CIMMYT, Addis Ababa, pp 253–263

Tewari S, Arora NK (2016) Soybean production under flooding stress and its mitigation using plant growth-promoting microbes. In: Mohammad M (ed) Environmental stresses in soybean production: soybean production, vol 2. Academic press, San Diego, pp 23–40

Turhan E, Ergin S (2012) Soluble sugars and sucrose-metabolizing enzymes related to cold acclimation of sweet cherry cultivars grafted on different rootstocks. Sci World J 2012:1–7

Turner NC, Wright GC, Siddique KHM (2001) Adaptation of grain legumes (pulses) to water-limited environments. Adv Agron 71:194–233

UNESCO (2018) Climate change and water security. https://en.unesco.org/themes/addressing-climate-change/climate-change-and-water-security. Accessed 12 Mar 2018

Vandoorne B, Descamps C, Mathieu AS, Van den Ende W, Vergauwen R, Javaux M, Lutts S (2014) Long term intermittent flooding stress affects plant growth and inulin synthesis of *Cichorium intybus* (var. sativum). Plant Soil 376:291–305

Wang Y, Zhang B, Jiang D, Chen G (2019) Silicon improves photosynthetic performance by optimizing thylakoid membrane protein components in rice under drought stress. Environ Exp Bot 158:117–124

Watts J (19 March 2018) Water shortages could affect 5bn people by 2050, UN report warns. The Guardian. https://www.theguardian.com/environment/2018/mar/19/water-shortages-could-affect-5bn-people-by-2050-un-report-warns. Accessed 31 Oct 2019

Wei H, Chen C, Ma X, Zhang Y, Han J, Mei H, Yu S (2017) Comparative analysis of expression profiles of panicle development among tolerant and sensitive rice in response to drought stress. Front Plant Sci 8:1–10

Williams G, Vanniarajan C, Vetriventhan M, Thiageshwari S, Anandhi K, Rajagopal B (2019) Genetic variability for seedling stage salinity tolerance in barnyard millet [*Echinochloa frumentacea* (Roxb.) Link]. Electron J Plant Breed 10:552–558

Wise RR, Naylor AW (1987) Chilling-enhanced photooxidation: the peroxidative destruction of lipids during chilling injury to photosynthesis and ultrastructure. Plant Physiol 83(2):272–277

Wise RR, Olson AJ, Schrader SM, Sharkey TD (2004) Electron transport is the functional limitation of photosynthesis in field-grown Pima cotton plants at high temperature. Plant Cell Environ 27:717–724

Wu G, Kang H, Zhang X, Shao H, Chu L, Ruan C (2010) A critical review on the bio-removal of hazardous heavy metals from contaminated soils: issues, progress, eco-environmental concerns and opportunities. J Hazard Mater 174:1–8

Wu GQ, Wang JL, Feng RJ, Li SJ, Wang CM (2018) iTRAQ-based comparative proteomic analysis provides insights into molecular mechanisms of salt tolerance in sugar beet (*Beta vulgaris* L.). Int J Mol Sci 19:1–20

Xiong ZT, Zhao F, Li MJ (2006) Lead toxicity in *Brassica pekinensis* Rupr.: effect on nitrate assimilation and growth. Environ Toxicol 21:147–153

Yadav P, Kaur R, Kanwar MK, Bhardwaj R, Sirhindi G, Wijaya L, Alyemeni MN, Ahmad P (2018) Ameliorative role of castasterone on copper metal toxicity by improving redox homeostasis in *Brassica juncea* L. J Plant Growth Regul 37:575–590

Yamauchi T, Colmer TD, Pedersen O, Nakazono M (2018) Regulation of root traits for internal aeration and tolerance to soil waterlogging-flooding stress. Plant Physiol 176:1118–1130

Yazaki K (2006) ABC transporters involved in the transport of plant secondary metabolites. FEBS Lett 580:1183–1191

Yin D, Chen S, Chen F, Guan Z, Fang W (2009) Morphological and physiological responses of two chrysanthemum cultivars differing in their tolerance to waterlogging. Environ Exp Bot 67:87–93

Yu B, Zhao CY, Li J, Li JY, Peng G (2015) Morphological, physiological, and biochemical responses of *Populus euphratica* to soil flooding. Photosynthetica 53:110–117

Zeng F, Ali S, Zhang H, Ouyang Y, Qiu B, Wu F, Zhang G (2011) The influence of pH and organic matter content in paddy soil on heavy metal availability and their uptake by rice plants. Environ Pollut 159:84–91

Zhang X, Wan S, Hao J, Hu J, Yang T, Zong X (2016) Large-scale evaluation of pea (*Pisum sativum* L.) germplasm for cold tolerance in the field during winter in Qingdao. Crop J 4:377–383

Zhang C, Bai T, Xie D, Wang Y, Zhang H, Chen Y, Ni Z (2019a) Effects of PEG-6000 simulation drought stress on seed germination of mango (*Mangiferca indica* L.). Nanfang Nongye Xuebao 50:600–606 (in Chinese)

Zhang W, Wang S, Yu F, Tang J, Shan X, Bao K, Yu L, Wang H, Fei Z, Li J (2019b) Genome-wide characterization and expression profiling of SWEET genes in cabbage (*Brassica oleracea* var. *capitata* L.) reveal their roles in chilling and clubroot disease responses. BMC Genomics 20:1–16

Zhao C, Yang R, Hong Y, Ren Z, Tang K, Zhang H, Zhu JK (2019) A role for PICKLE in the regulation of cold and salt stress tolerance in Arabidopsis. Front Plant Sci 10:1–11
Zheng C, Jiang D, Liu F, Dai T, Jing Q, Cao W (2009) Effects of salt and waterlogging stresses and their combination on leaf photosynthesis, chloroplast ATP synthesis, and antioxidant capacity in wheat. Plant Sci 176:575–582
Zhou M (2010) Improvement of plant waterlogging tolerance. In: Mancuso S, Shabala S (eds) Waterlogging signalling and tolerance in plants. Springer, Heidelberg, pp 267–285
Zhu JK (2001) Plant salt tolerance. Trends Plant Sci 6:66–71
Zinselmeier C, Jeong BR, Boyer JS (1999) Starch and the control of kernel number in maize at low water potentials. Plant Physiol 121:25–36
Zörb C, Geilfus CM, Dietz KJ (2019) Salinity and crop yield. Plant Biol 21:31–38

Chapter 6
Physiological and Molecular Aspects of Retrieving Environmental Stress in Plants by Microbial Interactions

Swati Rastogi, Sheel Ratna, Olfa Ben Said, and Rajesh Kumar

Abstract Environmental stress is the foremost limiting factor for agricultural productivity. It is essential to alleviate the distress caused by environmental and edaphic conditions in plants, failing to which would affect their growth, development, and productivity. Microorganisms have enormous metabolic capabilities to lessen the environmental stress and their interactions with plants provide a local and systemic defense under various biotic and abiotic stresses. Due to an increase in adverse external conditions, it is imperative to study plant–microbe relationships at the physiological and molecular level to provide deeper insights into the stress-mitigating mechanisms.

6.1 Introduction

Decline in crop yield and ultimately agricultural productivity has resulted from extreme climatic conditions which burden the land with environmental stress. The geographical land area that is unaffected by any biotic or abiotic constraint is diminishing with the advent of industrialization, pollution, and moderation in the environment. Drought, salinity, acidity, alkalinity, low/high temperature, and nutrient deprivation are some of the paramount abiotic stresses along with biotic stress which includes pest and diseases and show a huge impact on world's agriculture (Latef et al. 2016; Meena et al. 2017). Plants being sessile are generally exposed to these environmental stresses that cause morphological, physiological, biochemical, and molecular alterations in them and may lead to diminution in average yields by 50%. To fight back such constraints, plants have excogitated and developed several intricate specific mechanisms that provide them resistance and minimize damage

S. Rastogi · S. Ratna · R. Kumar (✉)
Rhizosphere Biology Laboratory, Department of Microbiology, Babasaheb Bhimrao Ambedkar University (A Central University), Lucknow, India

O. B. Said
Laboratory of Environment Biomonitoring, Coastal Ecology and Ecotoxicology Unit, Faculty of Sciences of Bizerte, University of Carthage, Zarzouna, Tunisia

A. Sharma (ed.), *Microbes and Signaling Biomolecules Against Plant Stress*, Rhizosphere Biology, https://doi.org/10.1007/978-981-15-7094-0_6

caused by stresses, thereby protecting invaluable resources for growth and development (Atkinson and Urwin 2012; Lenoir et al. 2016).

Though plants have their own internal mechanistic systems which activate specific and unique stress responses when subjected to multiple stress conditions but sometimes this mechanism cannot defy all the harmful stress factors. Complex surroundings where they flourish are comprised of numerous biotic and abiotic entities which tend to interact with them. Microorganisms are known to be ubiquitous in nature and plants are also found in many regions of the earth as they adapt to those environments that are otherwise not dwelled by humans and other larger animals (Coleman-Derr and Tringe 2014). Based on their presence and diversity, plants and microbes communicate with each other and have been observed to possess both negative and positive interactions. In order to cope up with antagonistic environment, plants and microbes have developed synergy to extract maximum benefit from one another. When exposed to environmental stress, such plant–microbe association shows mutual symbiotic relationship where plants may provide shelter, support, and nutrition to the microbes and microbes in return secrete exogenous metabolites and supplements such as primary nutrient sources, signaling molecules, chemo-attractants, etc. that would, otherwise, not be available to the plants (Babar et al. 2016).

The previous knowledge on plant–microbiome interactions at physiological, biochemical as well as at molecular level has provided significant support to establish and understand the complex integrated cellular paradox during such antagonistic variations and laid emphasis on more technological advancement towards plant–microbe interactions in relation to defense against environmental stresses (Nadeem et al. 2014; Ansari 2018). The present chapter summarizes the impact of environmental stress on plants and defense responses induced in plants due to plant–microbe interactions at the physiological and molecular level.

6.2 How Does Environmental Stress Affect Plants?

Plants need an optimum quantity of abiotic entity for their proper growth but they are said to be in a state of stress if certain determinants affect their growth and development, below or beyond optimum dosage or intensity. Plants are exposed to both abiotic and biotic stresses, and among them, abiotic stress is the main cause of abstaining agricultural productivity. These environmental stresses diminish the physiological processes of plants and bring about biochemical and molecular changes along with the generation of reactive oxygen species (ROS) (Ansari 2018).

The foremost effect of environmental stress in plants is the availability of low nutrient content which usually leads to poor growth and development. Drought and salt stress affect plant physiology by reducing soil water potential, leaf water potential (Doubková et al. 2013), stomatal conductance, photosynthetic rate, and chlorophyll pigment (Chakraborty et al. 2013), increased photorespiration, reduced seed germination index, leaf size, seed number, and viability (Aroca et al. 2013), root and shoot growth, decreased production of phytohormones (Palaniyandi et al. 2014; Shahzad et al. 2017), generation of free radicals as hydroxyl, superoxide, peroxide, etc. which disintegrate cell organelles membrane and interfere with metabolic signal pathways (Vurukonda et al. 2016; Lata et al. 2018). High salt concentration in arable soil causes a varying level of acidity or alkalinity, thereby increasing ion toxicity and burdening osmotic stress on plants, decreases crop yield, early senescence, production of oxidizing radicals (Brotman et al. 2013; Kim et al. 2014), affects the expression of metabolic genes (Nautiyal et al. 2013; Bharti et al. 2016), disturbs ion balance (Shukla et al. 2012). Heavy metal stress results in the production of ROS, affects cell's redox homeostasis and causing oxidative stress (Rastogi et al. 2019; De Souza et al. 2015; Fuentes et al. 2016). High or low temperature affects crop productivity by interfering with the biological activities of the soil, causes seasonal freeze-thaw effect leading to severe conditions, denaturation of membrane proteins (Subramanian et al. 2015; Lata et al. 2018). On the other hand, biotic stress is generally linked to the onset of diseases in plant and leads to severe morphological and internal damage, affects defense hormone levels, signaling pathways, and oxidative stress (Singh et al. 2013a, b; Waqas et al. 2015).

6.3 Plant–Microbe Interactions in Mitigating Environmental Stress: Beneficial Microorganisms

6.3.1 Arbuscular Mycorrhizal Symbiosis

The most communal kind of symbiosis is arbuscular mycorrhizal (AM) symbiosis where arbuscular mycorrhizal fungi (AMF) provide water and nutrients to the plants and in return take up carbon (organic matter) and their interaction involves the production of significant signal molecules which activate symbiotic genes and develop the symbiotic association with plants. These mycorrhizal fungi belong to phylum Glomeromycota and can show both ecto- and endo-mycorrhizal symbiosis with their host plants. The term "arbuscular" denotes association of fungal hyphae into the internal cortex of roots in the form of branched structures "arbuscules" where the exchange of nutrients takes place (Kim et al. 2012; Pozo et al. 2013).

6.3.1.1 Physiological Aspects of AMF

AMF form endosymbiotic relationships with plant's roots and help in amelioration of drought stress by direct or indirect mechanisms. Changes include underground and above the ground modifications in plants which may encounter changes in plant–water relations. Some examples of modifications are: changes in CO_2 assimilation, relative water content, growth and photosynthetic pigment parameters, stomatal conductance, leaf water potential, and improved efficiency of photosystem

II with enhanced photosynthetic rate and nutrient uptake (Rapparini and Penuelas 2014; Balestrini et al. 2017; Sun et al. 2017; Quiroga et al. 2017). It is proposed that physiological modifications in plants above the ground under drought stress by AM symbiosis largely depend on the type of host plant and the fungal species involved along with the involvement of hormone "abscisic acid" (ABA). An increased level of ABA is found responsible for bringing out such modifications. Chitarra et al. (2016) supported this fact and showed increased levels of ABA in non-mycorrhizal plants, which means that AM plants experience less drought stress. This theory was also supported by a study conducted on tomato plants. One of the indirect mechanisms by which AMF works in positive manner for plants is by improving soil structure in terms of soil porosity and organic matter. Drought stress is strongly linked with nutrient stress in plants and thus leads to poor soil structure under water deficit conditions and low nutrient availability (Kong et al. 2014; Bardi and Malusà 2012). Comparative studies on maize plants grown in two different types of coal mines under drought stress showed decreased plant dry weights, leaf moisture percentage, water utilization efficiency, nutrient content, and disturbed carbon:nitrogen:phosphorus (C:N:P) stoichiometry (Zhao et al. 2015). An underground association of AM with plant roots provides resistance by enhancing the absorption of water and nutrients from soil pores having low water potential through their hyphae (Hameed et al. 2014; Maiti and Ahirwal 2019). Fungi can form soil aggregates with the help of their hyphae and stabilize soil particles and enhance soil water retention potential. Besides this, extra mycelial growth along with AMF spores can contribute towards soil organic carbon and improve soil organic matter or structure. AMF colonization rate can range from 60 to 90% and can significantly increase the water use efficiency and leaf moisture percentage of fresh weight and improves C:N:P ratio (Zhao et al. 2015). A similar study where the presence of AMF "*Glomus constrictum* Trappe" in the rhizospheric region of marigold (*Tagetes erecta*) alleviated drought stress by improving phosphorus content and photosynthetic pigments (Asrar and Elhindi 2011). The growth of plants in arid or barren soil areas under drought stress can be improved by growing plants concomitantly with AMF which will solve the problem of water scarcity and will lead to restoration of ecosystem. Many studies have supported this fact and showed improved plant growth if inoculated with AMF. Plants generally have their own pathway for nutrient uptake but are depleted under drought stress and thus mycorrhizal uptake pathways come into action. High affinity uptake transporters of mycorrhiza absorb nutrients from soil and transfer them to plants and thus the nutrient level in plants for their growth is improved (Bücking and Kafle 2015). The role of AMF in phosphorus uptake is well established but recently many studies have also supported their role in nitrogen uptake, for example, in sunflower in the form of ammonium ions (Heidari and Karami 2014). Water deficit condition and its physiological impact and alleviation of water stress by AMF have been studied in many plants, *Robinia pseudoacacia* L. (Yang et al. 2015; He et al. 2016), *Zea mays* (Quiroga et al. 2017; Pavithra and Yapa 2018), *Vigna unguiculata* (Oyewole et al. 2017), *Rosa damascena* Mill. (Abdel-Salam et al. 2018).

Salinity is another environmental stress which affects the growth and development of plants. Though salt stress can negatively affect AMF but the constructive role of AMF in plants under salt stress is also well documented. Accumulation of excess salts in alkaline and saline soils generally found in arid and semi-arid geographical regions and also regions like estuaries and coastal fringes may lead to salinity. The physiological aspect of plant–mycorrhizal interaction results in high K^+/Na^+ content, improved photosynthetic and water status of the host plants, improved production of photosynthetic pigments, less ABA accumulation in AMF plants than non-mycorrhizal plants, other phytohormone production, nitrogen fixation, accumulation of compatible solutes and osmolytes, better host plant nutrition, improved antioxidant activity of enzymes to prevent generation of ROS (Porcel et al. 2012; Latef and Miransari 2014). Salinity affects nutrient uptake capacity of the plants (Miransari 2014) and AM fungi help in mitigating nutrient deficiency by assimilation of those nutrients in host plants. For example, elements like nitrogen are provided to plants by first assimilating it in the form of nitrate in their mycelia by using the activity of nitrate reductase enzyme that forms arginine which leads to the formation of other nitrogen containing substances along with reduction in toxic effects of sodium ions, thereby conserving the chlorophyll pigment of plant (Evelin et al. 2009; Latef and Miransari 2014). Latef and Miransari (2014) have also advocated that the other mineral elements like phosphorus, potassium, calcium, and magnesium are also provided to the plant by compartmentalization of sodium and chloride ions within vacuoles that would otherwise interfere with metabolic pathways in the plants and thus improved potassium-to-sodium ion ratio. Many studies have been conducted with crop plants like *Ocimum basilicum* L. (Elhindi et al. 2017), *Cucumis sativus* L. (Hashem et al. 2018), *Cicer arietinum* L. (Garg and Bhandari 2016), *Puccinellia tenuiflora* (Liu et al. 2018) and grass species like *Leymus chinensis* (Lin et al. 2016, 2017) and established the role of AMF in amelioration of salt stress.

6.3.1.2 Molecular Aspects of AM Symbiosis in Alleviating Environmental Stress

Plants respond to drought stress through signal transduction pathway which involves ABA production. Transcriptomic studies showed the involvement of responsive to dehydration genes (rd) and early response to dehydration genes (erd) in response to stress. In fact, ABA signaling pathway is often found to be linked with nitric oxide (NO) pathway as removal of NO is important; otherwise, it would inhibit stomatal closure. Other physiological responses involved during stress in plants are: accumulation of compatible solutes and expression of osmolyte genes as they imitate for water and provide resistance against drought stress (Kim et al. 2012; Lata et al. 2018). A study based on earlier microarray assay followed by reverse transcriptase polymerase chain reaction (RT-PCR) confirmed the expression of four responsive dehydration genes in *Poncirus trifoliata* inoculated with *Glomus mosseae* (AMF) that led to relevant physiological and biochemical changes (increased plant height, chlorophyll content, etc.) in the plant during drought stress. Gene probe 401 was involved in activation of antioxidant system and genes 15,555 and 3652 had similar activities as MIOX1 (myo-inositol oxygenase) and GLX1 (glyoxalase I), respectively, which are involved in ascorbic acid (an antioxidant) biosynthesis and glutathione (GSH) based detoxification of methylglyoxal, respectively. Gene 6927 coded for chalcone isomerase enzyme was involved in flavonoid synthesis which modulates ROS levels under water deficit conditions. Their mRNA transcripts were higher in AM inoculated *P. trifoliata* plants (Fan and Liu 2011). Besides this, studies on significant genes which regulate important membrane proteins involved in water transport and uptake could provide insight into drought tolerance mechanism by AMF. These membrane proteins are called aquaporins which channelize water and other molecules in roots and leaves of the plants (Wang et al. 2018). Studies have revealed the presence of an increased concentration of such membrane proteins in plants and fungi which help in providing drought resistance. They can control hydraulic conductivity of the roots, moisture potential of the leaves, and transpiration rate in mycorrhized plants (Lenoir et al. 2016; Srivastava et al. 2016). A study conducted by He et al. (2016) on *Robinia pseudoacacia* L. when inoculated with mycorrhizal fungus (*Rhizophagus irregularis*) under water stress revealed that AM regulated the expression of aquaporin genes in host plant and its own genes and thus improved plant physiological parameters. Another study on two cultivars of maize focused on water conservation status by AM plants showed that up- and down-regulation of some aquaporin genes took place to minimize water loss and maintenance of water homeostasis (Quiroga et al. 2017).

6.3.1.3 The Rhizo-Chemistry Between Plant and Plant Growth-Promoting Rhizobacteria (PGPR)

Beneficial plant–microbe interactions in the rhizosphere result in increased crop productivity. Plant growth-promoting rhizobacteria (PGPR) fall under such beneficial interactions. They can provide resistance to plants against biotic and abiotic stresses with the help of their characteristic properties like phosphate solubilization, production of 1-aminocyclopropane-1-carboxylic acid deaminase (ACC), biosurfactants, phytohormones, and exopolysaccharide, biological nitrogen fixation, siderophore production, production of many volatile compounds, and regulation of various stress-related genes (Vurukonda et al. 2016; Souza et al. 2015; Viscardi et al. 2016).

6.3.1.4 Potential of Exopolysaccharides (EPS) as Bioadsorbent

Rhizobacterial exopolysaccharides are generally high molecular weight complex organic macromolecules that can bind soil particles to form aggregates which can easily occupy specific locations on the plant roots or hyphae of fungi and thus stabilize the aggregates and thereby minimize water stress. They reduce saline stress by binding to sodium ions and also reduce heavy metal toxicity (Selvakumar et al. 2012; Salwan et al. 2019). Exopolysaccharide mediated working mechanism of

Bacillus amyloliquefaciens FZB42, a PGPR in imparting resistance to disease and salt stress is not much explored in *Arabidopsis thaliana* (Lu et al. 2018). So, they explored the role of exopolysaccharide gene (epsC) in FZB42 when inoculated in *A. thaliana* during drought stress and found that inoculation of epsC containing PGPR enhanced plant growth with increased root and shoot length and their dry weights along with increased expression of drought protection marker genes such as RD29A, LEA14, etc. than epsC FZB42 mutants (Lu et al. 2018; Lal et al. 2018). Another study supported the inoculation of exopolysaccharide (eps) producing PGPR *Burkholderia cepacia* in conferring resistance to *Capsicum annuum* during drought and salt stress through the production of about 4.893 ± 0.06 mg/mg eps protein (Maxton et al. 2018). The alternative role of EPS in heavy metal remediation could be suggested in the form of bioadsorbents. The focus needs to be shifted in utilizing negatively charged EPS such as hyaluronan (*Pseudomonas aeruginosa*), gellan (*Sphingomonas paucimobilis*), etc. in biosorption and biomineralization of positively charged heavy metal ions as their surfaces have abundant anionic groups that can also be chemically modified by acetylation, carboxymethylation, etc. thus enhanced applicability. EPS production is associated with processes such as biofilm formation that beneficiates these biosorption process (Ojuederie and Babalola 2017). EPS of *Rhizobium radiobacter* was used in biosorption of lead (Pb^{2+}) and zinc (Zn^{2+}) ions (Wang et al. 2013). In another study, the role of EPS as bioadsorbent from *Klebsiella* sp. J1, *Pseudomonas aeruginosa* CPSB1 and *Azotobacter chroococcum* CAZ3 was demonstrated in removal and reduction of chromium, Cr (VI) and lead (Pb^{2+}) ions (Wei et al. 2018; Rizvi and Khan 2019). Therefore, the inoculation of rhizobacterial EPS as bioadsorbent could be helpful in phytoremediation and suppression of heavy metal stress in plants.

6.3.1.5 Nitrogen Fixation, Phosphate Solubilization, and Production of Siderophore and Indole Acetic Acid

PGPR can work directly or indirectly for plant growth promotion and confer resistance against stress by fixing atmospheric nitrogen as done by a genus of *Rhizobium* or solubilize phosphates and chelate iron by secreting siderophores and provide nutrition to the host plant under stress environment (Dimkpa et al. 2009; Kumar et al. 2019). Rojas-Tapias et al. (2012) explored the effect of inoculation of *Azotobacter* strains in maize plants under salt stress. Under different salt concentrations they inoculated the plants with PGPR strains of *Azotobacter* and evaluated the changes in physiological parameters. Results indicated nitrogen fixation and phosphate solubilization irrespective of salt concentrations by two strains of *Azotobacter* C5 and C9 and auxin was also synthesized by C5 along with improved chlorophyll content and potassium-to-sodium ion ratio. A similar study was performed by Cardinale on barley (*Hordeum vulgare* L.) where PGPR improved physiological parameters of barley under salt stress. Their study also revealed that pure cultures might not alone show all PGP activities but hold plant growth promotion up to high levels and thus selection of strain should be done cautiously otherwise that would preclude new PGPR and novel plant growth mechanisms (Cardinale et al. 2015). The role of PGPR in providing protection against environmental stress has been studied by many researchers in many food and non-food crops to get insights of sustainable agriculture such as *Lavandula* and *Salvia* plants (Armada et al. 2014), *Bacopa monnieri* (Bharti et al. 2013), *Triticum aestivum* (Chakraborty et al. 2013; Kasim et al. 2013; Timmusk et al. 2014), *Capsicum annuum* (Del Amor and Cuadra-Crespo 2012), *Solanum tuberosum* (Gururani et al. 2013), *Helianthus* sp. (Shilev et al. 2012), *Glycine max* (Kang et al. 2014). Many studies revealed the inoculation of PGP bacteria having ACC deaminase was efficient in conferring resistance to plants against water stress (Choudhary et al. 2016). Salinity stress also results in declined production of phytohormones and use of IAA produced by PGP bacteria. PGPR can compensate for the decreased IAA levels from plants through their ACC deaminase activity, thereby improving root growth and absorption of nutrients in pea plants (Belimov et al. 2009; Wang et al. 2016). Inoculation of one of the common PGPR *Bacillus* sp. with characteristic features like siderophore, phosphate solubilization, and ACC deaminase activity can bring about modulation in gene profile and enzymes related to ROS scavenging (as found through mRNA profiling) and ethylene biosynthesis pathway thus improved physiological parameters during drought, salt, and heavy metal stress in potato plants (Gururani et al. 2013).

6.3.1.6 Biosurfactants

Another important metabolite produced by PGP bacteria is biosurfactant that can be isolated from hydrocarbon-contaminated soil. Biosurfactants show tremendous results in terms of remediation of heavy metal stress in plants. Their industrial usage has increased in recent years as they are biodegradable, amphiphilic in nature, highly effective, and less toxic (Rastogi and Kumar 2020; Etesami 2018). Identification of heavy metal resistant PGPR strains with biosurfactant production ability could provide insights into novel methods through which plants can escape antagonistic environments (Abou-Aly et al. 2019). A study conducted by Yaseen et al. (2019) showed the role of biosurfactant producing *Pseudochrobactrum lubricantis*, *Lysobacter novalis* and a fungus *Aspergillus niger* in mitigating lead ions stress in plants. Plants employ several internal mechanisms to alleviate heavy metal stress which can be enhanced with the inoculation of biosurfactant producing rhizobacteria (Gupta and Kumar 2017). Bioaugmentation assisted phytoremediation of heavy metal and petroleum contaminated soil with rhamnolipid (biosurfactant) producing *Pseudomonas aeruginosa* has been reported in a study. The study reported 56–105% increase in shoot and root growth, respectively, and other physiological parameters in alfalfa (Agnello et al. 2016).

6.3.1.7 Phytohormone Production

Bacteria use different mechanisms to produce phytohormones and contribute towards plant growth at root and shoot level (Egamberdieva 2013). Plant–microbe interaction does not involve any specific mechanism in coping up with environmental stress rather they work concomitantly involving multiple stress regulating mechanisms (Fahad et al. 2015). Khan et al. (2017) explored the role of *Sphingomonas* sp. LK11 and exogenous application of a phytohormone (jasmonic acid) in mitigating salt stress in *Solanum pimpinellifolium* and its non-isogenic mutant (Got-3). The improved physiological factors showed expression of glutathione-related genes in LK11 but stunted growth in Got-3 due to lack of glutamate oxaloacetate transaminase as its role lies in the assimilation of ammonia, aspartate, and in glutamate synthesis. Another study involving the co-inoculation of AMF (*Glomus etunicatum*) and PGP bacteria (*Methylobacterium oryzae* CBMB20) in maize plant has resulted in improved nutrient assimilation, phytohormone production, and decreased proline and sodium ion uptake during salinity stress (Lee et al. 2015). Symbiotic association of multiple microbes in the rhizospheric region of a plant provides a strong interactive environment to combat environmental stress. The interaction of AMF and *Bacillus subtilis* with *Acacia gerrardii* during salt stress resulted in greater root and shoot dry weight, improved nodulation for nitrogen fixation and nutrient acquisition (Hashem et al. 2016). In recent years, many studies were conducted to explain the interaction of valuable microorganisms with plants at physiological and molecular level to combat environmental stress as presented in Table 6.1.

6.3.1.8 Volatile Organic Compounds (VOCs)

Some PGPR strains release VOCs in the rhizosphere. During salt stress, bacterial VOCs activate high affinity potassium transporter 1 (HKT1) that prevents sodium ion accumulation in shoot region of the plants and also induces expression of salt overly sensitive 3 (SOS3) gene which controls post-transcriptional activity of proton/sodium ion (H^+/Na^+), antiporter SOS1 that regulates Na^+ homeostasis in VOC-treated plants and also leads to acidification of rhizosphere and thus generating a H^+ gradient which allows SOS1 mediated Na^+ exclusion from roots (Liu and Zhang 2015). VOCs by PGPR *Pseudomonas simiae* AU triggered induced systemic tolerance (IST) and plant growth in soybean (*Glycine max* L. Merrill) during salt stress (Vaishnav et al. 2015). Detailed proteomic analysis and western blotting revealed the suppression of salt stress by the accumulation of vegetative storage proteins (VSP), ϒ-glutamyl hydrolase (GGH), and ribulose-1,5-bisphosphate carboxylase oxygenase (RuBisCo) enzyme that also helped in sustaining plant growth (Vaishnav et al. 2015). During water stress, expression of transcripts for enzyme phosphoethanolamine N-methyltransferase (PEAMT) that facilitates tolerance to dehydration and expression of vacuolar H^+ pumps is known to rise in VOC-treated plants (Farag et al. 2013; Liu and Zhang 2015; Vigani et al. 2019).

Table 6.1 Plant–microbe interactions to alleviate environmental stress at physiological and molecular level

S. No.	Stress	Detrimental effects	Affected plant	Microorganism	Physiological and molecular mechanism involved	References
1	Drought	Reduced leaf and water potential, nutrient deprivation	*Fragaria* × *ananassa*	AMF (*Rhizophagus clarus*)	Increased antioxidant enzyme defense system, photosynthetic machinery, and water status	Moradtalab et al. (2019)
2	Drought	Oxidative stress	*Triticum durum* Desf; cv. Mongibello, *Triticum aestivum*; cv. Chinese spring	AMF (*Funneliformis mosseae*)	Metabolite regulation, stimulated brassinosteroids biosynthetic pathway, cross-talk between phytohormones	Bernardo et al. (2019)
3	Drought	Oxidative stress	*Cicer arietinum*	PGPR (*Bacillus subtilis*, *B. thuringiensis*, *B. megaterium*)	Enhanced accumulation of amino acids and other metabolites such as proline, L-arginine, etc.	Khan et al. (2019)
4	Drought	Affected plant growth parameters	*Triticum aestivum*	AMF (*Glomus mosseae*)	Improved physiological aspects in terms of plant biomass and yield	Rani et al. (2018)
5	Salinity	Ion imbalance, accumulation of Na^+ in leaves	*Sulla coronaria*	AMF (*Rhizophagus intraradices*), PGPR (*Pseudomonas* sp., *Bacillus subtilis*)	Enhanced physiological processes in terms of improved nutrient uptake, reduced electrolyte absorption, malondialdehyde levels	Hidri et al. (2019)
6	Salinity	Ion toxicity, hormonal imbalance, nutrient deficiency	*Cicer arietinum*	*Rhizobium* sp., *Stenotrophomonas maltophilia, mycorrhizal* fungi	Improve nodulation, nitrogen fixation, and nutritional status	Abd-alla et al. (2019)
7	Salinity	Altered osmotic pressure, toxic Na^+ ion accumulation	*Alfalfa, Arabidopsis thaliana*	*Enterobacter* sp. SA187	Induced systemic tolerance (IST) via ethylene signaling pathway by converting bacterial 2-keto-4-methylthiobutyric acid (KMBA) into ethylene	De Zélicourt et al. (2018)

8	Salinity	Disturbed osmotic, K^+/Na^+ ion ratio	*Arabidopsis thaliana*	*Serendipita indica* (formerly *Piriformospora indica*)	Na^+ detoxification via ENA ATPases	Lanza et al. (2019); Gill et al. (2016)
9	Drought and salinity	Disturbed water potential, electrolyte imbalance, reduced photosynthetic rate	*Medicago truncatula*	AMF (*Funneliformis mosseae, Rhizophagus irregularis*)	Induced resistance by 14-3-3-like protein expression via genes Fm_{201}, $Ri_{14\text{-}3\text{-}3}$ and $RiBMH_2$	Sun et al. (2018)
10	Nickel (Ni) and salinity	Electrolyte leakage, diminished chlorophyll, proline contents, etc.	*Helianthus annuus*	*Pseudomonas libanensis* TR1, AMF (*Claroideoglomus claroideum* BEG210)	Phytoremediation, improved physiological aspects	Ma et al. (2019)
11	Lead (Pb)	Decreased root and shoot dry weights, chlorophyll pigment, sugar content, etc.	*Lathyrus sativus*	PGPR consortium (*Rhizobium leguminosarum, Pseudomonas fluorescens, Luteibacter* sp., *Variovorax* sp. *Sinorhizobium meliloti*)	Mineral homeostasis, enhanced antioxidant enzyme activity, phenolic compounds biosynthesis, reduced oxidative damage	Abdelkrim et al. (2018)
12	Nutrient	Affected plant growth parameters	*Glycine max*	*Bradyrhizobium japonicum, AMF*	Nitrogen fixation and phosphate solubilization	Meena et al. (2018)
13	Allelopathic compounds	Effect on root and seedling growth	*Oryza sativa*	*Pseudomonas fluorescens* BRM-32111, *Burkholderia pyrrocinia* BRM-32113	Increased plant growth parameters such as carbon assimilation, chlorophyll a and b levels, etc.	Rego et al. (2018)
14	*Aspergillus niger*	Suppression of plant growth and ultimately senescence	*Zea mays*	*Bacillus megaterium, Pseudomonas aeruginosa*	IST via defense related enzyme production such as phenylalanine ammonia lyase (PAL) and β-1, 1, 3-glucanase	Jha (2018)
15	*Tomato spotted wilt virus* and *Potato virus Y*	Plant growth suppression	*Solanum lycopersicum*	*Bacillus amyloliquefaciens* strain MBI600	Induction of discrete gene expression patterns and genes related to salicylic acid (SA) or jasmonic acid (JA) signaling	Beris et al. (2018)

PGPR induced drought tolerance was recently reported in pepper plants (*Capsicum annuum* L.). PGPR inoculation not only reduced water stress but also increased plant's vacuolar pyrophosphatase (V-PPase) pump enhancing photosynthetic activity and larger root system (Vigani et al. 2019).

At the molecular level, bacteria can provide stress resistance by modulating the gene expression of stress-causing genes. Growth and physiological responses of the plant can be accelerated by using recombinant technologies along with inoculation of multiple microbes as reported by Calvo-Polanco et al. (2016). They used four recombinant inbred lines of tomato and inoculated them with AMF and PGPR to see the amelioration effect of drought. These recombinant lines showed better stress-mitigating properties than wild ones (Calvo-Polanco et al. 2016). Another study on soybean (*Glycine max* L. Merrill) reported modulation and up-regulation of transcription factors such as dehydration responsive element binding (DREB) and ethylene responsive element binding (EREB), water uptake transporters like plasma membrane intrinsic proteins (PIP), and tonoplast intrinsic proteins (TIP) and many others due to inoculation with PGPR *Pseudomonas simiae* strain AU (Vaishnav and Choudhary 2019). In order to have a sustainable agriculture in future, the focus is on developing technologies and insight into plant–microbiome interaction at the molecular level without affecting the yield or productivity (Barea 2015). Plant and bacteria grow concomitantly in the soil where secretions of bacteria can trigger a series of internal reactions and contribute towards plant growth. Many genes have been reported in bacteria to promote plant growth such as auxin-induced genes (At4g36110, At2g33830) and techniques like DNA microarray, SAGE (serial analysis of gene expression), cDNA-AFLP (cDNA-amplified fragment length polymorphism), etc. are available to monitor plant and pathogen interaction during biotic stress (Verma et al. 2018).

6.4 Conclusion

Environmental stress causes a significant diminution in the growth of plants by hampering them at morphological, biochemical, physiological, and molecular level. The molecular aspects of plant–microbe interaction are the current focus in assessing stress-mitigating mechanism as it lays the foundation for breeding technology to pool out requisite genes, provides insights into pathways involved, in the production of miscellaneous novel compounds and in the development of new methods to ameliorate environmental stress. The alternate approaches for use of novel microbial compounds that are produced during stress and omics-based research including genomics, meta-genomics, metabolomics, etc. are highly recommended.

Acknowledgement The authors sincerely acknowledge the facilities provided by the university and the research work of various authors cited in this chapter.

References

Abd-Alla MH, Nafady NA, Bashandy SR, Hassan AA (2019) Mitigation of effect of salt stress on the nodulation, nitrogen fixation and growth of chickpea (*Cicer arietinum* L.) by triple microbial inoculation. Rhizosphere 10:100148

Abdelkrim S, Jebara SH, Saadani O, Jebara M (2018) Potential of efficient and resistant plant growth-promoting rhizobacteria in lead uptake and plant defence stimulation in *Lathyrus sativus* under lead stress. Plant Biol 20:857–869

Abdel-Salam E, Alatar A, El-Sheikh MA (2018) Inoculation with arbuscularmycorrhizal fungi alleviates harmful effects of drought stress on damask rose. Saudi J Biol Sci 25:1772–1780

Abou-Aly HE, Youssef AM, El-Meihy RM, Tawfik TA, El-Akshar EA (2019) Evaluation of heavy metals tolerant bacterial strains as antioxidant agents and plant growth promoters. Biocatal Agric Biotechnol 19:101–110

Agnello AC, Bagard M, Van HED, Esposito G, Huguenot D (2016) Comparative bioremediation of heavy metals and petroleum hydrocarbons co-contaminated soil by natural attenuation, phytoremediation, bioaugmentation and bioaugmentation-assisted phytoremediation. Sci Total Environ 563:693–703

Ansari MI (2018) Plant microbiome and its functional mechanism in response to environmental stress. Int J Green Pharm 12:81–92

Armada E, Roldán A, Azcon R (2014) Differential activity of autochthonous bacteria in controlling drought stress in native *Lavandula* and *Salvia* plants species under drought conditions in natural arid soil. Microb Ecol 67:410–420

Aroca R, Ruiz-Lozano JM, Zamarreño ÁM, Paz JA, García-Mina JM, Pozo MJ, López-Ráez JA (2013) Arbuscular mycorrhizal symbiosis influences strigolactone production under salinity and alleviates salt stress in lettuce plants. J Plant Physiol 170:47–55

Asrar AWA, Elhindi KM (2011) Alleviation of drought stress of marigold (*Tagetes erecta*) plants by using arbuscular mycorrhizal fungi. Saudi J Biol Sci 18:93–98

Atkinson NJ, Urwin PE (2012) The interaction of plant biotic and abiotic stresses: from genes to the field. J Exp Bot 63:3523–3543

Babar MM, Khan SF, Zargaham MK, Gul A (2016) Plant-microbe interactions: a molecular approach. In: Hakeem KR, Akhtar MS (eds) Plant, soil and microbes. Springer, Cham, pp 1–22

Balestrini R, Chitarra W, Fotopoulos V, Ruocco M (2017) Potential role of beneficial soil microorganisms in plant tolerance to abiotic stress factors. In: Lukac M, Grenni P, Gamboni M (eds) Soil biological communities and ecosystem resilience. Springer, Cham, pp 191–207

Bardi L, Malusà E (2012) Drought and nutritional stresses in plant: alleviating role of rhizospheric microorganisms. In: Haryana N, Punj S (eds) Abiotic stress: new research. Nova Science Publishers, Hauppauge, pp 1–57

Barea JM (2015) Future challenges and perspectives for applying microbial biotechnology in sustainable agriculture based on a better understanding of plant-microbiome interactions. J Soil Sci Plant Nutr 15:261–282

Belimov AA, Dodd IC, Hontzeas N, Theobald JC, Safronova VI, Davies WJ (2009) Rhizosphere bacteria containing 1-aminocyclopropane-1-carboxylate deaminase increase yield of plants grown in drying soil via both local and systemic hormone signalling. New Phytol 181:413–423

Beris D, Theologidis I, Skandalis N, Vassilakos N (2018) *Bacillus amyloliquefaciens* strain MBI600 induces salicylic acid dependent resistance in tomato plants against Tomato spotted wilt virus and Potato virus Y. Sci Rep 8:10320

Bernardo L, Carletti P, Badeck FW, Rizza F, Morcia C, Ghizzoni R, Rouphael Y, Colla G, Terzi V, Lucini L (2019) Metabolomic responses triggered by arbuscular mycorrhiza enhance tolerance to water stress in wheat cultivars. Plant Physiol Biochem 137:203–212

Bharti N, Yadav D, Barnawal D, Maji D, Kalra A (2013) *Exiguobacterium oxidotolerans*, a halotolerant plant growth promoting rhizobacteria, improves yield and content of secondary metabolites in *Bacopa monnieri* (L.) Pennell under primary and secondary salt stress. World J Microbiol Biotechnol 29:379–387

Bharti N, Pandey SS, Barnawal D, Patel VK, Kalra A (2016) Plant growth promoting rhizobacteria *Dietzianatrono limnaea* modulates the expression of stress responsive genes providing protection of wheat from salinity stress. Sci Rep 6:34768

Brotman Y, Landau U, Cuadros-Inostroza Á, Tohge T, Fernie AR, Chet I, Viterbo A, Willmitzer L (2013) *Trichoderma*-plant root colonization: escaping early plant defense responses and activation of the antioxidant machinery for saline stress tolerance. PLoS Pathog 9:e1003221

Bücking H, Kafle A (2015) Role of arbuscular mycorrhizal fungi in the nitrogen uptake of plants: current knowledge and research gaps. Agronomy 5:587–612

Calvo-Polanco M, Sánchez-Romera B, Aroca R, Asins MJ, Declerck S, Dodd IC, Martinez-Andujar C, Albacete A, Ruiz-Lozano JM (2016) Exploring the use of recombinant inbred lines in combination with beneficial microbial inoculants (AM fungus and PGPR) to improve drought stress tolerance in tomato. Environ Exp Bot 131:47–57

Cardinale M, Ratering S, Suarez C, Montoya AMZ, Geissler-Plaum R, Schnell S (2015) Paradox of plant growth promotion potential of rhizobacteria and their actual promotion effect on growth of barley (*Hordeum vulgare* L.) under salt stress. Microbiol Res 181:22–32

Chakraborty U, Chakraborty BN, Chakraborty AP, Dey PL (2013) Water stress amelioration and plant growth promotion in wheat plants by osmotic stress tolerant bacteria. World J Microbiol Biotechnol 29:789–803

Chitarra W, Pagliarani C, Maserti B, Lumini E, Siciliano I, Cascone P, Schubert A, Gambino G, Balestrini R, Guerrieri E (2016) Insights on the impact of arbuscular mycorrhizal symbiosis on tomato tolerance to water stress. Plant Physiol 171:1009–1023

Choudhary DK, Kasotia A, Jain S, Vaishnav A, Kumari S, Sharma KP, Varma A (2016) Bacterial-mediated tolerance and resistance to plants under abiotic and biotic stresses. J Plant Growth Regul 35:276–300

Coleman-Derr D, Tringe SG (2014) Building the crops of tomorrow: advantages of symbiont-based approaches to improving abiotic stress tolerance. Front Microbiol 5:283

De Souza R, Meyer J, Schoenfeld RDA, Costa PB, Passaglia LM (2015) Characterization of plant growth-promoting bacteria associated with rice cropped in iron-stressed soils. Ann Microbiol 65:951–964

De Zélicourt A, Synek L, Saad MM, Alzubaidy H, Jalal R, Xie Y, Andres-Barrao C, Rolli E, Guerard F, Mariappan KG, Daur I, Colcombet J, Benhamed M, Depaepe T, Straeten DVD, Hirt H (2018) Ethylene induced plant stress tolerance by *Enterobacter* sp. SA187 is mediated by 2-keto-4-methylthiobutyric acid production. PLoS Genet 14:e1007273

Del Amor FM, Cuadra-Crespo P (2012) Plant growth-promoting bacteria as a tool to improve salinity tolerance in sweet pepper. Funct Plant Biol 39:82–90

Dimkpa CO, Merten D, Svatos A, Buchel G, Kothe E (2009) Siderophores mediate reduced and increased uptake of cadmium by *Streptomyces tendae* F4 and sunflower (*Helianthus annuus*), respectively. J Appl Microbiol 107:1687–1696

Doubková P, Vlasáková E, Sudová R (2013) Arbuscular mycorrhizal symbiosis alleviates drought stress imposed on *Knautia arvensis* plants in serpentine soil. Plant Soil 370:149–161

Egamberdieva D (2013) The role of phytohormone producing bacteria in alleviating salt stress in crop plants. In: Miransari M (ed) Biotechnological techniques of stress tolerance in plants. Stadium Press, Auburn, pp 21–39

Elhindi KM, El-Din AS, Elgorban AM (2017) The impact of arbuscular mycorrhizal fungi in mitigating salt-induced adverse effects in sweet basil (*Ocimum basilicum* L.). Saudi J Biol Sci 24:170–179

Etesami H (2018) Bacterial mediated alleviation of heavy metal stress and decreased accumulation of metals in plant tissues: mechanisms and future prospects. Ecotoxicol Environ Saf 147:175–191

Evelin H, Kapoor R, Giri B (2009) Arbuscular mycorrhizal fungi in alleviation of salt stress: a review. Ann Bot 104:1263–1280

Fahad S, Hussain S, Bano A, Saud S, Hassan S, Shan D, Khan FA, Khan F, Chen Y, Wu C, Tabassum MA, Chun MX, Afzal M, Jan A, Jan MT, Huang J (2015) Potential role of phytohormones and plant growth-promoting rhizobacteria in abiotic stresses: consequences for changing environment. Environ Sci Pollut Res 22:4907–4921

Fan QJ, Liu JH (2011) Colonization with arbuscular mycorrhizal fungus affects growth, drought tolerance and expression of stress-responsive genes in *Poncirus trifoliata*. Acta Physiol Plant 33:1533–1542

Farag MA, Zhang H, Ryu CM (2013) Dynamic chemical communication between plants and bacteria through airborne signals: induced resistance by bacterial volatiles. J Chem Ecol 39:1007–1018

Fuentes A, Almonacid L, Ocampo JA, Arriagada C (2016) Synergistic interactions between a saprophytic fungal consortium and *Rhizophagus irregularis* alleviate oxidative stress in plants grown in heavy metal contaminated soil. Plant Soil 407:355–366

Garg N, Bhandari P (2016) Silicon nutrition and mycorrhizal inoculations improve growth, nutrient status, K+/Na+ ratio and yield of *Cicer arietinum* L. genotypes under salinity stress. Plant Growth Regul 78:371–387

Gill SS, Gill R, Trivedi DK, Anjum NA, Sharma KK, Ansari MW, Ansari AA, Johri AK, Prasad R, Pereira E, Varma A, Tuteza N (2016) *Piriformospora indica*: potential and significance in plant stress tolerance. Front Microbiol 7:332

Gupta P, Kumar V (2017) Value added phytoremediation of metal stressed soils using phosphate solubilizing microbial consortium. World J Microbiol Biotechnol 33:9

Gururani MA, Upadhyaya CP, Baskar V, Venkatesh J, Nookaraju A, Park SW (2013) Plant growth-promoting rhizobacteria enhance abiotic stress tolerance in *Solanum tuberosum* through inducing changes in the expression of ROS-scavenging enzymes and improved photosynthetic performance. J Plant Growth Regul 32:245–258

Hameed A, Wu QS, Abd-Allah EF, Hashem A, Kumar A, Lone HA, Ahmad P (2014) Role of AM fungi in alleviating drought stress in plants. In: Miransari M (ed) Use of microbes for the alleviation of soil stresses. Springer, New York, pp 55–75

Hashem A, Abd_Allah EF, Alqarawi AA, Al-Huqail AA, Wirth S, Egamberdieva D (2016) The interaction between arbuscular mycorrhizal fungi and endophytic bacteria enhances plant growth of *Acacia gerrardii* under salt stress. Front Microbiol 7:1089

Hashem A, Alqarawi AA, Radhakrishnan R, Al-Arjani ABF, Aldehaish HA, Egamberdiev D, Abd_Allah EF (2018) Arbuscular mycorrhizal fungi regulate the oxidative system, hormones and ionic equilibrium to trigger salt stress tolerance in *Cucumis sativus* L. Saudi J Biol Sci 25:1102–1114

He F, Zhang H, Tang M (2016) Aquaporin gene expression and physiological responses of *Robinia pseudoacacia* L. to the mycorrhizal fungus *Rhizophagus irregularis* and drought stress. Mycorrhiza 26:311–323

Heidari M, Karami V (2014) Effects of different mycorrhiza species on grain yield, nutrient uptake and oil content of sunflower under water stress. J Saudi Soc Agric Sci 13:9–13

Hidri R, Mahmoud OMB, Farhat N, Cordero I, Pueyo JJ, Debez A, Barea JM, Abdelly C, Azcon R (2019) Arbuscular mycorrhizal fungus and rhizobacteria affect the physiology and performance of *Sulla coronaria* plants subjected to salt stress by mitigation of ionic imbalance. J Plant Nutr Soil Sci 182:451–462

Jha Y (2018) Induction of anatomical, enzymatic, and molecular events in maize by PGPR under biotic stress. In: Meena VS (ed) Role of rhizospheric microbes in soil. Springer, Singapore, pp 125–141

Kang SM, Radhakrishnan R, Khan AL, Kim MJ, Park JM, Kim BR, Shin DH, Lee IJ (2014) Gibberellin secreting rhizobacterium, *Pseudomonas putida* H-2-3 modulates the hormonal and stress physiology of soybean to improve the plant growth under saline and drought conditions. Plant Physiol Biochem 84:115–124

Kasim WA, Osman ME, Omar MN, El-Daim IAA, Bejai S, Meijer J (2013) Control of drought stress in wheat using plant-growth-promoting bacteria. J Plant Growth Regul 32:122–130

Khan AL, Waqas M, Asaf S, Kamran M, Shahzad R, Bilal S, Khan MA, Kang SM, Kim YH, Yun BW, Al-Rawahi A, Al-Harrasi A, Lee IJ (2017) Plant growth-promoting endophyte

Sphingomonas sp. LK11 alleviates salinity stress in *Solanum pimpinellifolium*. Environ Exp Bot 133:58–69

Khan N, Bano A, Rahman MA, Guo J, Kang Z, Babar MA (2019) Comparative physiological and metabolic analysis reveals a complex mechanism involved in drought tolerance in chickpea (*Cicer arietinum* L.) induced by PGPR and PGRs. Sci Rep 9:2097

Kim YC, Glick BR, Bashan Y, Ryu CM (2012) Enhancement of plant drought tolerance by microbes. In: Aroca R (ed) Plant responses to drought stress. Springer, Berlin, pp 383–413

Kim K, Jang YJ, Lee SM, Oh BT, Chae JC, Lee KJ (2014) Alleviation of salt stress by *Enterobacter* sp. EJ01 in tomato and *Arabidopsis* is accompanied by up-regulation of conserved salinity responsive factors in plants. Mol Cells 37:109

Kong J, Pei Z, Du M, Sun G, Zhang X (2014) Effects of arbuscular mycorrhizal fungi on the drought resistance of the mining area repair plant Sainfoin. Int J Min Sci Technol 24:485–489

Kumar A, Patel JS, Meena VS, Ramteke PW (2019) Plant growth-promoting rhizobacteria: strategies to improve abiotic stresses under sustainable agriculture. J Plant Nutr 42:1402–1415

Lal S, Ratna S, Said OB, Kumar R (2018) Biosurfactant and exopolysaccharide-assisted rhizobacterial technique for the remediation of heavy metal contaminated soil: an advancement in metal phytoremediation technology. Environ Technol Innov 10:243–263

Lanza M, Haro R, Conchillo LB, Benito B (2019) The endophyte Serendipita indica reduces the sodium content of Arabidopsis plants exposed to salt stress: fungal ENA ATPases are expressed and regulated at high pH and during plant co-cultivation in salinity. Environ Microbiol 21:3364–3378

Lata R, Chowdhury S, Gond SK, White JF Jr (2018) Induction of abiotic stress tolerance in plants by endophytic microbes. Lett Appl Microbiol 66:268–276

Latef AAHA, Miransari M (2014) The role of arbuscular mycorrhizal fungi in alleviation of salt stress. In: Miransari M (ed) Use of microbes for the alleviation of soil stresses. Springer, New York, pp 23–38

Latef AAHA, Hashem A, Rasool S, Abd_Allah EF, Alqarawi AA, Egamberdieva D, Sumira J, Naser AA, Ahmad P (2016) Arbuscular mycorrhizal symbiosis and abiotic stress in plants: a review. J Plant Biol 59:407–426

Lee Y, Krishnamoorthy R, Selvakumar G, Kim K, Sa T (2015) Alleviation of salt stress in maize plant by co-inoculation of arbuscular mycorrhizal fungi and *Methylobacterium oryzae* CBMB20. J Korean Soc Appl Biol Chem 58:533–540

Lenoir I, Fontaine J, Sahraoui ALH (2016) Arbuscular mycorrhizal fungal responses to abiotic stresses: a review. Phytochemistry 123:4–15

Lin J, Yu D, Shi Y, Sheng H, Li C, Wang Y, Yingnan W, Chungsheng M, Li X (2016) Salt-alkali tolerance during germination and establishment of *Leymus chinensis* in the Songnen Grassland of China. Ecol Eng 95:763–769

Lin J, Wang Y, Sun S, Mu C, Yan X (2017) Effects of arbuscular mycorrhizal fungi on the growth, photosynthesis and photosynthetic pigments of *Leymus chinensis* seedlings under salt-alkali stress and nitrogen deposition. Sci Total Environ 576:234–241

Liu XM, Zhang H (2015) The effects of bacterial volatile emissions on plant abiotic stress tolerance. Front Plant Sci 6:774

Liu C, Dai Z, Cui M, Lu W, Sun H (2018) Arbuscular mycorrhizal fungi alleviate boron toxicity in *Puccinellia tenuiflora* under the combined stresses of salt and drought. Environ Pollut 240:557–565

Lu X, Liu SF, Yue L, Zhao X, Zhang YB, Xie ZK, Wang RY (2018) Epsc involved in the encoding of exopolysaccharides produced by *Bacillus amyloliquefaciens* FZB42 act to boost the drought tolerance of *Arabidopsis thaliana*. Int J Mol Sci 19:3795

Ma Y, Rajkumar M, Oliveira RS, Zhang C, Freitas H (2019) Potential of plant beneficial bacteria and arbuscular mycorrhizal fungi in phytoremediation of metal-contaminated saline soils. J Hazard Mater 379:120813

Maiti SK, Ahirwal J (2019) Ecological restoration of coal mine degraded lands: topsoil management, pedogenesis, carbon sequestration, and mine pit limnology. In: Pandey VM, Bauddh K (eds) Phytomanagement of polluted sites. Elsevier, Amsterdam, pp 83–111

Maxton A, Singh P, Masih SA (2018) ACC deaminase-producing bacteria mediated drought and salt tolerance in Capsicum annum. J Plant Nutr 41:574–583

Meena KK, Sorty AM, Bitla UM, Choudhary K, Gupta P, Pareek A, Singh DP, Prabha R, Sahu PK, Gupta VK, Krishanani KK, Singh HB (2017) Abiotic stress responses and microbe-mediated mitigation in plants: the omics strategies. Front Plant Sci 8:172

Meena RS, Vijayakumar V, Yadav GS, Mitran T (2018) Response and interaction of *Bradyrhizobium japonicum* and arbuscular mycorrhizal fungi in the soybean rhizosphere. Plant Growth Regul 84:207–223

Miransari M (2014) Mycorrhizal fungi to alleviate salinity stress on plant growth. In: Miransari M (ed) Use of microbes for the alleviation of soil stresses. Springer, New York, pp 77–86

Moradtalab N, Hajiboland R, Aliasgharzad N, Hartmann TE, Neumann G (2019) Silicon and the association with an arbuscular-mycorrhizal fungus (*Rhizophagus clarus*) mitigate the adverse effects of drought stress on strawberry. Agronomy 9:41

Nadeem SM, Ahmad M, Zahir ZA, Javaid A, Ashraf M (2014) The role of mycorrhizae and plant growth promoting rhizobacteria (PGPR) in improving crop productivity under stressful environments. Biotechnol Adv 32:429–448

Nautiyal CS, Srivastava S, Chauhan PS, Seem K, Mishra A, Sopory SK (2013) Plant growth-promoting bacteria *Bacillus amyloliquefaciens* NBRISN13 modulates gene expression profile of leaf and rhizosphere community in rice during salt stress. Plant Physiol Biochem 66:1–9

Ojuederie O, Babalola O (2017) Microbial and plant-assisted bioremediation of heavy metal polluted environments: a review. Int J Environ Res Public Health 14:1504

Oyewole BO, Olawuyi OJ, Odebode AC, Abiala MA (2017) Influence of Arbuscular mycorrhiza fungi (AMF) on drought tolerance and charcoal rot disease of cowpea. Biotechnol Rep 14:8–15

Palaniyandi SA, Damodharan K, Yang SH, Suh JW (2014) *Streptomyces* sp. strain PGPA39 alleviates salt stress and promotes growth of 'Micro Tom' tomato plants. J Appl Microbiol 117:766–773

Pavithra D, Yapa N (2018) Arbuscular mycorrhizal fungi inoculation enhances drought stress tolerance of plants. Groundw Sustain Dev 7:490–494

Porcel R, Aroca R, Ruiz-Lozano JM (2012) Salinity stress alleviation using arbuscular mycorrhizal fungi: a review. Agron Sustain Dev 32:181–200

Pozo MJ, Jung SC, Martínez-Medina A, Lopez-Raez JA, Azcon-Aguilar C, Barea JM (2013) Root allies: arbuscular mycorrhizal fungi help plants to cope with biotic stresses. In: Aroca R (ed) Symbiotic endophytes. Springer, Berlin, pp 289–307

Quiroga G, Erice G, Aroca R, Chaumont F, Ruiz-Lozano JM (2017) Enhanced drought stress tolerance by the arbuscular mycorrhizal symbiosis in a drought-sensitive maize cultivar is related to a broader and differential regulation of host plant aquaporins than in a drought-tolerant cultivar. Front Plant Sci 8:1056

Rani B, Madan S, Pooja SKD, Kumari N, Kumar A (2018) Mitigating the effect of drought stress on yield in wheat (*Triticum aestivum*) using arbuscular mycorrhiza fungi (*Glomus mosseae*). Indian J Agric Sci 88:95–100

Rapparini F, Penuelas J (2014) Mycorrhizal fungi to alleviate drought stress on plant growth. In: Miransari M (ed) Use of microbes for the alleviation of soil stresses, volume 1. Springer, New York, pp 21–42

Rastogi S, Kumar R (2020) Remediation of heavy metals using non-conventional adsorbents and biosurfactant-producing bacteria. In: Kumar V, Singh J, Kumar P (eds) Environmental degradation: causes and remediation strategies, pp 133–153. https://doi.org/10.26832/aesa-2020-edcrs-010

Rastogi S, Kumar J, Kumar R (2019) An investigation into the efficacy of fungal biomass as a low cost bio-adsorbent for the removal of lead from aqueous solutions. Int Res J Eng Technol 6:7144–7149

Rego MC, Cardoso AF, Ferreira T d C, de Filippi MC, Batista TF, Viana RG, da Silva GB (2018) The role of rhizobacteria in rice plants: growth and mitigation of toxicity. J Integr Agr 17:2636–2647

Rizvi A, Khan MS (2019) Putative role of bacterial biosorbent in metal sequestration revealed by SEM–EDX and FTIR. Indian J Microbiol 59:246–249

Rojas-Tapias D, Moreno-Galvan A, Pardo-Diaz S, Obando M, Rivera D, Bonilla R (2012) Effect of inoculation with plant growth-promoting bacteria (PGPB) on amelioration of saline stress in maize (*Zea mays*). Appl Soil Ecol 61:264–272

Salwan R, Sharma A, Sharma V (2019) Microbes mediated plant stress tolerance in saline agricultural ecosystem. Plant Soil 442:1–22

Selvakumar G, Panneerselvam P, Ganeshamurthy AN (2012) Bacterial mediated alleviation of abiotic stress in crops. In: Maheshwari DK (ed) Bacteria in agrobiology: stress management. Springer, Berlin, pp 205–224

Shahzad R, Khan AL, Bilal S, Waqas M, Kang SM, Lee IJ (2017) Inoculation of abscisic acid-producing endophytic bacteria enhances salinity stress tolerance in *Oryza sativa*. Environ Exp Bot 136:68–77

Shilev S, Sancho ED, Benlloch-Gonzalez M (2012) Rhizospheric bacteria alleviate salt-produced stress in sunflower. J Environ Manag 95:S37–S41

Shukla PS, Agarwal PK, Jha B (2012) Improved salinity tolerance of *Arachis hypogaea* (L.) by the interaction of halotolerant plant-growth-promoting rhizobacteria. J Plant Growth Regul 31:95–206

Singh A, Jain A, Sarma BK, Upadhyay RS, Singh HB (2013a) Rhizosphere microbes facilitate redox homeostasis in *Cicer arietinum* against biotic stress. Ann Appl Biol 163:33–46

Singh A, Sarma BK, Upadhyay RS, Singh HB (2013b) Compatible rhizosphere microbes mediated alleviation of biotic stress in chickpea through enhanced antioxidant and phenylpropanoid activities. Microbiol Res 168:33–40

Souza RD, Ambrosini A, Passaglia LM (2015) Plant growth-promoting bacteria as inoculants in agricultural soils. Genet Mol Biol 38:401–419

Srivastava AK, Penna S, Nguyen DV, Tran LSP (2016) Multifaceted roles of aquaporins as molecular conduits in plant responses to abiotic stresses. Crit Rev Biotechnol 36:389–398

Subramanian P, Mageswari A, Kim K, Lee Y, Sa T (2015) Psychrotolerant endophytic *Pseudomonas* sp. strains OB155 and OS261 induced chilling resistance in tomato plants (*Solanum lycopersicum* Mill.) by activation of their antioxidant capacity. Mol Plant-Microbe Interact 28:1073–1081

Sun X, Shi J, Ding G (2017) Combined effects of arbuscular mycorrhiza and drought stress on plant growth and mortality of forage sorghum. Appl Soil Ecol 119:384–391

Sun Z, Song J, Xin XA, Xie X, Zhao B (2018) Arbuscular mycorrhizal fungal 14-3-3 proteins are involved in arbuscule formation and responses to abiotic stresses during AM symbiosis. Front Microbiol 9:1–17

Timmusk S, Abd El-Daim IA, Copolovici L, Tanilas T, Kannaste A, Behers L, Nevo E, Seisenbaeva G, Stenstrom E, Niinemets U (2014) Drought-tolerance of wheat improved by rhizosphere bacteria from harsh environments: enhanced biomass production and reduced emissions of stress volatiles. PLoS One 9:e96086

Vaishnav A, Choudhary DK (2019) Regulation of drought-responsive gene expression in Glycine max l. Merrill is mediated through *Pseudomonas simiae* strain AU. J Plant Growth Regul 38:333–342

Vaishnav A, Kumari S, Jain S, Varma A, Choudhary DK (2015) Putative bacterial volatile-mediated growth in soybean (*Glycine max* L. Merrill) and expression of induced proteins under salt stress. J Appl Microbiol 119:539–551

Verma RK, Sachan M, Vishwakarma K, Upadhyay N, Mishra RK, Tripathi DK, Sharma S (2018) Role of PGPR in sustainable agriculture: molecular approach toward disease suppression and growth promotion. In: Meena VS (ed) Role of rhizospheric microbes in soil. Springer, Singapore, pp 259–290

Vigani G, Rolli E, Marasco R, Dell'Orto M, Michoud G, Soussi A, Raddadi N, Borin S, Sorlini C, Zocchi G, Daffonchio D (2019) Root bacterial endophytes confer drought resistance and enhance expression and activity of a vacuolar H+-pumping pyrophosphatase in pepper plants. Environ Microbiol 21:3212–3228

Viscardi S, Ventorino V, Duran P, Maggio A, De Pascale S, Mora ML, Pepe O (2016) Assessment of plant growth promoting activities and abiotic stress tolerance of *Azotobacter chroococcum* strains for a potential use in sustainable agriculture. J Soil Sci Plant Nutr 16:848–863

Vurukonda SSKP, Vardharajula S, Shrivastava M, SkZ A (2016) Enhancement of drought stress tolerance in crops by plant growth promoting rhizobacteria. Microbiol Res 184:13–24

Wang L, Yang J, Chen Z, Liu X, Ma F (2013) Biosorption of Pb (II) and Zn (II) by extracellular polymeric substance (Eps) of *Rhizobium radiobacter*: equilibrium, kinetics and reuse studies. Arch Environ Prot 39:129–140

Wang Q, Dodd IC, Belimov AA, Jiang F (2016) Rhizosphere bacteria containing 1-aminocyclopropane-1- carboxylate deaminase increase growth and photosynthesis of pea plants under salt stress by limiting Na+ accumulation. Funct Plant Biol 43:161–172

Wang R, Wang M, Chen K, Wang S, Mur LAJ, Guo S (2018) Exploring the roles of aquaporins in plant–microbe interactions. Cell 7:267

Waqas M, Khan AL, Hamayun M, Shahzad R, Kim YH, Choi KS, Lee IJ (2015) Endophytic infection alleviates biotic stress in sunflower through regulation of defence hormones, antioxidants and functional amino acids. Eur J Plant Pathol 141:803–824

Wei W, Li A, Ma F, Pi S, Yang J, Wang Q, Ni BJ (2018) Simultaneous sorption and reduction of Cr (VI) in aquatic system by microbial extracellular polymeric substances from *Klebsiella* sp. J1. J Chem Technol Biotechnol 93:3152–3159

Yang Y, Han X, Liang Y, Ghosh A, Chen J, Tang M (2015) The combined effects of arbuscular mycorrhizal fungi (AMF) and lead (Pb) stress on Pb accumulation, plant growth parameters, photosynthesis, and antioxidant enzymes in *Robinia pseudoacacia* L. PLoS One 10:e0145726

Yaseen RY, El-Aziz SA, Eissa DT, Abou-Shady AM (2019) Application of biosurfactant producing microorganisms to remediate heavy metal pollution in El-Gabal El-Asfar area. Alex Sci Exch J 39:17–34

Zhao R, Guo W, Bi N, Guo J, Wang L, Zhao J, Zhang J (2015) Arbuscular mycorrhizal fungi affect the growth, nutrient uptake and water status of maize (*Zea mays* L.) grown in two types of coal mine spoils under drought stress. Appl Soil Ecol 8:41–49

Chapter 7
Plant–Microbe Interactions: An Insight into the Underlying Mechanisms to Mitigate Diverse Environmental Stresses

Asifa Mushtaq and Seema Rawat

Abstract Plants, being sessile, are often exposed to a range of biotic and abiotic stresses, which can negatively impact their growth and development causing humongous losses to the agriculture across the globe. Plants have evolved to sense and mitigate the adverse environmental challenges with an array of protective mechanisms, among which the plant–microbe interactions play a remarkable role in conferring tolerance to multiple stresses. The response of plant to a particular stress is governed by highly complex signaling pathways influenced by the associated microbes. As the abrupt climatic changes are posing a significant threat to the crop productivity, it becomes imperative to elucidate the plant–microbe interactions with respect to protection against environmental hostilities. Essentially, an exhaustive knowledge of the underlying mechanisms of microbe-mediated stress mitigation in plants can be translated into the development of resistant crop varieties, eventually paving the way for sustainable agriculture.

7.1 Introduction

Plants are often exposed to the diverse environmental stresses which affect their vital activities causing a substantial loss in crop quality as well as productivity. Even though a precise estimation of agricultural loss due to various stresses cannot be made, it has been very well documented that such stresses wallop large areas of land and consequently affect crop quality and quantity (Cramer et al. 2011). The varied stresses that a plant can experience have been broadly categorized into two groups, viz, biotic and abiotic. The biotic stresses are a manifestation of the interaction of a plant with pathogenic bacteria, fungi, or viruses as well as insects, and nematodes while the abiotic stresses are posed by drought, temperature extremes (high/low),

A. Mushtaq
Department of Botany and Microbiology, H.N.B. Garhwal (A Central) University, Srinagar, Uttarakhand, India

S. Rawat (✉)
School of Life Sciences. Central University of Gujarat, Gandhinagar, Gujarat, India

A. Sharma (ed.), *Microbes and Signaling Biomolecules Against Plant Stress*, Rhizosphere Biology, https://doi.org/10.1007/978-981-15-7094-0_7

salinity, light intensity, nutrient starvation, presence of heavy metals, and anaerobiosis. Although both biotic and abiotic stresses can impact the plant growth on various levels, some of the usual repercussions of these stresses on plants encompass hormonal and nutritional disparity, physiological disorders such as abscission, distorted growth, epinasty, senescence, and increased susceptibility to diseases (Nadeem et al. 2014). Plants initially perceive the effect of any hostile condition at the cellular level and the physiological symptoms become apparent with the passage of time. In a state of multiple biotic and abiotic adversities the plant metabolism can be put out of homeostasis which can stimulate the plant to make necessary advancements in genetic and metabolic mechanisms (Apel and Hirt 2004; Foyer and Noctor 2005; Gill and Tuteja 2010; Meena et al. 2017). Over the course of time plants have evolved to sense and mitigate the adverse environmental challenges with an array of protective mechanisms. The ability of plants to endure the challenging environmental conditions depends on a combination of processes comprising a wide spectrum of molecular and regulatory mechanisms (Duque et al. 2013; Mosa et al. 2017).

Plants interact with a myriad of soil microorganisms that bestow several benefits on the host plant, besides playing a quintessential role in quelling the burden of environmental stresses (Marulanda et al. 2006; Turner et al. 2013; Ngumbi and Kloepper 2014). Under the influence of several biotic and abiotic challenges, the consequence of plant–microbe interplay is regulated by the prioritization of physiological pathways. This cross-talk between plants and microbes is governed by a series of complex network of signaling pathways (Fujita et al. 2006; Pieterse et al. 2012; Mushtaq and Rawat 2017) which induce heterogeneous local and systemic responses that revamp metabolic capabilities of the plants to combat various forms of biotic and abiotic stresses (Kavamura et al. 2013; Nguyen et al. 2016). A variety of mechanisms have been proposed for underlying microbe-induced stress tolerance in plants (Fig. 7.1). The bacteria belonging to genera *Achromobacter*, *Azospirillum*, *Bacillus*, *Burkholderia*, *Enterobacter*, *Klebsiella*, *Methylobacterium*, *Microbacterium*, and *Serratia* have been reported to assist the host plants to adapt in different biotic and abiotic adversities through a wide spectrum of fine level molecular machinery (Joseph et al. 2007; Grover et al. 2011; Saharan and Nehra 2011). Goswami et al. (2016) reported that the plant-associated microbes employ either direct or indirect strategies to foster adaptive capacity in plants. Direct mechanisms comprise those in which plant–microbes enable an even distribution of plant growth regulators either by releasing the growth regulators themselves which are incorporated into the plant or by acting as a reservoir of hormones released by the plant itself and also by refining plant acclimatization via stimulating their metabolic activities (Govindasamy et al. 2011; Glick 2014). Indirect mechanisms involve the assistance of plant defensive metabolic processes primarily in the form of induced systemic resistance to plant pathogens (Choudhary et al. 2016). The main focus of this chapter is to get an insight into the underlying mechanisms which are employed by plant-associated microbes to impart resistance in plants against some of the major environmental constraints.

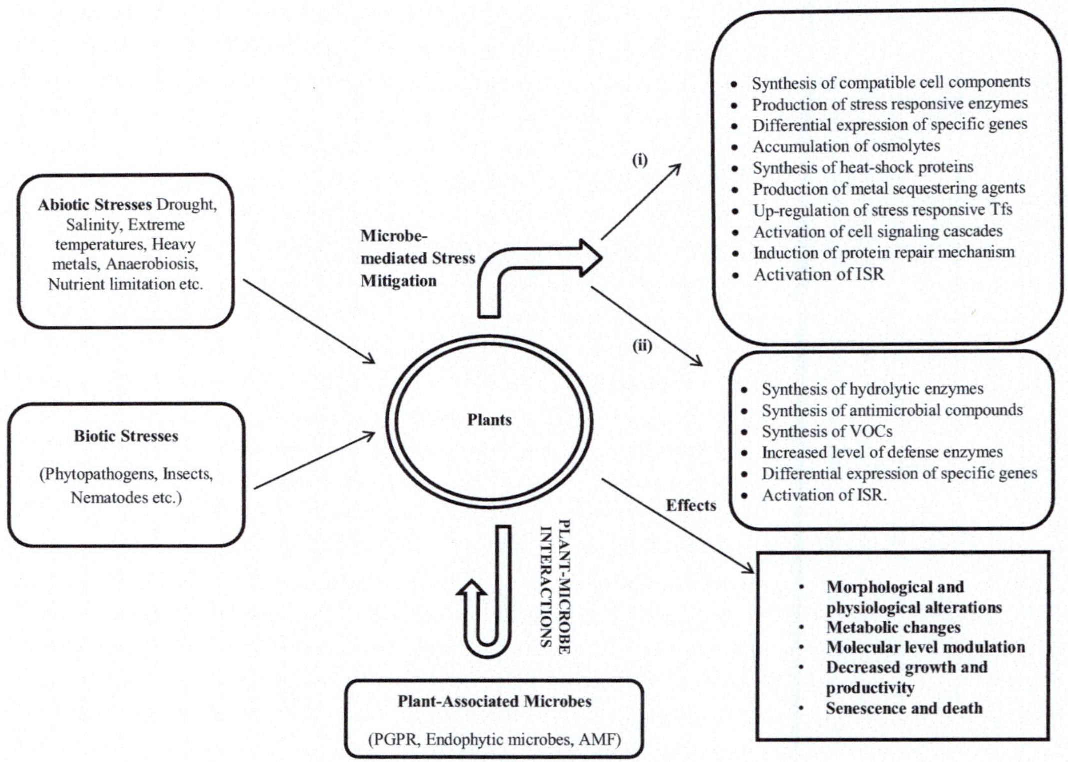

Fig. 7.1 Schematic illustration of biotic and abiotic stresses encountered by plants, their effect, and the microbe-mediated alleviation of stress induced symptoms. (i) Abiotic stress responses. (ii) Biotic stress responses

7.2 Mechanisms Underlying Microbe-Mediated Stress Tolerance in Plants

7.2.1 Abiotic Stresses

Hostile environmental conditions are becoming an abiding threat to agricultural crops across the globe. Abiotic stresses manifest a sequence of biochemical, physiological, morphological, and molecular alterations hampering plant growth and productivity (Wang et al. 2001). Over the years, role of connate metabolic and genetic capabilities of microorganisms to alleviate abiotic stresses in plants has become the intriguing area of research (Nadeem et al. 2014; Gopalakrishnan et al. 2015; Souza et al. 2015) (Table 7.1). This section will comprehend the mechanisms underlying microbe-mediated stress management in plants against the environmental hostilities that cause humongous losses to agriculture annually.

7.2.1.1 Drought Stress Management

Drought being one of the primary abiotic stresses directly affects all the aspects of plant growth and development. The arid conditions induce a metabolic and osmotic variance which causes decline in cell turgor and stomata closure (Chaves et al. 2009) imposing a restriction on the uptake of CO_2 and thus resulting in suppressed cell growth and reduced photosynthesis (Shinozaki and Yamaguchi-Shinozaki 2000). The extended exposure of plants to these conditions can cause irreversible damage leading to plant death.

In order to survive the deleterious effects of drought, plants must respond in an appropriate manner. Plant-associated microbes are known to have a striking contribution in alleviating the drought stress symptoms. The role of plant growth promoting bacteria in eliciting the plant tolerance under arid and semiarid conditions has been widely recognized. It has been reported that plant growth promoting (PGP) bacteria confer Induced Systemic Tolerance (IST) in plants to cope with water deficit through diverse mechanisms. Plants are known to produce uncurbed levels of ethylene, when exposed to drought stress which modulates plant homeostasis internally leading to decreased root and shoot growth (Mahajan and Tuteja 2005). One of the defined properties of PGP bacteria is their ability to produce 1-aminocyclopropane-1-carboxylase deaminase, an enzyme that catalyzes the conversion of 1-aminocyclopropane-1-carboxylate (precursor of ethylene biosynthesis) to ammonia and α-ketobutyrate (Qin et al. 2016). Degradation of ethylene precursor by ACC deaminase releases plant stress and restores standard plant growth (Mayak et al. 2004; Naveed et al. 2014a, b). Plants, in order to overcome the drought-induced effects, employ a strategy of lowering the ethylene levels by associating with ACC deaminase producing microorganisms. The role of ACC deaminase producing PGPR strains to endow numerous plants with IST under drought stress has been thoroughly studied (Guo et al. 2015). Kohler et al. (2008) demonstrated that ACC

Table 7.1 Microbe-mediated abiotic stress tolerance in plants

Stress	Plant	Microorganism	Underlying mechanism	References
Drought	*Triticum aestivum*	*Pantoea alhagi*	Modulated accumulation of soluble sugars, proline, and malondialdehyde	Chen et al. (2017)
	Triticum aestivum	*Burkholderia phytofirmans* PsJN	Increased activity of glutathione reductase and CAT Increased lipid peroxidase and malondialdehyde. Enhanced accumulation of osmolytes	Naveed et al. (2014a, b)
	Cucumis sativus	*Bacillus cereus*, *B. subtilis*, *Serratia*	Efficient photosynthetic systems, antioxidases	Wang et al. (2012)
	Phaseolus vulgaris	*Rhizobiumetli*	Overexpression of trehalose-6-phosphatase synthase gene	Grover et al. (2011)
	Phaseolus vulgaris	*Rhizobium tropici* and *Paenibacillus polymyxa*	Upregulation of specific genes involved in stress tolerance	Figueiredo et al. (2008)
	Solanum lycopersicum	*Achromobacter piechaudii* ARV8	ACC deaminase production	Kohler et al. (2008)
Heavy metals	*Triticum aestivum*	*Stenotrophomonas maltophilia*	Decreased proline and malondialdehyde, elevated activity of SOD, CAT, and POX	Singh and Jha (2017)
	Brassica juncea	*Pseudomonas brassicacearum*, *rhizobium leguminosarum*	Production of metal chelating molecules	Adediran et al. (2016)
	Phragmites australis	*Photobacterium* sp.	Increased activity of mercury reductase, IAA production	Mathew et al. (2015)
	Vicia faba	*Enterobacter cloacae* and *Pseudomonas* sp.	Modulated activity of SOD and CAT	Fatnassi et al. (2015)
	Zea mays	*Glomus* spp.	Increased SOD activity, high lead concentration in mycorrhizal roots	Zhang et al. (2010)

(continued)

Table 7.1 (continued)

Stress	Plant	Microorganism	Underlying mechanism	References
Salinity	*Solanum lycopersicum*	*Enterobacter* sp. EJ01	Rapid upregulation of conserved salt stress responsive genes	Kim et al. (2014)
	Oryza sativa	*Bacillus amyloliquefaciens*	Differential expression of specific set of genes	Nautiyal et al. (2013)
	Glycine max	*Metarhizium anisopliae*	Enhanced levels of endogenous JA, reduced malondialdehyde, elevated SOD activity	Khan et al. (2012)
	Zea mays	*Glomus mosseae*	Thicker root system and efficient root activity	Sheng et al. (2009)
	Solanum lycopersicum	*Glomus mosseae*	Elevated levels of SOD, POX, CAT, and ascorbate peroxidase	He et al. (2007)
Temperature	*Triticum aestivum*	*Enterobacter cloacae ZNP-3*	Synthesis of ACC deaminase, IAA, elevated levels of compatible solutes	Singh et al. (2017)
	Triticum aestivum	*Bacillus amyloliquefaciens*, *Azospirillum brasilense*	Limited regeneration of ROS, preactivation of heat shock Tfs, modified metabolome	El-Daim et al. (2014)
	Lycopersicum esculentum L. (seedlings)	*Trichoderma harzianum*	Reduced accumulation of ROS	Mastouri et al. (2010)
	Vitis vinifera	*Burkholderia phytofirmans*	ISR activation	Barka et al. (2006)

deaminase producing *Achromobacter piechaudii* ARV8 curtails the ethylene production in tomato seedlings exposed to conditions of drought and also shows a remarkable increase in the fresh and dry weight of tomato and pepper seedlings in water-deficit environment. Likewise, a consortium of PGPR species, namely, *Bacillus cereus* AR156, *Bacillus subtilis* SM21, and *Serratia* sp.XY21 has been shown to impart induced systemic tolerance in cucumber plants under dry spell via retaining efficiency of photosynthetic systems, maintaining root vigor as well as increased activity of antioxidases, evading the involvement of ACC deaminase to reduce the levels of ethylene (Wang et al. 2012). Naveed et al. (2014a, b) have demonstrated that the bacterial inoculation of maize seedlings with endophytic strains *Burkholderia phytofirmans* PsJN and *Enterobacter* sp. FD17 minimizes the drought-imposed inhibitory effects besides strikingly increasing root biomass, shoot biomass, leaf area, chlorophyll content, and photochemical efficiency of Photosystem II.

Plants, under water-deficit conditions, elevate the osmotic potential by increasing the levels of osmolytes within the cells (Farooq et al. 2009). Dimkpa et al. (2009) reported that rhizobacteria produce osmolytes such as glycine betaine which might act in cooperation with glycine betaine produced by the plants in response to drought stress. Yuwono et al. (2005) demonstrated that rice plants inoculated with osmolyte-producing rhizobacteria on exposure to water-deficit conditions, exhibited more shoot and root dry weight along with number of tillers as compared to uninoculated plants. The roots in the inoculated rice plants proliferated vigorously (evidently to absorb more water) which could possibly be the effect of indole acetic acid (IAA) synthesized by osmotolerant bacteria, a hormone that stimulates the plant growth.

Plant root-associated bacteria have been acknowledged to influence various phenomena taking place at cell membrane interfaces. These processes are essential in shaping the physiological status of a plant including root cell membrane elasticity which plays a crucial role during the inadequate water availability. There are several studies which suggest rhizobacteria colonization directs plant root cell membranes to make necessary modifications in phospholipid composition modulating the cell membrane elasticity under water deficit. It has also been concluded that rhizobacteria-mediated modifications in the elasticity of root cell membranes could be one of the strategies implemented by plants in developing resistance against drought (Dimkpa et al. 2009). Sueldo et al. (1996) reported that wheat plants inoculated with *Azospirillum* restrain the rise in phosphatidylcholine and decrease in phosphatidylethanolamine content. Similar observations were reported by Pereyra et al. (2006).

Besides making significant metabolic and physiologic changes, plants also undergo various modifications at transcriptional level under water-deficit conditions. Timmusk and Wagner (1999) reported that the drought resistance in *Arabidopsis thaliana* was influenced by the PGPR strain *Paenibacillus polymyxa* which triggered transcription of drought-response gene ERD15 (Early Responsive to Dehydration) and Abscisic acid (ABA)-responsive gene RAB18 in the host plant. Likewise, another study carried by Rocha et al. (2007) on drought treated sugarcane highlights the differential expression of as many as 93 genes, encompassing conventional drought-responsive genes encoding MRB and WRKY transcription factors. Similarly, inoculation of *Rhizobium etli* (overexpressing trehalose-6-phosphatase synthase gene) was found to stimulate nodule formation with elevated nitrogenase activity, more biomass and drought resistance in *Phaseolus vulgaris* in contrast to the plants inoculated with wild type *Rhizobium etli*. Microarray analysis of 7200 expressed sequence tags of nodules from test plants divulged upregulation of genes entailed in stress resilience, suggesting the involvement of trehalose signal transduction (Grover et al. 2011).

Apart from plant-associated bacteria, arbuscular mycorrhizal fungi have also been reported to play a significant role in conferring drought resistance in the host plants through alteration of plant physiology and differential expression of specific genes (Ruiz-Lozano and Azcon 2000). Porcel et al. (2006) identified a 14-3-3 protein encoding gene from *Glomus intraradices* subjected to in vitro water deficit. The proteins regulating both signaling pathways as well as effector proteins were proposed to act in conferring drought resistance in the host plants. Aroca et al. (2007) reported the regulation of root hydraulic properties, in substantial correlation with plasma membrane intrinsic proteins (PIP) gene expression and protein phosphorylation by AM fungus *Glomus intraradices* in symbiotic association with *Phaseolus vulgaris* under arid conditions. Under inadequate water availability, amount of glutathione and ascorbate accumulation plays a critical role in maintaining the plant metabolism. Colonization of autochthonous drought resistant *Glomus intraradices* and *Glomus* sp. in lavender plants resulted in reduced accumulation of glutathione and ascorbate and increased tolerance against drought (Marulanda et al. 2007). A recent study conducted by Chen et al. (2017) highlighted the impact of colonization of a novel endophyte *Pantoea alhagi* in the drought exposed wheat plants. It was observed that the endophyte had a pronounced effect on the accumulation of soluble sugars, proline, and malondialdehyde (MDA) as well as on the degradation of the chlorophyll, thereby leading to the enhanced drought tolerance in wheat plants.

7.2.1.2 Temperature Stress Management

Since temperature plays a critical role in the plant growth and development, any fluctuation in the optimal growth temperature (both below and above the optimum) might lead to severe repercussions in terms of plant growth and productivity (Koini et al. 2009; Pareek et al. 2010). Higher temperatures affect plant water relations, nutrient uptake, seed germination, membrane permeability, respiration and photosynthetic rate, and vigor of the organs involved in procreation and pollination (limiting fertilization), while extremely high temperatures can result in heat shock. Phytohormone levels may be elevated as a result of temperature fluctuations, revamping primary and secondary metabolites causing over expression of heat shock proteins, and generating reactive oxygen species (ROS) (Iba 2002). Prolonged exposure to such increased temperatures leads to excessive transpiration from the stomata leading to wilting and even death of the plant. At molecular level, heat shock can stimulate expression of otherwise unexpressed proteins altering DNA transposition recurrence and protein denaturation (Peterson and Higley 2000). On the other hand, excessive drop in optimal temperatures can cause intense mechanical and physical damages to the plants. Drop in temperatures below 0 °C leads to the formation of ice within the intercellular spaces which in turn increases the intracellular fluid accumulation resulting in increased cell membrane pressure and hence cell disruption (Sazzad 2007). Low temperatures also stimulate enhanced production of ROS within the cells causing oxidative damage to proteins, lipids, and DNA (Choudhury et al. 2013).

In order to cope with the challenging temperature fluctuations, plants undergo diverse physiological changes such as increased starch deposition, proline biosynthesis, and phenol accumulation which are the classic indicators of Induced Systemic Resistance (ISR) activation. Additionally, plants activate potential mechanisms that assist in upholding the cell membrane stability as well as induce MAPK (Mitogen-activated Protein Kinase) and CDPK (Calcium-dependent Protein Kinase) signaling pathways (Wang and Li 2006; Janska et al. 2010). Furthermore, increase in the level of antioxidants and appropriate solutes, synthesis of molecular chaperones, and activation of transcription factors are some of the prevalent phenomena occurring within the plants to cope up with harmful effects of temperature (Wang and Li 2006). Since chilling temperatures lead to reduced plant biomass and cell membrane injury resulting in electrolyte leakage, plants essentially accumulate cellular metabolites like sugars, proline (protects membranes and proteins), and anthocyanin in considerable amount to withstand the hostilities of chilling temperatures. Simultaneously, a variety of signaling pathways involved in the production of cold responsive proteins are also triggered, which include antifreeze proteins, detoxification enzymes, enzymes catalyzing biosynthesis of osmoprotectants, late embryogenesis abundant (LEA) proteins, lipid-transfer proteins, molecular chaperones, mRNA binding proteins, proteinase inhibitors, and transporter proteins (Shinozaki and Yamaguchi-Shinozaki 2000; Xiong and Zhu 2001; Wang et al. 2003; Grennan 2006; Nakashima and Yamaguchi-Shinozaki 2006; Yamaguchi-Shinozaki and Shinozaki 2006).

Plenty of literature accentuating the ability of plant-associated microbes to ameliorate the temperature induced stress tolerance in plants is available. Various studies advocating a positive correlation between fluctuating temperatures and microbe-mediated ISR have been put forth (Choudhary et al. 2016).

Colonization of bacteria with ice-nucleating activity can be an efficacious means of managing the cold temperature stress in plants. Lindow and Leveau (2002) reported that the epiphytic bacteria such as *Erwinia herbicola* and *Pseudomonas syringae* impeded the supercooling of plants at freezing temperatures by raising the ice-nucleation temperature, thereby, protecting the plants from frost damage. A plant growth promoting bacteria *Burkholderia phytofirmans* strain PsJN was reported to induce tolerance in *Vitis vinifera* exposed to heat stress (Compant et al. 2005) and cold temperature (Barka et al. 2006). The inoculation of *Burkholderia phytofirmans* PsJN strain into *Vitis vinifera* subjected to cold treatment lowered the biomass reduction rate as well as electrolyte leakage and also assisted the host plant in post-chilling recovery. The plants also showed increased accumulation of carbohydrates as compared to control plants. The marked increase observed in the rate of photosynthesis, starch deposition as well as higher levels of proline and phenols was suggestive of ISR activation.

Ali et al. (2009) reported the potential of a thermotolerant *Pseudomonas* sp. to mitigate the heat stress in sorghum seedlings by synthesizing high molecular weight proteins, improving biomass and increasing amino acid, sugar, proline, and chlorophyll content of the host plant. Ali et al. (2011) investigated the effect of inoculation of thermotolerant PGP *Pseudomonas putida* strain AKMP7 on the growth of wheat plants subjected to heat stress. Inoculated plants showed a marked difference in terms of root and shoot length, quantity of tillers, dry biomass, spike length, and grain formation as compared to uninoculated plants. Inoculated plants also had improved levels of cellular metabolites such as sugars, starch, amino acids, proline, chlorophyll, and essential proteins. These plants were also found to have minimum membrane injury with increased activity of antioxidases such as SOD, APX, and CAT to scavenge ROS. Selvakumar et al. (2012) demonstrated the ability of novel cold resistant PGPR, namely, *Exiguobacterium acetylicum*, *Pantoea dispersa*, *Pseudomonas fragi*, *Pseudomonas lurida*, and *Serratia marcescens* promoting growth of plants under extremely low temperatures (Tewari and Arora 2013). Similarly, inoculation of plant growth promoting *Azospirillum brasilense* and *Bacillus amyloliquefaciens* to *Triticum aestivum* plant seedlings was found to impart heat stress tolerance to the plants by causing a reduced regeneration of ROS, changes in the metabolome, and preactivation of heat shock transcription factors (El-Daim et al. 2014).

7.2.1.3 Salinity Stress Management

Increasing soil salinity is a challenging environmental issue that affects the agricultural crops worldwide (Coleman-Derr and Tringe 2014). There occurs a three-fold effect of the salt stress reducing the water potential, affecting cellular metabolism, and causing ion imbalance and toxicity (Singh 2014). Some vital physiological processes including seed germination, germination time, Na^+/K^+ ratio in root and shoot, and root/shoot dry weight which are pivotal in the plant growth and development are also affected (Parida and Das 2005). As a result of the untimely exposure to salinity, ion toxicity occurs within the plant cells causing imbalance in the osmoregulation especially during the prolonged stress conditions (Munns and Tester 2008). Such deleterious conditions induce the generation of ROS which are detrimental for cell viability, thereby, limiting plant growth and development.

Ideally, both ionic and osmotic homeostasis need to be upheld within the plant cells to sustain the salinity stress. Plants usually combat the salinity stress by keeping susceptible tissues less exposed to the saline environment either by exuding ions from the roots or compartmentalizing ions distant from the cytoplasm of physiologically vital cells (Silva et al. 2010). At the molecular level, plants respond to salt stress via conserved signal transduction pathways triggering several stress forbearance and endurance processes (Xiong et al. 2002) encompassing the role of a stress phytohormone Abscisic acid (ABA). ABA is pivotal in such processes as it regulates the expression of genes whose promoter regions have ABA-responsive elements.

Several plant-associated bacteria have been reported to enhance salt resistance of their host. Such resistance is primarily explained by employing multiple mechanisms such as maintaining water and hormone homeostasis, accumulation of appropriate osmolytes, synthesis of antioxidants, production of volatile compounds and ACC deaminase as well as altering Na^+ uptake/transport (Yang et al. 2009; Dodd and Perez-Alfocea 2012). Interference in the production of ABA has been chiefly ascribed to boost the tolerance under saline conditions (Aroca et al. 2006; Jahromi et al. 2008). Similarly, the mechanisms employed by arbuscular mycorrhizal fungi to alleviate the adverse effects caused by salinity stress include alteration in physiological and enzymatic activities of the host plant, modifying root architecture to assist in water absorption, revamping the plant nutrition, and revising Na^+ and K^+ uptake (Evelin et al. 2009; Gamalero et al. 2010; Zhang et al. 2011; Zolfaghari et al. 2013; Krishnamoorthy et al. 2016).

Bano and Fatima (2009) reported that *Zea mays* plants inoculated with *Pseudomonas* sp. and *Rhizobium* sp. developed tolerance to salinity stress by increasing proline accumulation, reducing the electrolyte leakage, stimulating selective uptake of K^+ ions, thereby, retaining the water content of the plant leaves. ABA producing PGPRs isolated from halophytic weeds were shown to alleviate in vitro salt stress in *Glycine max* (Naz et al. 2009). *Pseudomonas* sp. with its desirable root colonizing ability along with its potential to produce exopolysaccharides has been demonstrated to confer increased tolerance to rice plants exposed to saline conditions (Sen and Chandrasekhar 2014). Similarly, *Enterobacter* sp. EJ01 was found to facilitate the growth in *Arabidopsis* and tomato subjected to salinity stress via activation of osmolytes accumulation, ROS scavenging, LEA (Late embryogenesis abundant protein) genes governing the restoration and conservation of proteins (Kim et al. 2014). The bacteria exhibited the property of in vitro ACC deaminase and IAA production. It was also observed that the short-term treatment of *Arabidopsis* with *Enterobacter* sp. EJ01 enhanced the expression of salt stress responsive genes such as DREB2b, RD29A, RD29B, and RAB18. The expression of genes involved in proline biosynthesis (P5CS1 and P5CS2) and priming processes (MPK3 and MPK6) was also up-regulated in response to stress stimulus. *Bacillus amyloliquefaciens* RWL-1 has been reported to confer tolerance in *Oryza sativa* against salinity stress via production of ABA and auxin phytohormones (Shahzad et al. 2017). Likewise, phytohormones producing endophytic bacteria along with exogenous application of jasmonic acid have been reported to enhance the growth of *Solanum pimpinellifolium* under saline environment (Khan et al. 2017). A promising soil bacterium *Pseudomonas frederiksbergensis* strain OS261 has been reported to mitigate salt stress in red pepper plants via reduction of ethylene production and regulating the activity of antioxidant enzymes (ascorbate peroxidase, superoxide dismutase, and catalase) (Chatterjee et al. 2017).

7.2.1.4 Heavy Metal Stress Management

Heavy metals are metallic elements with relatively higher density (>4 g/cm^3) and high relative atomic weight (>20). They are non-degradable and even toxic at low concentrations (Duruibe et al. 2007; Chibuike and Obiora 2014; Ma et al. 2016a, b). Co, Cu, Fe, Mn, Mo, Ni, V, and Zn are some of the heavy metals which are essential for the organisms in trace amounts (Wintz et al. 2002) but uncurbed quantities of these elements can have a negative impact on the organisms. Pb, Cd, Hg, and As, on the other hand, are the main cause of concern as they are highly toxic to the plants and animals, even in smaller amounts. Although heavy metals are naturally present in the soil, but continued industrialization, intensive agricultural practices, and anthropogenic activities have led to an increased concentration of heavy metals to the level where they become pernicious to the plants and animals. Although plants uptake certain heavy metals from the surrounding soil to facilitate their metabolic activities, accumulation of metal concentrations beyond the optimal levels can have adverse effects on plant growth and development such as inhibition of cytoplasmic enzymes, oxidative damage to the cellular structures (Jadia and Fulekar 2009), metabolism alteration, reduced biomass, chlorosis, decreased rate of photosynthesis, and senescence (Nagajyoti et al. 2010).

Plant growth promoting bacteria have been reported to exhibit bioremediation properties through various mechanisms such as detoxification of the metal pollutants via oxidation-reduction, biosorption and bioaccumulation at the cell wall, precipitation of heavy metals, transport of metals across the cytoplasm, and metal entrapment in the bacterial capsule (Zubair et al. 2016; Tiwari and Lata 2018). Many rhizobacteria have the ability to release metal chelators such as siderophores which form stable complexes with heavy metals that negatively impact the plant health such as Al, Cd, Cu, Ga, In, Pb, Ni, and Zn (Neubauer et al. 2000; Egamberdiyeva 2007; Dimkpa et al. 2008; Ahemad and Kibret 2013). Siderophore binding thus increases the concentration of soluble metal (Rajkumar et al. 2010) influencing bioavailability of toxic heavy metals, thereby, aiding the plant to overcome the constraints of excessive levels of heavy metals in the soil. Pishchik et al. (2002) observed the significant effects of PGPR *Klebsiella mobilis* CIAM inoculation on barley plants grown on Cd-contaminated soil. The inoculated plants showed higher grain yield and two-fold decrease in Cd content in the grain. It was observed that the ability of bacteria to bind free Cd ions seldom makes these ions available for the plants. Furthermore, the test bacterial strains also possessed the properties of nitrogen fixation, IAA, or ethylene production possibly playing a vital role in conferring stress tolerance to the plants. Interestingly, cell wall components of some of the heavy metal resistant PGPR (such as *Pseudomonas*) possess metal binding properties which might aid in intracellular accumulation of metals such as copper and cadmium. Association of such bacteria can limit the metal uptake by plants (Ganesan 2008; Sinha and Mukherjee 2008; Pabst et al. 2010), thereby helping the plant to acclimatize in hostile conditions. Fatnassi et al. (2015) demonstrated that the combined inoculation of copper accumulating *Rhizobium* and PGPR (*Enterobacter cloacae* and *Pseudomonas* sp.) was able to reduce the inhibitory effects of copper stress on *Vicia faba* plants. The copper tolerance in the plant was linked to the modulated activity of antioxidases (both superoxide dismutase and catalase) influenced by the co-inoculation. Kamran et al. (2016) demonstrated the potential of PGPR *Pseudomonas putida* to enhance the growth and tolerance of *Eruca sativa* under nickel stress.

AMF have also been reported to play a critical role in phytoremediation, especially in phytostabilization (Leyval et al. 2002; Orlowska et al. 2002; Regvar et al. 2003; Barea et al. 2005). AMF possibly sequesters heavy metals either by releasing chelators or by directly absorbing the heavy metals. Recently the role of AMF to hyper-accumulate heavy metals has generated a lot of interest among the researchers (Turnau et al. 2005). A recent study conducted by Hashem et al. (2016) has shown the role of AMF in mitigating the adverse effects of cadmium stress in *Solanum lycopersicum* by reducing malondialdehyde and hydrogen peroxide production, thereby palliating the oxidative stress as well as by increasing the production of proline and phenol content of the cells. Another study by Franchi et al. (2017) has revealed that the addition of thiosulfate with metal resistant PGPR improves the mobilization and uptake of arsenic and mercury in *Brassica juncea* and *Lupinus albus* fostering bioavailability and phytoextraction. Similarly, very recently, a metal resistant *Azotobacter chroococcum* strain CAZ3 has been shown to ameliorate the heavy metal (Cu and Pb) induced oxidative damage in *Zea mays*. The strain produced significant amounts of IAA, ACC deaminase, ammonia, and siderophores under metal pressure. The isolate also showed the metal chelating properties. Furthermore, the inoculation of the strain led to the decreased levels of proline, malondialdehyde, and antioxidases in plant leaves (Rizvi and Khan 2018).

7.2.2 *Biotic Stress Management*

In addition to varying abiotic stresses, plants are also inflicted by the attack of pathogens (bacteria, fungi, viruses, protozoa, and nematodes) as well as by other plant pests (Atkinson and Urwin 2012). Abiotic stresses may also contribute to weakening the defense mechanisms of plants and thus making them susceptible to phytopathogens (Suzuki et al. 2014), for example elevated temperatures are known to facilitate pathogen transmission (Madgwick et al. 2011; Nicol et al. 2011). Likewise, chilling temperatures have been reported to compromise gene silencing, an efficient defense mechanism against plant viruses (Szittya et al. 2003; Suzuki et al. 2014). Biotic stresses are known to cause massive destruction of crop productivity annually (Sazzad 2007; Ramegowda and Senthil-Kumar 2015). Phytopathogens have been recognized as a main threat challenging the food productivity and ecosystem stability across the globe. Some of the common manifestations of these biotic factors are nutrient imbalance, imbalanced hormonal regulation, and various physiological disorders.

Table 7.2 Microbe-mediated biotic stress tolerance in plants

Plants	Pathogen	Biocontrol agent	Resistance mechanism	References
Cucumis sativus and *Triticum aestivum*	*Fusarium oxysporum*, *Botryosphaeria ribis*, *Alternaria* (Nees), *Rhizoctonia solani*	*Paenibacillus jamilae*HS-26	Synthesis of hydrolytic enzymes, antifungal metabolites and VOC	Wang et al. (2019)
Persea americana	*Phytophthora cinnamomi*, *Colletotrichum gloeosporioides*	*Serratia* sp.	Synthesis of antifungal secondary metabolites	Granada et al. (2018)
Triticum aestivum	*Fusarium graminearum*	*Stenotrophomonas maltophilia*	Increased level of defense enzymes such as PAL, PO, PPO, β-1,3 glucanase	Singh and Jha (2017)
Piper nigrum	*Stemphylium lycopersici*	*Brevibacterium iodinum*	Increased expression of pathogen-related protein genes	Son et al. (2014)
Cucumis sativus	Cucumber mosaic cucumovirus	*Azotobacter* chroococcum, *Bacillus subtilis*, *Pseudomonas fluorescens*	Increased activities of POX and β-1,3 glucanase Secretion of pathogen-related proteins	El-Borollosy and Oraby (2012)
Oryza sativa	*Xanthomonas oryzae*	*Bacillus* sp.	Enhanced accumulation of POX, phenylalanine lyase, and polyphenol oxidase	Chithrashree et al. (2011)
Arabidopsis thaliana	*Pseudomonas syringae*	*Bacillus subtilis*	Activation of induced systemic resistance	Rudrappa et al. (2008)

Large body of literature eliciting the potential of plant-associated microbes in ameliorating the stress symptoms imposed by the attack of phytopathogens is available (Dobbelaere et al. 2003; Domenech et al. 2006; Saleem et al. 2007) (Table 7.2). The widely recognized mechanisms through which these microbes impart tolerance to the plants include activation of cellular component including cellular burst, accumulation of secondary metabolites, production of inhibitory allelochemicals, cell wall reinforcement, expression of genes encoding pathogenesis-related proteins, and induced systemic resistance (ISR) (Nurnberger et al. 2004; Compant et al. 2005; Lugtenberg and Kamilova 2009). The hormones that play a vital role in plant defense include ethylene, jasmonic acid (JA), and salicylic acid (SA) (Bari and Jones 2009; Verhage et al. 2010). *Bacillus* and *Pseudomonas* species have been widely recognized to potentiate ISR against varied phytopathogens via diverse mechanisms (Choudhary and Johri 2009; Bhattacharyya and Jha 2012). Majority of PGPR have been reported to elicit ISR potentially in the host plant via multiple signal transductions like SA-independent JA-ethylene-dependent (Pettersson and Bååth 2004) and NPR 1-dependent signaling pathways (Niu et al. 2011) as well as stimulating ISR-related gene expression.

PGPR strains have been reported to elicit ISR in a range of plant hosts via the activation of various defense related enzymes such as β-1,3-glucanase, chitinases, peroxidases, phenylalanine ammonia lyase, and polyphenol oxidase (Bharathi 2004). Many studies have highlighted the ability of rhizobacteria to contribute in the antibiosis of the host plant through the production of peptide antibiotics, siderophores, volatiles, and other organic metabolites (Maksimov et al. 2011). The ability to release a wide spectrum of lytic enzymes such as cellulases, chitinases, glucanases, and proteases to curb the growth of filamentous fungal pathogens by *Bacillus*, *Pseudomonas*, and other antagonistic bacteria makes them the appropriate biocontrol agents (Chernin and Chet 2002; Maksimov et al. 2011). Various reports are suggestive of ethylene synthesis inhibitors having an antagonistic effect against plant pathogens. It has also been reported that co-inoculation of PGPR with AMF curbs the repercussions of phytopathogens through stimulating growth attributes and increasing the resistance against plant diseases (Dohroo and Sharma 2012). PGPB have also been reported to evoke a series of other defensive strategies such as regulating quorum sensing phenomenon, activation of antioxidant enzymes, activation of phenylpropanoid pathway triggering the phenolic production, lignification of cell walls, and transgenerational defense response to overcome the hostilities of the parasite (Mishra et al. 2015).

Compant et al. (2005) reported that PGPM activate defense mechanisms in *Dianthus caryophyllus* against wilt causing *Cucumis sativus* and *Fusarium* sp. against foliar disease causing *Colletotrichum orbicular* pathogens. Rudrappa et al. (2008) reported that *Bacillus subtilis* FB17 protects the host plant against foliar pathogen *Pseudomonas syringae* pv. tomato Pst DC3000 through ISR. Many *Bacillus* strains have been shown to elicit ISR in rice against *Xanthomonas oryzae* pv. oryzae, the causative agent of bacterial leaf blight (Chithrashree et al. 2011). Similarly, Lee et al. (2015) reported the activation of ISR in *Panax ginseng* against *Phytophthora cactorum* by *Bacillus amyloliquefaciens* strain HK34. *Paenibacillus* P16 has been reported to be an effective biocontrol agent in cabbage against black rot caused by *Xanthomonas campestris*, simultaneously inducing systemic resistance (Ghazalibigla et al. 2016). Recently, *Serratia* sp. isolated from avocado plants has been demonstrated to produce secondary metabolites like serratamolide, haterumalide NC, prodigiosin with a significant inhibitory activity against *Phytophthora cinnamomi* and *Colletotrichum gloeosporioides* (Granada et al. 2018). A very recent study by Wang et al. (2019) has shown the biocontrol ability of *Paenibacillus jamilae* HS-26 isolated from cucumber rhizosphere against several fungal pathogens. The bacteria has been shown to produce hydrolytic enzymes and antifungal metabolites as well as volatile organic compounds essentially, N, N-diethyl-1, 4-phenylenediamine.

7.3 Conclusion

Plants share an intriguing relationship with the associated microbes. This interaction plays a quintessential role in maintaining the plant health both under biotic and abiotic stresses. Plant-associated microbes assist in reprogramming of plant physiology and metabolism via multiple and intricate signaling cascades for perceiving and responding to a diversity of biotic and abiotic adversities. Microbe-mediated stress mitigation provides an excellent model for apprehending the mechanisms underlying stress management. A comprehensive understanding of these interactions can be underpinning in developing biotechnological tools for a sustainable agriculture. Development of stress-resistant microbial bioformulations can be an efficient way to improve crop quality in challenging environmental conditions. As our understanding on the mechanisms governing plant responses to diverse environmental adversities expands, we will be able to get closer in comprehending the functioning of natural ecosystems where plants interact with biotic and abiotic factors. The use of OMICs in understanding the precise mechanisms of microbe-mediated stress mitigation in plants will strengthen the scope of developing crops capable of enduring extreme environmental conditions which have been anticipated to worsen in future.

Acknowledgement Authors sincerely acknowledge the facilities provided by the Institutes and the work of various researchers cited in this chapter.

References

Adediran GA, Ngwenya BT, Mosselmans JFW, Heal KV (2016) Bacteria–zinc co-localization implicates enhanced synthesis of cysteine-rich peptides in zinc detoxification when *Brassica juncea* is inoculated with *Rhizobium leguminosarum*. New Phytol 209:280–293

Ahemad M, Kibret M (2013) Mechanisms and applications of plant growth promoting rhizobacteria: current perspective. J King Saud Univ Sci 26(1):1–20

Ali SKZ, Sandhya V, Grover M, Kishore Linga VR, Bandi V (2009) *Pseudomonas* sp. strain AKM-P6 enhances tolerance of sorghum seedlings to elevated temperatures. Biol Fert Soils 46:4555

Ali SKZ, Sandhya V, Grover M, Kishore Linga VR, Bandi V (2011) Effect of inoculation with a thermotolerant plant growth promoting *Pseudomonas putida* strain AKMP7 on growth of wheat (*Triticum* spp.) under heat stress. J. Plant Interact 6(4):239–246

Apel K, Hirt H (2004) Reactive oxygen species: metabolism, oxidative stress, and signal transduction. Ann Rev Plant Biol 55:373–399

Aroca R, Ferrante A, Vernieri P, Chrispeels MJ (2006) Drought, abscisic acid and transpiration rate effects on the regulation of PIP aquaporin gene expression and abundance in *Phaseolus vulgaris* plants. Ann Bot 98:1301–1310

Aroca R, Porcel R, Ruiz-Lozano JM (2007) How does arbuscular mycorrhizal symbiosis regulate root hydraulic properties and plasma membrane aquaporins in *Phaseolus vulgaris* under drought, cold or salinity stresses? New Phytol 173:808–816

Atkinson NJ, Urwin PE (2012) The interaction of plant biotic and abiotic stresses: from genes to the field. J Exp Bot 63:3523–3543

Bano A, Fatima M (2009) Salt tolerance in *Zea mays* (L) following inoculation with *Rhizobium* and *Pseudomonas*. Biol Fert Soils 45:405413

Barea JM, Pozo MJ, Azcón R, Azcón-Aguilar C (2005) Microbial co-operation in the rhizosphere. J Exp Bot 56(417):1761–1778

Bari R, Jones JD (2009) Role of plant hormones in plant defence responses. Plant Mol Biol 69:473–488

Barka EA, Nowak J, Clement C (2006) Enhancement of chilling resistance of inoculated grapevine plantlets with a plant growth-promoting rhizobacterium, *Burkholderia phytofirmans* strain PsJN. Appl Environ Microbiol 72:7246–7252

Bharathi S (2004) Development of botanical formulations for the management of major fungal diseases of tomato and onion. PhD thesis, Tamil Nadu Agricultural University, Coimbatore, India, p 152

Bhattacharyya PN, Jha DK (2012) Plant growth-promoting rhizobacteria (PGPR): emergence in agriculture. World J Microbiol Biotechnol 28(4):1327–1350

Chatterjee P, Samaddar S, Anandham R, Kang Y, Kim K, Selvakumar G, Sa T (2017) Beneficial soil bacterium *Pseudomonas frederiksbergensis* OS261 augments salt tolerance and promotes red pepper plant growth. Front Plant Sci 8:705

Chaves MM, Flexas J, Pinheiro C (2009) Photosynthesis under drought and salt stress: regulation mechanisms from whole plant to cell. Ann Bot 103(4):551–560

Chen C, Xin K, Liu H, Cheng J, Shen X, Wang Y, Zhang L (2017) *Pantoea alhagi*, a novel endophytic bacterium with ability to improve growth and drought tolerance in wheat. Sci Rep 7:41564

Chernin L, Chet I (2002) Microbial enzymes in biocontrol of plant pathogens and pests. In: Burns RG, Dick RP (eds) Enzymes in the environment: activity, ecology, and applications. MarcelDekker, New York, pp 171–225

Chibuike GU, Obiora SC (2014) Heavy metal polluted soils: effect on plants and bioremediation methods. Appl Environ Soil Sci 2014:752708

Chithrashree, Udayashankar AC, Nayaka SC, Reddy MS, Srinivas C (2011) Plant growth-promoting rhizobacteria mediate induced systemic resistance in rice against bacterial leaf blight caused by *Xanthomonas oryzae* pv. oryzae. Biol Control 59:114–122

Choudhary DK, Johri BN (2009) Interactions of *Bacillus* spp. and plants—with special reference to induced systemic resistance (ISR). Microbiol Res 164:493–513

Choudhary DK, Kasotia A, Jain S, Vaishnav A, Kumari S, Sharma KP, Varma A (2016) Bacterial-mediated tolerance and resistance to plants under abiotic and biotic stresses. J Plant Growth Regul 35:276–300

Choudhury S, Panda P, Sahoo L, Panda SK (2013) Reactive oxygen species signaling in plants under abiotic stress. Plant Signal Behav 4:e2368

Coleman-Derr D, Tringe SG (2014) Building the crops of tomorrow: advantages of symbiont-based approaches to improving abiotic stress tolerance. Front Microbiol 5:283

Compant S, Reiter B, Sessitsch A, Nowak J, Clement C, Ait Bakra E (2005) Endophytic colonization of *Vitis vinifera* L. by plant growth promoting bacterium *Burkholderia* sp. strain PsJN. Appl Environ Microbiol 71:1685–1693

Cramer GR, Urano K, Delrot S, Pezzotti M, Shinozaki K (2011) Effects of abiotic stress on plants: a systems biology perspective. BMC Plant Biol 11:163

Dimkpa C, Svatoš A, Merten D, Büchel G, Kothe E (2008) Hydroxamate siderophores produced by *Streptomyces acidiscabies* E13 bind nickel and promote growth in cowpea (*Vigna unguiculata* L.) under nickel stress. Can J Microbiol 54:163–172

Dimkpa C, Weinand T, Asch F (2009) Plant–rhizobacteria interactions alleviate abiotic stress conditions. Plant Cell Environ 32:1682–1694

Dobbelaere S, Vanderleyden J, Okon Y (2003) Plant growth-promoting effects of diazotrophs in the rhizosphere. Crit Rev Plant Sci 22:107–149

Dodd IC, Perez-Alfocea F (2012) Microbial amelioration of crop salinity stress. J Exp Bot 63:3415–3428

Dohroo A, Sharma DR (2012) The role of plant growth promoting rhizobacteria, arbuscular mycorrhizal fungi and their helper bacteria on growth parameters and root rot of apple. Walailak J Sci Technol 2:35–38

Domenech J, Reddy MS, Kloepper JW, Ramos B, GutierrezMañero J (2006) Combined application of the biological product LS213 with *Bacillus*, *Pseudomonas* or *Chryseobacterium* for growth promotion and biological control of soil-borne diseases in pepper and tomato. BioControl 51:245–258

Duque AS, de Almeida AM, da Silva AB, da Silva JM, Farinha AP, Santos D, Fevereiro P, Araujo S (2013) Abiotic stress responses in plants: unraveling the complexity of genes and networks to survive. In: Vahdati K, Leslie C (eds) Abiotic stress- plant responses and applications in agriculture. In Tech, Rijeka. ISBN 978-953-51-1024-8

Duruibe JO, Ogwuegbu MOC, Egwurugwu JN (2007) Heavy metal pollution and human bio toxic effects. Int J Physical Sci 2:112–118

Egamberdiyeva D (2007) The effect of plant growth promoting bacteria on growth and nutrient uptake of maize in two different soils. Appl Soil Ecol 36:184–189

El-Borollosy AM, Oraby MM (2012) Induced systemic resistance against cucumber mosaic cucumovirus and promotion of cucumber growth by some plant growth-promoting rhizobacteria. Ann. Agric Sci 57:91–97

El-Daim IAA, Bejai S, Meijer J (2014) Improved heat stress tolerance of wheat seedlings by bacterial seed treatment. Plant Soil 379:337–350

Evelin H, Kapoor R, Giri B (2009) Arbuscular mycorrhizal fungi in alleviation of salt stress: a review. Ann Bot 104:1263–1280

Farooq M, Wahid A, Kobayashi N, Fujita D, Basra SMA (2009) Plant drought stress: effects, mechanisms and management. Agron Sustain Dev 29:185–212

Fatnassi IC, Chiboub M, Saadani O, Jebara M, Jebara SH (2015) The impact of dual inoculation with *Rhizobium* and PGPR on growth and antioxidant status of *Vicia faba* L. under copper stress. C R Biol 338(4):241–254

Figueiredo MVB, Burity HA, Martinez CR, Chanway CP (2008) Alleviation of drought stress in common bean (*Phaseolus vulgaris* L.) by coinoculation with *Paenibacillus polymyxa* and *Rhizobium tropici*. Appl Soil Ecol 40:182–188

Foyer CH, Noctor G (2005) Oxidant and antioxidant signalling in plants: a re-evaluation of the concept of oxidative stress in a physiological context. Plant Cell Environ 28:1056–1071

Franchi E, Rolli E, Marasco R, Agazzi G, Borin S, Cosmina P, Pedron F, Rosellini I, Barbafieri M, Petruzzelli G (2017) Phytoremediation of a multi contaminated soil: mercury and arsenic phytoextraction assisted by mobilizing agent and plant growth promoting bacteria. J Soils Sediments 17:1224–1236

Fujita M, Fujita Y, Noutoshi Y, Takahashi F, Narusaka Y, Yamaguchi-Shinozaki K, Shinozaki K (2006) Crosstalk between abiotic and biotic stress responses: a current view from the points of convergence in the stress signaling networks. Curr Opin Plant Biol 9:436–442

Gamalero E, Berta G, Massa N, Glick BR, Lingua G (2010) Interactions between *Pseudomonas putida*UW4 and *Gigaspora rosea* BEG9 and their consequences on the growth of cucumber under salt stress conditions. J Appl Microbiol 108:236–245

Ganesan V (2008) Rhizoremediation of cadmium soil using a cadmium-resistant plant growth-promoting rhizopseudomonad. Curr Microbiol 56(4):403–407

Ghazalibigla H, Hampton JG, Jong EZ, Holyoake A (2016) Is induced systemic resistance the mechanism for control of black rot in *Brassica oleracea* by a *Paenibacillus* sp.? Biol Control 92:195–201

Gill SS, Tuteja N (2010) Reactive oxygen species and antioxidant machinery in abiotic stress tolerance in crop plants. Plant Physiol Biochem 48:909–930

Glick BR (2014) Bacteria with ACC deaminase can promote plant growth and help to feed the world. Microbiol Res 169:30–39

Gopalakrishnan S, Sathya A, Vijayabharathi R, Varshney RK, Gowda CL, Krishnamurthy L (2015) Plant growth promoting rhizobia: challenges and opportunities. Biotech 5:355–377

Goswami D, Thakker JN, Dhandhukia PC (2016) Portraying mechanics of plant growth promoting rhizobacteria (PGPR): a review. Cogent Food Agric 2:1127500

Govindasamy V, Senthilkumar M, Magheshwaran V, Kumar U, Bose P, Sharma V, Annapurna K (2011) *Bacillus* and *Paenibacillus* spp.: potential PGPR for sustainable agriculture. In: Maheshwari DK (ed) Plant growth and health promoting bacteria. Springer, Berlin, pp 333–364

Granada SD, Ramírez-Restrepo S, López-Luján L, Peláez-Jaramillo CA, Bedoya-Pérez JC (2018) Screening of a biological control bacterium to fight avocado diseases: from agroecosystem to bioreactor. Biocatal Agric Biotechnol 14:109–115

Grennan AK (2006) Abiotic stress in rice. An "omic" approach. Plant Physiol 140:1139–1141

Grover M, Ali SKZ, Sandhya V, Rasul A, Venkateswarlu B (2011) Role of microorganisms in adaptation of agriculture crops to abiotic stresses. W J Microbiol Biotechnol 27:1231–1240

Guo JH, Jiang CH, Xie P, Huang ZY, Fa ZH (2015) The plant healthy and safety guards plant growth promoting rhizobacteria (PGPR). Transcriptomics 3:109

Hashem A, Abd_Allah EF, Alqarawi AA, Al Huqail AA, Egamberdieva D, Wirth S (2016) Alleviation of cadmium stress in *Solanum lycopersicum* L. by arbuscular mycorrhizal fungi via induction of acquired systemic tolerance. Saudi J Biol Sci 23(2):272–281

He Z, He C, Zhang Z, Zou Z, Wang H (2007) Changes of antioxidative enzymes and cell membrane osmosis in tomato colonized by arbuscular mycorrhizae under NaCl stress. Colloids Surf B: Biointerfaces 59:128–133

Iba K (2002) Acclimative response to temperature stress in higher plants: approaches of gene engineering for temperature tolerance. Ann Rev Plant Biol 53:225–245

Jadia CD, Fulekar MH (2009) Phytoremediation of heavy metals: recent techniques. Afr J Biotechnol 8(6):921–928

Jahromi F, Aroca R, Porcel R, Ruiz-Lozano JM (2008) Influence of salinity on the *In vitro* development of *Glomus intraradices* and on the *In vivo* physiological and molecular responses of mycorrhizal lettuce plants. Microb Ecol 55:45–53

Janska A, Marsik P, Zelenkova S, Ovesna J (2010) Cold stress and acclimation – what is important for metabolic adjustment? Plant Biol 12:395–405

Joseph B, Patra RR, Lawrence R (2007) Characterization of plant growth promoting rhizobacteria associated with chickpea (*Cicer arietinum* L). Int J Plant Product 1:141–152

Kamran MA, Eqani SAMAS, Bibi S, Xu RK, Monis MFH, Katsoyiannis A, Bokhari H, Chaudhary HJ (2016) Bioaccumulation of nickel by *E. sativa* and role of plant growth promoting rhizobacteria (PGPRs) under nickel stress. Ecotoxicol Environ Saf 126:256–263

Kavamura VN, Santos SN, da Silva JL, Parma MM, Ávila LA, Visconti A, Zucchi TD, Taketani RG, Andreote FD, de Melo IS (2013) Screening of *Brazilian cacti* rhizobacteria for plant growth promotion under drought. Microbiol Res 168:183–191

Khan AL, Hamayun M, Khan SA, Kang SM, Shinwari ZK, Kamran M, Rehman S, Kim JG, Lee IJ (2012) Pure culture of *Metarhizium anisopliae* LHL07 reprograms soybean to higher growth and mitigates salt stress. World J Microbiol Biotechnol 28(4):1483–1494

Khan AL, Waqas M, Asaf S, Kamran M, Shahzad R, Bilal S, Khan MA, Kang SM, Kim YH, Yun BW, Al-Rawahi A (2017) Plant growth-promoting endophyte *Sphingomonas* sp: LK11 alleviates salinity stress in *Solanum pimpinellifolium*. Environ Exp Bot 133:58–69

Kim K, Jang YJ, Lee SM, Oh BT, Chae JC, Lee KJ (2014) Alleviation of salt stress by *Enterobacter* sp. EJ01 in tomato and arabidopsis is accompanied by up-regulation of conserved salinity responsive factors in plants. Mol Cells 37(2):109–117

Kohler J, Hernandez JA, Caravaca F, Roldán A (2008) Plant-growth-promoting rhizobacteria and arbuscular mycorrhizal fungi modify alleviation biochemical mechanisms in water-stressed plants. Funct Plant Biol 35:141–151

Koini MA, Alvey L, Allen T, Tilley CA, Harberd NP, Whitelam GC, Franklin KA (2009) High temperature-mediated adaptations in plant architecture require the bHLH transcription factor PIF4. Curr Biol 19:408–413

Krishnamoorthy R, Kim K, Subramanian P, Senthilkumar M, Anandham R, Sa T (2016) Arbuscular mycorrhizal fungi and associated bacteria isolated from salt-affected soil enhance the tolerance of maize to salinity in coastal reclamation soil. Agric Ecosyst Environ 1:231–239

Lee BD, Dutta S, Ryu H, Yoo SJ, Suh DS, Park K (2015) Induction of systemic resistance in *Panax ginseng* against *Phytophthora cactorum* by native *Bacillus amyloliquefaciens* HK34. J Ginseng Res 39(3):213–220

Leyval C, Joner EJ, del Val C, Haselwandter K (2002) Potential of arbuscular mycorrhizal fungi for bioremediation. In: Gianinazzi S, Schuepp H, Barea JM, Haselwandter K (eds) Mycorrhiza technology in agriculture: from genes to bioproducts. Birkhäuser Verlag, Basel, pp 175–186

Lindow SE, Leveau JHJ (2002) Phyllosphere microbiology. Curr Opin Biotechnol 13:238–243

Lugtenberg B, Kamilova F (2009) Plant-growth-promoting rhizobacteria. Ann Rev Microbiol 63:541–556

Ma Y, Rajkumar M, Zhang C, Freitas H (2016a) Inoculation of *Brassica oxyrrhina* with plant growth promoting bacteria for the improvement of heavy metal phytoremediation under drought conditions. J Hazard Mater 320:36–44

Ma Y, Rajkumar M, Zhang C, Freitas H (2016b) Beneficial role of bacterial endophytes in heavy metal phytoremediation. J Environ Manag 1(174):14–25

Madgwick J, West J, White R, Semenov M, Townsend J, Turner J, Fitt BL (2011) Impacts of climate change on wheat anthesis and fusarium ear blight in the UK. Eur J Plant Pathol 130:117–131

Mahajan S, Tuteja N (2005) Cold, salinity and drought stresses: an overview. Arch Biochem Biophys 444:139–158

Maksimov IV, Abizgil'dina RR, Pusenkova LI (2011) Plant growth promoting rhizobacteria as alternative to chemical crop protectors from pathogens. Appl Biochem Microbiol 47:333–345

Marulanda A, Barea JM, Azcon R (2006) An indigenous drought-tolerant strain of *Glomus intraradices* associated with a native bacterium improves water transport and root development in *Retama sphaerocarpa*. Microb Ecol 52:670–678

Marulanda A, Porcel R, Barea JM, Azcon R (2007) Drought tolerance and antioxidant activities in lavender plants colonized by native drought tolerant or drought sensitive *Glomus* species. Microb Ecol 54(3):543–552

Mastouri F, Björkman T, Harman GE (2010) Seed treatment with *Trichoderma harzianum* alleviates biotic, abiotic, and physiological stresses in germinating seeds and seedlings. Phytopathology 100(11):1213–1221

Mathew DC, Ho YN, Gicana RG, Mathew GM, Chien MC, Huang CC (2015) A rhizosphere-associated symbiont, *Photobacterium* spp. strain MELD1, and its targeted synergistic activity for phytoprotection against mercury. PLoS One 10:e0121178

Mayak S, Tirosh T, Glick BR (2004) Plant growth-promoting bacteria that confer resistance in tomato to salt stress. Plant Physiol Biochem 42:565–572

Meena KK, Sorty AM, Bitla UM, Choudhary K, Gupta P, Pareek A, Singh DP, Prabha R, Sahu PK, Gupta VK, Singh HB, Krishanani KK, Minhas PS (2017) Abiotic stress responses and microbe-mediated mitigation in plants: the omics strategies. Front Plant Sci 8:172

Mishra S, Singh A, Keswani C, Saxena A, Sarma BK, Singh HB (2015) Harnessing plant-microbe interactions for enhanced protection against phytopathogens. In: Arora N (ed) Plant microbes symbiosis: applied facets. Springer, Berlin, pp 111–125

Mosa KA, Ismail A, Helmy M (2017) Introduction to plant stresses. In: Plant stress tolerance. Springer, Cham, pp 1–19

Munns R, Tester M (2008) Mechanisms of salinity tolerance. Ann Rev Plant Biol 59:651–681

Mushtaq A, Rawat S (2017) Rhizosphere microbiome metagenomics: elucidating the abditive microflora. In: Kalia VC, Shouche Y, Purohit HJ, Rahi P (eds) Mining of microbial wealth and meta genomics. Springer, Singapore, pp 11–27

Nadeem SM, Ahmad M, Zahir ZA, Javaid A, Ashraf M (2014) The role of mycorrhizae and plant growth promoting rhizobacteria (PGPR) in improving crop productivity under stressful environments. Biotechnol Adv 32:429–448

Nagajyoti PC, Lee KD, Sreekanth TVM (2010) Heavy metals, occurrence and toxicity for plants: a review. Environ Chem Lett 8:199–216

Nakashima K, Yamaguchi-Shinozaki K (2006) Regulons involved in osmotic stress-responsive and cold stress-responsive gene expression in plants. Physiol Plant 126:62–71

Nautiyal CS, Srivastava S, Chauhan PS, Seema K, Mishra A, Sopory SK (2013) Plant growth-promoting bacteria *Bacillus amyloliquefaciens* NBRISN13 modulates gene expression profile of leaf and rhizosphere community in rice during salt stress. Plant Physiol Biochem 66:1–9

Naveed M, Hussain MB, Zahir ZA, Mitter B, Sessitsch A (2014a) Drought stress amelioration in wheat through inoculation with *Burkholderia phytofirmans* strain PsJN. Plant Growth Regul 73 (2):121–131

Naveed M, Mittera B, Reichenauer TG, Wieczorek K, Sessitsch A (2014b) Increased drought stress resilience of maize through endophytic colonization by *Burkholderia phytofirmans* PsJN and *Enterobacter* sp. FD17. Environ Exp Bot 97:30–39

Naz I, Bano A, Ul-Hassan T (2009) Isolation of phytohormones producing plant growth promoting rhizobacteria from weeds growing in Khewra salt range, Pakistan and their implication in providing salt tolerance to *Glycine max* L. Afr J Biotechnol 8:57–62

Neubauer U, Furrer G, Kayser A, Schulin R (2000) Siderophores, NTA, and citrate: potential soil amendments to enhance heavy metal mobility in phytoremediation. Int J Phytoremediation 2:353–368

Ngumbi E, Kloepper J (2014) Bacterial-mediated drought tolerance: current and future prospects. Appl Soil Ecol 105:109–125

Nguyen D, Rieu I, Mariani C, van Dam NM (2016) How plants handle multiple stresses: hormonal interactions underlying responses to abiotic stress and insect herbivory. Plant Mol Biol 91:727–740

Nicol JM, Turner SJ, Coyne DL, Nijs LD, Hockland S, Maafi ZT (2011) Current nematode threats to world agriculture. In: Jones J, Gheysen G, Fenoll C (eds) Genomics and molecular genetics of plant–nematode interactions. Springer, Amsterdam, pp 21–43

Niu DD, Liu HX, Jiang CH, Wang YP, Wang QY, Jin HL, Guo JH (2011) The plant growth promoting rhizobacterium *Bacillus cereus* AR156 induces systemic resistance in *Arabidopsis thaliana* by simultaneously activating salicylate-and jasmonate/ethylene-dependent signaling pathways. Mol Plant-Microbe Interact 24:533–542

Nurnberger T, Brunner F, Kemmerling B, Piater L (2004) Innate immunity in plants and animals: striking similarities and obvious differences. Immunol Rev 198:249–266

Orlowska E, Zubek SZ, Jurkiewicz A, Szarek-Łukaszewska G, Turnau K (2002) Influence of restoration on arbuscular mycorrhiza of *Biscutella laevigata* L. (Brassicaceae) and *Plantago lanceolata* L. (Plantaginaceae) from calamine spoil mounds. Mycorrhiza 12:153–160

Pabst MW, Miller CD, Dimkpa CO, Anderson AJ, McLean JE (2010) Defining the surface adsorption and internalization of copper and cadmium in a soil bacterium, *Pseudomonas putida*. Chemosphere 81(7):904–910

Pareek A, Sopory SK, Bohnert HK, Govindjee (eds) (2010) Abiotic stress adaptation in plants: physiolgical, molecular and genomic foundation. Springer, Dordrecht, p 526

Parida AK, Das AB (2005) Salt tolerance and salinity effects on plants: a review. Ecotoxicol Environ Saf 60:324–349

Pereyra MA, Zalazar CA, Barassi CA (2006) Root phospholipids in *Azospirillum*-inoculated wheat seedlings exposed to water stress. Plant Physiol Biochem 44:873–879

Peterson RKD, Higley LG (2000) Biotic stress and yield loss, 1st edn. CRC Press, Boca Raton

Pettersson ME, Bååth (2004) Effects of the properties of the bacterial community on pH adaptation during recolonization of a humus soil. Soil Biol Biochem 36:1383–1388

Pieterse CMJ, van der Does D, Zamioudis C, Leon-Reyes A, vanWees SCM (2012) Hormonal modulation of plant immunity. Ann Rev Cell Dev Biol 28:489–521

Pishchik VN, Vorobyev NI, Chernyaeva II, Timofeeva SV, Kozhemyakov AP, Alexeev YV, Lukin SM (2002) Experimental and mathematical simulation of plant growth promoting rhizobacteria and plant interaction under cadmium stress. Plant Soil 243:173–186

Porcel R, Aroca R, Cano C, Bago A, Ruiz-Lozano JM (2006) Identification of a gene from the arbuscular mycorrhizal fungus *Glomus intraradices* encoding for a 14-3-3 protein that is upregulated by drought stress during the AM symbiosis. Microb Ecol 52(3):575–582

Qin Y, Druzhinina IS, Pan X, Yuan Z (2016) Microbially mediated plant salt tolerance and microbiome-based solutions for saline agriculture. Biotechnol Adv 34(7):1245–1259

Rajkumar M, Ae N, Prasad MNV, Freitas H (2010) Potential of siderophore-producing bacteria for improving heavy metal phytoextraction. Trends Biotechnol 28:142–149

Ramegowda V, Senthil-Kumar M (2015) The interactive effects of simultaneous biotic and abiotic stresses on plants: mechanistic understanding from drought and pathogen combination. J Plant Physiol 176:47–54

Regvar M, Vogel K, Irgel N, Wraber T, Hildebrandt U, Wilde P, Bothe H (2003) Colonization of pennycresses (Thlaspi spp.) of the Brassicaceae by arbuscular mycorrhizal fungi. J. Plant Physiol 160:615–626

Rizvi A, Khan MS (2018) Heavy metal induced oxidative damage and root morphology alterations of maize (*Zea mays* L.) plants and stress mitigation by metal tolerant nitrogen fixing *Azotobacter chroococcum*. Ecotoxicol Environ Saf 157:9–20

Rocha FR, Papini-Terzi FS, Nishiyama MY, Vêncio RZ, Vicentini R, Duarte RD, de Rosa VE, Vinagre F, Barsalobres C, Medeiros AH, Rodrigues FA (2007) Signal transduction-related responses to phytohormones and environmental challenges in sugarcane. BMC Genomics 8 (1):71

Rudrappa T, Czymmek KJ, Paré PW, Bais HP (2008) Root-secreted malic acid recruits beneficial soil bacteria. Plant Physiol 148:1547–1556

Ruiz-Lozano JM, Azcon R (2000) Symbiotic efficiency and infectivity of an autochthonous arbuscular mycorrhizal *Glomus* sp. from saline soils and *Glomus deserticola* under salinity. Mycorrhiza 10:137–143

Saharan BS, Nehra V (2011) Plant growth promoting rhizobacteria a critical review. Life Sci Med Res 21:1–10

Saleem M, Arshad M, Hussain S, Bhatti AS (2007) Perspective of plant growth promoting rhizobacteria (PGPR) containing ACC deaminase in stress agriculture. J Ind Microbiol Biotechnol 34(10):635–648

Sazzad K (2007) Exploring plant tolerance to biotic and abiotic stresses. Swedish University of Agricultural Sciences, Uppsala

Selvakumar G, Panneerselvam P, Ganeshamurthy AN (2012) Bacterial mediated alleviation of abiotic stress in crops. In: Maheshwari DK (ed) Bacteria in agrobiology: stress management. Springer, Berlin, pp 205–224

Sen S, Chandrasekhar CN (2014) Effect of PGPR on growth promotion of rice (*Oryza sativa* L.) under salt stress. Asian J Plant Sci Res 4:62–67

Shahzad R, Khan AL, Bilal S, Waqas M, Kang SM, Lee IJ (2017) Inoculation of abscisic acid-producing endophytic bacteria enhances salinity stress tolerance in *Oryza sativa*. Environ Exp Bot 1(136):68–77

Sheng M, Tang M, Chen H, Yang BW, Zhang FF, Huang YH (2009) Influence of arbuscular mycorrhizae on the root system of maize plants under salt stress. Can J Microbiol 55:879–886

Shinozaki K, Yamaguchi-Shinozaki K (2000) Molecular responses to dehydration and low temperature: differences and cross-talk between two stress signaling pathways. Curr Opin Plant Biol 3:217–223

Silva EN, Ribeiro RV, Ferreira-Silva SL, Viégas RA, Silveira JAG (2010) Comparative effects of salinity and water stress on photosynthesis, water relations and growth of *Jatropha curcas* plants. J Arid Environ 74:1130–1137

Singh S (2014) A review on possible elicitor molecules of cyanobacteria: their role in improving plant growth and providing tolerance against biotic or abiotic stress. J Appl Microbiol 117:1221–1244

Singh RP, Jha PN (2017) The PGPR *Stenotrophomonas maltophilia* SBP-9 augments resistance against biotic and abiotic stress in wheat plants. Front Microbiol 8:1945

Singh RP, Jha P, Jha PN (2017) Bio-inoculation of plant growth-promoting rhizobacterium *Enterobacter cloacae* ZNP-3 increased resistance against salt and temperature stresses in wheat plant (*Triticum aestivum* L.). J Plant Growth Regul 36(3):783–798

Sinha S, Mukherjee SK (2008) Cadmium–induced siderophore production by a high Cd-resistant bacterial strain relieved Cd toxicity in plants through root colonization. Curr Microbiol 56 (1):55–60

Son JS, Sumayo M, Hwang YJ, Kim BS, Ghim SY (2014) Screening of plant growth-promoting rhizobacteria as elicitor of systemic resistance against gray leaf spot disease in pepper. Appl Soil Ecol 73:1–8

Souza RD, Ambrosini A, Passaglia LMP (2015) Plant growth-promoting bacteria as inoculants in agricultural soils. Genet Mol Biol 38:401–419

Sueldo RJ, Invernati A, Plaza SG, Barassi CA (1996) Osmotic stress in wheat seedlings: effects on fatty acid composition and phospholipid turnover in coleoptiles. Cereal Res Commun 24:77–84

Suzuki N, Rivero RM, Shulaev V, Blumwald E, Mittler R (2014) Abiotic and biotic stress combinations. New Phytol 203:32–43

Szittya G, Silhavy D, Molnar A, Havelda Z, Lovas A, Lakatos L, Banfalvi Z, Burgyan J (2003) Low temperature inhibits RNA silencing-mediated defence by the control of siRNA generation. EMBO J 22:633–640

Tewari S, Arora NK (2013) Plant growth promoting rhizobacteria for ameliorating abiotic stresses triggered due to climatic variability. Climate Chang Environ Sustain 1(2):95–103

Timmusk S, Wagner EGH (1999) The plant-growth-promoting rhizobacterium *Paenibacillus polymyxa* induces changes in *Arabidopsis thaliana* gene expression a possible connection between biotic and abiotic stress responses. Mol Plant-Microbe Interact 12:951–959

Tiwari S, Lata C (2018) Heavy metal stress, signaling, and tolerance due to plant-associated microbes: an overview. Front Plant Sci 9:452

Turnau K, Jurkiewicz A, Lingua G, Barea JM, Gianinazzi Pearson V (2005) Role of arbuscular mycorrhiza and associated micro-organisms in phytoremediation of heavy metal polluted sites. In: Prasad MNV, Sajwan D, Ravi S (eds) Trace elements in the environment. Biogeochemistry, biotechnology and bioremediation. CRC Taylor & Francis, Boca Raton, London, New York, pp 235–252

Turner TR, James EK, Poole PS (2013) The plant microbiome. Genome Biol 14:209

Verhage A, van Wees SC, Pieterse CM (2010) Plant immunity: it's the hormones talking, but what do they say? Plant Physiol 154:536–540

Wang LJ, Li SL (2006) Salicylic acid-induced heat or cold tolerance in relation to Ca2C homeostasis and antioxidant systems in young grape plants. Plant Sci 170:685–694

Wang WX, Vinocur B, Shoseyov O, Altman A (2001) Biotechnology of plant osmotic stress tolerance: physiological and molecular considerations. Acta Hortic 560:285–292

Wang W, Vinocur B, Altman A (2003) Plant responses to drought, salinity and extreme temperatures: towards genetic engineering for stress tolerance. Planta 218:1–14

Wang CJ, Yang W, Wang C, Gu C, Niu DD, Liu HX, Wang YP, Guo JH (2012) Induction of drought tolerance in cucumber plants by a consortium of three plant growth promoting rhizobacterium strains. PLoS One 7:e52565

Wang X, Li Q, Sui J, Zhang J, Liu Z, Du J, Xu R, Zhou Y, Liu X (2019) Isolation and characterization of antagonistic bacteria *Paenibacillus jamilae* hs-26 and their effects on plant growth. Biomed Res Int 2019:3638926

Wintz H, Fox T, Vulpe C (2002) Responses of plants to iron, zinc and copper deficiencies. Biochem Soc Trans 30:766–768

Xiong L, Zhu JK (2001) Abiotic stress signal transduction in plants. Molecular and genetic perspectives. Physiol Plant 112:152–166

Xiong LM, Schumaker KS, Zhu JK (2002) Cell signaling during cold, drought, and salt stress. Plant Cell 14:S165–S183

Yamaguchi-Shinozaki K, Shinozaki K (2006) Transcriptional regulatory networks in cellular responses and tolerance to dehydration and cold stress. Ann Rev Plant Biol 57:781–803

Yang J, Kloepper JW, Ryu CM (2009) Rhizosphere bacteria help plants tolerate abiotic stress. Trends Plant Sci 14:1–4

Yuwono T, Handayani D, Soedarsono J (2005) The role of osmotolerant rhizobacteria in rice growth under different drought conditions. Aust J Agric Res 56(7):715

Zhang HH, Tang M, Chen H, Zheng C, Niu Z (2010) Effect of inoculation with AM fungi on lead uptake, translocation and stress alleviation of *Zea mays* L. seedlings planting in soil with increasing lead concentrations. Eur J Soil Biol 46:306–311

Zhang YF, Wang P, Yang YF, Bi Q, Tian SY, Shi XW (2011) Arbuscular mycorrhizal fungi improve reestablishment of *Leymus chinensis in* bare saline–alkaline soil: implication on vegetation restoration of extremely degraded land. J Arid Environ 75:773–778

Zolfaghari M, Nazeri V, Sefidkon F, Rejali F (2013) Effect of arbuscular mycorrhizal fungi on plant growth and essential oil content and composition of *Ocimum basilicum* L. Iranian J Plant Physiol 3:643–650

Zubair M, Shakir M, Ali Q, Rani N, Fatima N, Farooq S, Shafiq S, Kanwal N, Ali F, Nasir IA (2016) Rhizobacteria and phytoremediation of heavy metals. Environ Technol Rev 5 (1):112–119

Chapter 8
Omics Technology: Role and Future in Providing Biotic and Abiotic Stress Tolerance to Plants

Ashish Kumar Singh, Kuldeep Kumar, Lal Bahadur Singh, Amit Ahuja, Sharani Choudhury, M. L. Manoj, N. Sreeshma, Manoj Parihar, B. Jeewan, Amit U. Paschapur, Priyanka Khati, and K. K. Mishra

Abstract Development and productivity of the plants are constantly threatened by abiotic and biotic stresses. Plants have in turn developed various strategies to combat these stressed conditions. Technological advancement has led to the integration of biology and computer science which has helped mankind significantly with various analytical advancements to understand plant system biology more precisely. Genomics, transcriptomics, proteomics and metabolomics are collectively known as 'omics'. These molecular tools help to integrate biological information encoded within the molecular system of the plants. These strategies act as powerful tools in deciphering the intricate, well-coordinated signalling pathways and understanding of plant phenotypic responses at cellular and molecular level towards various environmental stimuli. This chapter briefly addresses the current omics strategies used for deciphering the mechanisms involved in plant stress tolerance.

8.1 Introduction

Plants being confined to a fixed area are exposed to several biotic and abiotic stresses throughout their life cycles. This attack is just not over with the completion of their life cycles; even the produce obtained after harvesting is threatened by several storage pests. There is a need to understand the molecular mechanisms of biotic and abiotic stresses in plants in order to define a strategy to cope against them. Many conventional methods like agronomic managements, development of tolerant varieties through plant breeding practices have helped the agricultural scientists

A. K. Singh (✉) · M. Parihar · B. Jeewan · A. U. Paschapur · P. Khati · K. K. Mishra
ICAR-Vivekananda Parvatiya Krishi Anusandhan Sansthan, Almora, Uttarakhand, India

K. Kumar · L. B. Singh · S. Choudhury · M. L. Manoj · N. Sreeshma
ICAR-NIPB, New Delhi, India

A. Ahuja
ICAR-IARI, New Delhi, India

A. Sharma (ed.), *Microbes and Signaling Biomolecules Against Plant Stress*, Rhizosphere Biology, https://doi.org/10.1007/978-981-15-7094-0_8

enormously in the past. But there is a continuous evolution in the agents causing biotic stress or increase in exposure to the degree of abiotic stresses on the plants. So, there is a need to adopt new technologies against these. Omics technologies have provided a way forward to biological science in this regard.

Omics are the emerging technologies to understand the behaviour and response of cells, tissues, organs and the whole organism at the cellular and molecular level using different modern-day technologies like genomics, proteomics, system biology, bioinformatics, as well as the computational tools needed to analyse and make sense of the data. *Omics* refers to the collective technologies used to explore the roles, relationships and actions of the various biomolecules that make up the cells of an organism. These technologies have potential to develop and integrate the available in vivo (phenotypic and other data such as inheritance pattern, etc.) and in vitro data (developed by several computational methods applied on living system like transcriptome or proteome data by using bioinformatics and molecular biology interventions).

These *omics* technologies measure the changes in the entire genome, gene, RNA, protein, metabolites or even at ionic level occurring at cellular level. Several fields of *omics* have evolved accordingly like the field of *omics* dealing with study at genome level is called as genomics and that deals with transcriptome is called as transcriptomics and so on. The field of *omics* includes: Genomics: study of structure, function of whole genome as well as comparison of genomes; Transcriptomics: study of change in gene expression profiling in different tissues; Proteomics: study of entire compendium of proteins; and Metabolomics: study of molecules involved in cellular metabolites

8.1.1 Genomics

'Genome' term was first given by Hans Winkler and it stands for haploid DNA content in a cell including the representation from organelle genome. In simpler term 'Genomics' is the field of omics which deals with the study of genome (Keller 2011). It involves molecular characterization and evaluation of whole genome in order to study its structure, function and comparison with related organism. Structural genomics, functional genomics and comparative genomics are different subsidiaries of genomics (Rubin et al. 2000). (1) Structural genomics is the field of genomics which deals with the study of genome and protein structure. Genome structure analysis mainly includes gene mapping, gene density, genome complexity, repeat analysis, cot curve analysis and other such studies, while in protein structure analysis it mainly includes high throughput methods for determination of protein structure by using a combination of experimental and modelling approaches (Baker and Sali 2001). (2) Functional genomics is the field of genomics which deals with experimental strategies adapted in order to reveal the function of different genes which genomes possess. It is further classified into two groups: one is forward genetics approaches, adapted to identify the causal gene when the only phenotype is known. It is usually done by creating random mutation in the genome either through chemical mutagens (EMS is mainly used for chemical mutagenesis), physical mutagens (γ radiation are main physical mutagens) or biological mutagens (transposable elements and T-DNA insertions are mainly used, they create insertion mutagenesis) followed by mapping of mutation. Classical plant breeding such as QTL mapping is also a forward genetics approach. Reverse genetics approach is adapted to identify the function and phenotype produced by a gene when its sequence is known. In order to learn the influence a sequence has on the phenotype, or to discover its biological function, researchers can engineer a change or disrupt the DNA sequences specifically or alter the expression of the concern sequences followed by tracking the changes occurring in the phenotype. There are several approaches available for reverse genetics, e.g. site directed mutagenesis, target induced local lesions in genome, RNAi, virus induced gene silencing, etc. (Robinson et al. 2011). (3) Comparative genomics approaches are used to compare and contrast the genome or genes to study their evolution pattern. Synteny and collinearity analysis, consensus maps and comparative maps, etc. are some of the approaches used in comparative genomics (Hardison 2003).

8.1.2 *Transcriptomics*

Gene expression inside a cell is highly spatio-temporal dependent. It is highly regulated by several environmental stimuli, for example, same tissue when uninfected and infected with a particular pathogen will have different gene expression profile which can be exploited to identify the causal genes (Geiger et al. 1996). Entire compendium of mRNA and small RNA expressing inside a cell constitutes transcriptome and a strategy adopted to study transcriptome is called as transcriptomics approach (Wang et al. 2009). Microarray analysis, serial analysis of gene expression, massive parallel signature sequencing, RNA-seq approaches are utilized for transcriptome analysis.

8.1.3 *Proteomics*

Entire compendium of protein expression inside the cell including its post-translational modification, interactome profile and splicing products constitutes proteome (Blackstock and Weir 1999). Microarray, 2-dimensional gel electrophoresis, mass spectrometry, etc. are some of the strategies adopted to study proteomics. With the advancement in genomics several discoveries are being made day by day. These fields have been extensively exploited in order to develop resistance in plants against several biotic and abiotic stresses which they encounter in their life cycles. Changes at genome level, alteration in gene expression profiling (in terms of transcriptome and proteome) are the keys to find the genetic reasons for the harmful metabolic drift and to develop strategies to cope up with them (Mittler and Blumwald 2010). In this chapter we are presenting a comprehensive compilation of several discoveries made in regard to develop biotic and abiotic resistance, identifying their causes using different genomics approaches.

8.2 Application of Different Omics Technique in Biotic Stress

When plants are affected by environmental factors like extreme temperatures, light, salinity, etc., it is called abiotic stress (Mittler 2006) and when stress is caused by living organisms it is called biotic stress (Suzuki et al. 2014). The components of biotic stress are fungi, bacteria, virus, parasites, insects, weeds, etc. Upon recognizing the pathogens, plants also modulate their defence pathway accordingly in an effort to cope up those stresses. Hence, gene expression pattern, proteome dynamics and metabolite pool get altered from those at natural condition. To address how plants use their intercellular and intracellular mechanisms to combat these challenges, multi-omics techniques are the best way to take up (Seneviratne et al. 2017).

The omic approaches are extensively engaged in elucidating biological pathways which are actively involved in triggering responses of host and disease development. When plant senses pathogen attack and recognizes pathogen associated molecules it rapidly alters its behaviour and results into different signalling event. The components of basal defence line up get triggered (Mehta et al. 2008). On the other hand, pathogen also continues its effort to make the infection successful by interfering plant defence line. In the course of this confrontation, plant resets the level, form and activity of its macromolecules in order to better withstand the pathogen intervention. The system level understanding to pursue all those interrelated changes in plant cells can be achieved through the application of different omic techniques. In this portion we will focus on transcriptomics, proteomics and metabolomics techniques deployed to explore molecules involved in resistance development in different crop under biotic stress.

8.2.1 Transcriptome Analysis

Transcriptomics study can provide a major understanding about why the same plant varies in resistance capability against the same pathogen at different ages or why within a genus some species can protect themselves whereas some are vulnerable to infection to a particular strain of pathogen (Dodds and Rathjen 2010). This also provides a fair idea about how genes are regulated at a particular time in a particular cell or tissue of an organism. Nowadays, transcriptome study is on high demand to screen out actively operating genes, pathways and interactions under any stress

Table 8.1 Some examples of RNA-seq deployed to identify the underlying gene and pathway

Crop	Pathogen	References
Poplar	*Marssonina brunnea*	Zhang et al. (2018)
Tomato	*Phytophthora parasitica*	Naveed and Ali (2018)
Nicotiana benthamiana	*Pseudomonas* spp.	Pombo et al. (2019)
Apple	Fungal pathogen Apple stem grooving virus *Erwinia amylovora* (bacteria)	Balan et al. (2018)
Rice (*Oryza sativa*)	*M. oryzae* Guy11 *M. oryzae* ZB13	Cohen et al. (2019)
Sunflower	*Plasmopara halstedii*	Ramu et al. (2016)

situation as it captures snapshot data on gene expression dynamics at a given time which becomes gradually instrumental to identify the underlying major gene(s). There are two main techniques in the field: microarray, which gives a relative quantitative value of the amount of transcript and identification based on a set of predefined sequences, and RNA-seq, which is proceeded through sequencing of transcript and mapping them in the reference, be it de novo or already existing Table 8.1. As the latter one provides very precise and absolute data about transcript read count and also works for exploring new transcript, it is adapted as a more popular technology in different molecular biology experiments (Tseng et al. 2012).

8.2.2 Proteomics Techniques

As and when pathogen entry gets initiated, many biological processes are triggered as well as many become suppressed. Functional genomics has been a boon to identify protein and corresponding gene by detecting those changes in transcript and protein level. Moreover, all the changes in transcriptome do not correlate with the proteome due to various factors like post-transcriptional modification, degradation, instability of the transcript, etc. Therefore, proteomics study has gained more focus to reveal the pathways and network of functional molecules in host plant.

Many proteomics tools and techniques have been exploited to date to detect pathogens and to shed light on host mechanisms under different biotic and abiotic stresses. These techniques have the potential to detect the proteins as well as define the underlying pathways and protein–protein interaction network. With the advent of enzyme linked immunosorbent assay (ELISA), detection of phytopathogen was remarkably easy (Lopez et al. 2003). 2D gel electrophoresis followed by mass spectrometry (MS) has long been exploited as a popular technique to identify the protein from stressed (biotic/abiotic/chemical) samples. Four defence-related protein (namely 14-3-3: master regulator active in biotic stress, Rab11: GTPase, RPN7: 26S proteasome subunit and an mRNA binding protein) and a ubiquitin pathway related protein were successfully detected and identified from *Capsicum annuum*

Table 8.2 Proteomics utilized against different pathogens

Plant–pathogen interaction	Strategy	References
Brassica napus–Piriformospora indica	LC-MS/MS	Shrivastava et al. (2018)
Brassica napus–Alternaria brassicae	2DE and MS	Sharma et al. (2007)
Wheat–*Puccinia triticina*	2DE and MS	Rampitsch et al. (2006)
Wheat–Fusarium graminearum	2DE and LC-MS/MS	Zhou et al. (2005)
Rice–*Xanthomonas oryzae*	2DE and MS	Mahmood et al. (2006)
Rice–*X. oryzae* pv. *Oryzae*		Chen et al. (2007)
Nicotiana benthamiana–pepper mild mottle virus	2DE and N-terminal sequencing	Pérez-Bueno et al. (2004)

cv. Bugang inoculated with Tobacco mosaic virus (TMV) (Lee et al. 2006). In rice, infected with yellow mosaic virus (RYMV), Brizard and co-workers found the significant presence of 223 proteins which were later well categorized among 3 classes: defence category, metabolism category, protein synthesis category (Brizard et al. 2006).

IP-MS is a technique which can address protein–protein interaction (PPI) between host and pathogen. When pathogens come in contact with their hosts, they try to turn down the host defence mechanism. In this course of action direct or indirect PPIs take place at different levels. Identification and characterization of PPIs can give scientists a lead to discover drugs against that pathogen. Not only that, IP-MS can also be exploited to pursue how disease progression has taken place. Detection and analysis of different mechanisms sensitized upon attack of pathogen HCMV was also performed by exploring IP-MS technique (Beltran et al. 2017).

Proteomics can be deployed to decipher the type of plant–pathogen interaction. A study looking for the differential protein profile in resistant and susceptible *Pinus lambertiana* infected with fungus *Cronartium ribicola* (biotrophic fungi) unravelled the significant downregulation of two acidic proteins similar to ß-1,3-glucanase in susceptible plants whereas chitinase level got increased in resistant plant (Ekramoddoullah and Hunt 1993). In another study, SDS-PAGE, peptide mass fingerprinting and sequence determination by MS were utilized to get an insight on the change in the protein profile of the xylem sap of tomato (*Lycopersicon esculentum*) in response to *Fusarium oxysporum* (necrotroph) infection (Rep et al. 2002). To identify the differentially expressed proteins in rice upon the infection of *Magnaporthe grisea* (hemibiotroph), 2DE and MALDI-TOF analysis were sought in another study (Kim et al. 2004) (Table 8.2).

Proteomics approach has the ability to study post-translational modification (PTM) also. When pathogen interacts with host, many PTMs take place in both host and pathogen proteome as a signature of its dynamic nature under selection pressure. Studying those PTMs may lead to discover biomarker with respect to that specific host–pathogen interaction. In one study on animal system this concept has been proved (Zhang and Ge 2011).

8.2.3 Metabolomics Technique

In spite of tremendous advancements in proteome strategies, many questions remain unanswered about the holistic changes in plant system that took place over pathogen attack. These questions can be addressed through the study of metabolome. Metabolome being more complex than proteome and metabolomic study to capture the changes in metabolite pool is still considered a tedious task due to its extremely dynamic nature albeit the evolution of more sensitive tools (Fessenden 2016). Other difficulties that come in the way of flawless insight into metabolome are: complex, sensitive, high level of fine-regulation and the presence of chemical contaminants (Lindon et al. 2011). Though current metabolomics techniques have their own advantages and disadvantages but scientists prefer to have deep insight on change in plant cell metabolome pool which is not possible with these techniques.

A study proved by metabolome analysis using LG-hybrid MS that in *Fusarium graminearum* infected yellow barley plant, resistance is directly correlated with the decrease in pathogen virulence factor and increase in detoxified product of virulence factor which is not the case for black barley (Kumaraswamya et al. 2011). In another study, in order to unravel the underlying biochemical activities in *Rhizoctonia solani* infected soybean orbitrap mass spectrometry coupled with GC/MS was carried out. For this experiment at first metabolite library of soybean was generated for quick reference of identified molecules. They detected movement of carbohydrates, disturbance of amino acid pools and elevation in the synthesis of various secondary metabolites like phenylpropanoid, a-linolenate and isoflavonoid, which in turn help plant to deter pathogen to colonize (Aliferis et al. 2014).

Pang et al. (2018) implemented another step to effectively differentiate plant metabolites from those of microbes before going for metabolic analysis to avoid the risk of contamination from pathogen metabolites. To study the effect of *Pseudomonas syringae* pv. tomato on *Arabidopsis thaliana*, the metabolome of microbes was labelled with C^{13} and targeted analysis of plant metabolome was carried out. Their study unveiled that the alteration in primary metabolic pathways and plant signalling finally led to stomatal movement restriction prior to pathogen entry into the plant (Pang et al. 2018). Biotic stresses also have an impact on plant lipidome. Upon sensitization of pathogen attack plants start producing reactive oxygen species (ROS) which leads to lipid peroxidation (Zoeller et al. 2012). The accurate changes what this lipid peroxidation brings to limit pathogen invasion may be taken up by the scientists as a topic to explore in the near future.

The phytonematodes are responsible for the huge crop yield losses worldwide (Dhaliwal et al. 2015). These threatening crop insect pests cause vast economic losses which are projected around 10.8% globally, in many cases these losses are unpredictable and uncountable. These tiny creatures possess a challenge to the crop protectionists as most often they remain undiagnosed and their management seems to be a cumbersome task with available resources. To overcome these issues and to devise the new management strategies, one should have to reveal the hidden genetic information of these organisms. The omics studies of nematodes provide a deeper insight into the crop protectionist to understand their exact regulatory mode of life (Mosa et al. 2017). Once the secrets of genomes are unveiled, then the particular regulatory genes can be targeted to disturb nematode's life cycle; thus, the crop can be prevented from the potential harm. The sedentary nematodes which include the genus *Meloidogyne*, *Heterodera*, *Nacobbus*, *Tylenchulus and Rotylenchulus* are considered the most dangerous to the crops as they alter the host physiology (Nicol et al. 2011), while few other ectoparasites like *Xiphinema*, *Trichodorus*, *Paratrichodorus*, *Longidorus* and *Paralongidorus* are responsible for spreading crop-damaging viruses (Taylor and Brown 1997). In the world of nematodes, the genomic databases are available for the free-living model nematode *Caenorhabditis elegans* which help and outstretch its applicability towards the understanding of the basic biology and lifestyle of other nematodes. Unlike free-living nematodes, research on biological aspects of phytonematodes is a tough job because many of them form feeding structures inside their hosts and due to their obligatory nature, they cannot be reared on artificial media in the laboratory. However, the genomic research progress will enhance the databases and expand the scope of target-specific management strategies (Abad and McCarter 2011).

8.3 Nematomics Application in the Field of Nematology

The whole genomes sequencing of an organism helps in identifying the total gene pool of that species and in later stages, the work functions of those genes can also be predicted by bioinformatics analysis. Recently, the next generation sequencing technologies are making the whole genome sequencing economical and feasible. Sequencing platforms like Illumina, 454 sequencings, SOLiD, and Nanopore, all have certain advantages and limitations over one another and are used today invariably by the researchers (Berglund et al. 2011). In the late 1990s, the whole genome sequencing of free-living model nematode *Caenorhabditis elegans* boosted the research for future nematode genomics (Hodgkin et al. 1995). The smallest and largest genomic sequence is reported for the *Pratylenchus* spp. (19 Mb) (Burke et al. 2015) and animal parasitic nematode *Ascaris suum* (0.5 Gb) (Jex et al. 2011), respectively. The huge variation in the length of genomic sequences among the species under phylum Nematoda puts a challenging question to the molecular biologists and nematologists to understand its nature, molecular, biological and ecological aspects. Within the genus of root-knot nematode, two species like *Meloidogyne incognita* and *Meloidogyne hapla* exhibit huge genomic variation, which is attributed to its size and reproduction physiology? The first one is expert reproducing via parthenogenesis mechanism and exhibits polyploidy while the later one reproduces by amphimixis. Today the sequences of economically important sedentary endoparasitic nematode *Meloidogyne incognita* (Abad et al. 2008), *Meloidogyne hapla* (Opperman et al. 2008) and migratory endoparasite pine wilt nematode and *Bursaphelenchus xylophilus* are available. Many sequencing projects of other important plant-parasitic nematodes are under progress in various parts of the world. These available whole genome sequences help in documentation of novel proteins involved in parasitism, reproduction, feeding behaviour and other aspects of biology. Upon identification of these functional genes and protein, some of them can be chosen as effective targets for drawing management perspectives. Study of transcriptome (RNA profile) helps indirectly in identifying the expressed genes as all the genes of an organism are not functional (Berglund et al. 2011). The transcriptome data of some important nematodes are available on the internet like WormBase (Harris et al. 2009) or NEMBASE (Parkinson et al. 2004), which can be utilized to understand the transcriptomes of closely related other important nematodes.

The overall interaction mechanism between nematode and its host depends upon the protein–protein interaction at the target site and the secretion of specific proteins is affected by the environmental factors (Chen et al. 2011). Numerous effecter proteins for the sedentary and migratory endoparasitic nematodes have been documented and the proteome data is generated. The mass spectrometry analysis of proteins also helps in rapid detection of the unknown species, based on data comparison with available data resources Ahmad and Babalola 2014).

In an attempt to feed on plant or for the establishment of a permanent feeding site inside their hosts, the phytonematodes exhibit an interesting pattern of movement and behavioural adaptations (Lewis et al. 2002). These nematodes move randomly in soil and upon sensing the host's cues, they exhibit a directed movement towards the gradient of particular stimulant. Then they invade a preselected site on the roots and feed by extracorporeal digestions and initiate reproduction. Once the host becomes devoid of sufficient food, then the nematode leaves the host and moves in the soil in search of a host and the same pattern of the life cycle is repeated. Based on the life cycle, whole pattern is categorized into various stages like host search, invasion, secretion, digestion, multiplication and host evasion. For successful completion of each stage, a gene or specific sets of the gene are responsible. The identification of these essential genes is involved in plant–host interaction (Gheysen and Fenoll 2011) and their functions by omics and knocking them down could disrupt the normal physiology of the nematodes and indirectly inhibit their advancements. So, omics knowledge is essential here for devising the target-specific management approaches to prevent the menace of phytonematodes on commercially important agricultural crops.

The only way to manage nematodes in the field is to prevent them to get established above certain threshold limits. The correct and timely identification of problem causing nematodes is the first step to apply any management practices. Omics provides a way for correct and fast identification of the exact nematodes based on the availability of sequence databases. Further omics helps in identifying the resistant genes in the host against nematodes, effector targets of nematode or genes utmost needed for the survival of nematodes. Thus these genes can be targeted by gene silencing methods which may help in devising alternative management tactics. Knocking down the genes involved in pathogenesis by nematodes helps in preventing losses to plant (Kaletta and Hengartner 2006). The necessity of omics in the field of nematology cannot be underestimated and there is an immediate need to develop nematode's genomics databases.

8.4 Applications of Different Omics Techniques in Abiotic Stress

Any change in optimal conditions of non-living factors in the environment adversely affects plant growth and development and is called as abiotic stress (Jenks and Hasegawa 2005). Major abiotic factors causing stress to plants are drought, salinity, temperature (cold and heat), flooding, heavy metals, etc. (Jenks and Hasegawa 2005). Stress signalling to plant response includes a lot of players from receptors, secondary messengers, transcriptional actors, regulatory enzymes, stress responsive proteins, ROS, different organic and inorganic molecules, etc. (Knight and Knight 2001). Conventional breeding approaches are not always useful exercises owing to the complex nature of the stress response by plants. The difference in the genome favours different plants to respond differentially to abiotic stress even of the same species. This is basically attributed to changes in genome sequence of different genotypes. Abiotic stress is a complex trait. Omics approaches are considered better strategies to take over. Genomics, transcriptomics, proteomics, metabolomics and ionomics provide several tools to understand and utilize the variations for crop improvement and breeding (Tripathi et al. 2012; Deshmukh et al. 2014).

8.4.1 Genomics

The complex nature of stress response by plants is not much efficient in improving the traits via conventional breeding practices (Varshney et al. 2011). The complex interactions occurring during abiotic response need to be understood in a holistic way to utilize that for crop improvement. Study of genome structure and function is necessary to gain a basic understanding of the molecular mechanisms involved in stress response of plants (Tripathi et al. 2012). Sequencing enables us to understand the whole genome of the organism and it paves way for more techniques to understand functional aspects of the genome and its evolution. Whole genome sequencing and resequencing enable identification of genome wide SNPs and SSRs and easy development of high density molecular maps (Gupta et al. 2008). Whole genome resequencing is a potential tool to find variations to develop molecular markers like SSR, SNPs and can be later used for genotyping by sequencing (Varshney et al. 2009). These techniques can be used for developing climate resilient crops (Henry 2014). QTL mapping for abiotic stress is a strategy to find chromosome regions contributing for a quantitative trait. Easy QTL mapping can be done using the genome wide high density markers developed previously using NGS technologies (Davey et al. 2011). Meta QTL analysis is done by combining many QTLs identified in a particular trait by using statistical tools in a single genetic map to predict the presence of QTLs (Goffinet and Gerber 2000). Genome wide association studies making use of genome wide SNP markers give more allelic diversity and increased resolution to QTL maps compared to biparental mapping methods (Myles et al. 2009). Genomic selection is another approach of crop improvement in which breeding values of different lines are predicted by using molecular markers (Desta and Ortiz 2014). This enables us to bring in all favourable QTLs even of minor effects together to develop plants tolerant to abiotic stress. Microarrays based SNP genotyping is another genomic tool which is less costly than NGS based short gun sequencing (Gunderson et al. 2005). SNP array is a hybridization-based technique and facilitates genome wide analysis and screening of a large number of genotypes easily. Sequencing and low-cost genome wide genotyping methods will fasten discovery of genetic variation, identification of genes and underlying molecular mechanisms and speed up crop breeding for stress tolerance.

8.4.2 *Transcriptomics*

Analysing the genes expressed in response to stress in tolerant and susceptible plants will give a way to study the tolerance mechanisms involved in the process (Clement et al. 2008). Earlier expressed sequence tags (ESTs) developed by sequencing of cDNAs were used for preparing spotted arrays for the study of a large number of transcripts. But later, microarray was developed which is a high throughput technology where labelled RNA samples hybridized with thousands of synthetic probes on a chip (which is previously developed) represent different transcripts which is useful to profile expression patterns of RNA under various spatial temporal and developmental stages (Table 8.3). This is used as an effective way for genome wide

Table 8.3 Some arrays developed in crop plants

Crop	Size (technology)	References
Apple	480 K (Affymetrix axiom)	Bianco et al. (2016)
Maize	600 K (Affymetrix axiom)	Unterseer et al. (2014)
Rice	1 M (Affymetrix)	McCouch et al. (2010)
Rice	50 K (Affymetrix axiom)	Singh et al. (2015)
Soybean	180 K (Affymetrix axiom)	Lee et al. (2014)
Wheat	820 K (Affymetrix axiom)	Winfield et al. (2015)
De novo (applicable to multiple crops)	50–300 K (GBS genotype-by-sequencing)	Elshire et al. (2011)

Table 8.4 RNA-seq in studying abiotic stress

Crop: stress	Findings	References
Chickpea: drought	261 and 169 drought stress responsive genes identified	Mashaki et al. (2018)
Sea buckthorn: drought	1644 unigenes were differentially expressed, 519 unigenes were upregulated and 1125 were downregulated	Ye et al. (2018)
Mustard: high temperature and drought	Differentially regulated 19,110 transcripts	Bhardwaj et al. (2015)
Bread wheat: salinity	5128 genes were differentially expressed due to salt stress	Amirbakhtiar et al. (2019)

expression profiling under stress or any other specific conditions. Affymetrix chip of 61 k probes in soybean is one example of microarray (Le et al. 2012).

RNA-Seq: RNA sequencing is an advanced technology for transcriptome profiling. The entire transcript from a sample is deeply sequenced in NGS tools like ILLUMINA. It gives accurate information on the genes, expressed in a plant sample and its quantity which gives a detailed transcript profile by comparing RNA-seq data between tolerant and susceptible plants under stress conditions. It can be used to understand the genes involved in abiotic stress responses (Table 8.4). RNA-seq data also helps to understand molecular regulation, small RNA expression, alternative splicing, etc. (Deshmukh et al. 2014).

8.4.3 Proteomics Techniques

Stress reception, signal transduction, changes in gene expression, etc. involve many proteins including enzymes and regulation of these proteins is done by many mechanisms such as post-translation modification, protein–protein interactions, targeting, degradation, etc. (Marks and Klingmüller 2008). Profiling the proteins which are being synthesized, modified, targeted and made functional and non-functional according to the stress perceived by the plant will provide a better account on stress tolerance level of plants (Table 8.5).

8.4.4 Metabolomics Technique

Identification and quantification of metabolome (entire metabolites) at a particular stage in a tissue enable us to understand the biochemical pathways involved and the underlying molecular mechanisms (Hall 2006). Techniques like gas chromatography mass spectrometry (GC-MS), liquid chromatography mass spectrometry (LC-MS), capillary electrophoresis mass spectrometry (CE-MS), Fourier transform ion

Table 8.5 Application of proteomic techniques in abiotic stress

Crop: stress	Techniques	Remarks	References
Tomato: thermotolerance	LC-MS fraction-ation protocol	Identified proteins involved in systemic and fundamental components of pollen thermotolerance	Jegadeesan et al. (2018)
Spinach: heat stress	iTRAQ Label-ling, LC-20A HPLC system	Proteomic and phosphoproteomics of heat tolerant line were studied	Zhao et al. (2018)
Wheat: heat stress	2DE	Characterization of heat tolerant and susceptible cultivars via flag leaf proteomics	Nandha et al. (2018)
Sugarcane: salt stress	ESI-LC-MS/MS	Analyse the proteomic effects of salt stress in micro-propagated shoots of two sugarcane cultivars	Passamani et al. (2017)
Spring wheat: drought	2D-PAGE, LC-ESI-MS/MS	Effect of drought in proteome of susceptible vs. tolerant cultivar	Michaletti et al. (2018)

Table 8.6 Application of metabolomics tools in abiotic stress research

Crop: stress	Techniques	Remarks	References
Spring wheat: drought	LC-MS	Effect of drought in metabolome of susceptible vs. tolerant cultivar	Michaletti et al. (2018)
Wheat: drought	Gas chromatography–time-of-flight/mass spectrometry	Mechanisms involved in wheat seedling drought tolerance	Guo et al. (2018)
Rice: salinity	GC-MS	Metabolomic responses of seedlings of tolerant varieties and susceptible varieties	Gupta and De (2017)

cyclotron resonance mass spectrometry (FT-ICR-MS) and nuclear magnetic resonance (NMR) are generally used in plant sciences (Putri et al. 2013) (Table 8.6).

8.4.5 *Ionomics*

Ionomics deals with the identification and quantification of elemental composition of a tissue or an organism (Singh et al. 2013). Plants growing under different soil types have developed mechanisms to take up different elements. Ionomics studies will help to improve the understanding of functions of elements in biochemistry, physiology and molecular mechanisms of plants (Salt 2004). Availability of elements, uptake, transport, evapotranspiration, etc. makes the ionome profile very sensitive to the prevailing conditions. Exposure to various abiotic stresses may drastically change the elemental conditions in various plant tissues; so the study of ionomics in responses to abiotic stress conditions can give us a better understanding of underlying mechanisms. Some techniques used in ionomics studies are neutron activation analysis (NAA), X-ray fluorescence (XRF), inductively coupled plasma-mass spectrometry (ICP-MS) and inductively coupled plasma-atomic/optical emission spectrometry (ICP-AES/OES) (Wu et al. 2013).

8.5 Conclusion

Integration of computer science and biology has resulted in the development of high throughput methods with fast data generation which help to identify, measure and quantify cellular molecules like DNA, RNA and proteins. These approaches are dependent on computational tools and have made an immense impact on our understanding of plant stress tolerance responses and signalling involved with the process. Omics measures change at gene, RNA, protein and even at ionic level within the cell. Further advancement and novel tools and computational models need to be developed in order to find the ultimate missing parts to decipher the complete mechanism of stress tolerance in plants and improve our understanding of plant system biology.

Acknowledgement We sincerely thank the facilities and support provided by our Institute.

References

Abad P, McCarter JP (2011) Genome analysis of plant parasitic nematodes. In: Genomics and molecular genetics of plant-nematode interactions. Springer, Dordrecht, pp 103–117

Abad P, Gouzy J, Aury JM, Castagnone-Sereno P, Danchin EG, Deleury E, Caillaud MC (2008) Genome sequence of the metazoan plant-parasitic nematode Meloidogyne incognita. Nat Biotechnol 26(8):909

Ahmad F, Babalola OO (2014) Application of mass spectrometry as rapid detection tool in plant nematology. Appl Spectrosc Rev 49(1):1–10

Aliferis KA, Faubert D, Jabaji S (2014) A metabolic profiling strategy for the dissection of plant defense against fungal pathogens. PLoS One 9(11):e111930

Amirbakhtiar N, Ismaili A, Ghaffari MR, Nazarian Firouzabadi F, Shobbar ZS (2019) Transcriptome response of roots to salt stress in a salinity-tolerant bread wheat cultivar. PLoS One 14(3):e0213305. https://doi.org/10.1371/journal.pone.0213305

Baker D, Sali A (2001) Protein structure prediction and structural genomics. Science 294 (5540):93–96

Balan B, Marra FP, Caruso T, Martinelli F (2018) Transcriptomic responses to biotic stresses in *Malus x domestica*: a meta-analysis study. Sci Rep 8:1970

Beltran PMJ, Federspiel JD, Sheng X, Cristea IM (2017) Proteomics and integrative omic approaches for understanding host–pathogen interactions and infectious diseases. Mol Syst Biol 13:922

Berglund EC, Kiialainen A, Syvänen AC (2011) Next-generation sequencing technologies and applications for human genetic history and forensics. Investig Genet 2(1):23

Bhardwaj AR, Joshi G, Kukreja B, Malik V, Arora P, Pandey R, Shukla RN, Bankar KG, Katiyar-Agarwal S, Goel S, Jagannath A, Kumar A, Agarwal M (2015) Global insights into high temperature and drought stress regulated genes by RNA-Seq in economically important oilseed crop *Brassica juncea*. BMC Plant Biol 15

Bianco L, Cestaro A, Linsmith G et al (2016) Development and validation of the Axiom(®) Apple480K SNP genotyping array. Plant J 86(1):62–74. https://doi.org/10.1111/tpj.13145

Blackstock WP, Weir MP (1999) Proteomics: quantitative and physical mapping of cellular proteins. Trends Biotechnol 17(3):121–127

Brizard JP, Carapito C, Delalande F, Van Dorsselaer A, Brugidou C (2006) Proteome analysis of plant–virus interactome: comprehensive data for virus multiplication inside their hosts. Mol Cell Proteomics 5:2279–2297

Burke M, Scholl EH, Bird DM, Schaff JE, Colman SD, Crowell R, Windham E (2015) The plant parasite *Pratylenchus coffeae* carries a minimal nematode genome. Nematology 17(6):621–637

Chen F, Yuan Y, Li Q, He Z (2007) Proteomic analysis of rice plasma membrane reveals proteins involved in early defense response to bacterial blight. Proteomics 7:1529–1539

Chen X, MacDonald MH, Khan F, Garrett WM, Matthews BF, Natarajan SS (2011) Two-dimensional proteome reference maps for the soybean cyst nematode *Heterodera glycines*. Proteomics 11(24):4742–4746

Clement M, Lambert A, Herouart D, Boncompagni E (2008) Identification of new up-regulated genes under drought stress ins soybean nodules. Gene 426:15–22. https://doi.org/10.1016/j.gene.2008.08.016

Cohen SP, Leach JE (2019) Abiotic and biotic stresses induce a core transcriptome response in rice. Sci Rep 9:6273. https://doi.org/10.1038/s41598-019-42731-8

Davey JW, Hohenlohe PA, Etter PD, Boone JQ, Catchen JM, Blaxter ML (2011) Genome-wide genetic marker discovery and genotyping using next-generation sequencing. Nat Rev Genet 12 (7):499

Deshmukh R, Sonah H, Patil G, Chen W, Prince S, Mutava R, Vuong T, Valliyodan B, Nguyen HT (2014) Integrating omic approaches for abiotic stress tolerance in soybean. Front Plant Sci 5:244. https://doi.org/10.3389/fpls.2014.00244

Desta ZA, Ortiz R (2014) Genomic selection: genome-wide prediction in plant improvement. Trends Plant Sci 19(9):592–601

Dhaliwal GS, Jindal V, Mohindru B (2015) Crop losses due to insect pests: global and Indian scenario. Indian J Entomol 77(2):165–168

Dodds PN, Rathjen JP (2010) Plant immunity: towards an integrated view of plant–pathogen interactions. Nat Rev Genet 11(8):539

Ekramoddoullah AKM, Hunt RS (1993) Changes in protein profile of susceptible and resistant sugar-pine foliage infected with the white pine blister rust fungus *Cronartium ribicola*. Can J Plant Path 15:259–264

Elshire RJ, Glaubitz JC, Sun Q, Poland JA, Kawamoto K, Buckler ES et al (2011) A robust, simple genotyping-by-sequencing (GBS) approach for high diversity species. PLoS One 6(5):e19379. https://doi.org/10.1371/journal.pone.0019379

Fessenden M (2016) Metabolomics: small molecules, single cells. Nature 540(7631):153–156

Geiger DR, Koch KE, Shieh WJ (1996) Effect of environmental factors on whole plant assimilate partitioning and associated gene expression. J Exp Bot 47:1229–1238

Gheysen G, Fenoll C (2011) In: Jones J (ed) Genomics and molecular genetics of plant-nematode interactions (no. PA 632.6 G33.). Springer, Dordrecht

Goffinet B, Gerber S (2000) Quantitative trait loci: a meta-analysis. Genetics 155(1):463–473

Gunderson KL, Steemers FJ, Lee G, Mendoza LG, Chee MS (2005) A genome-wide scalable SNP genotyping assay using microarray technology. Nat Genet 37(5):549

Guo R, Shi L, Jiao Y et al (2018) Metabolic responses to drought stress in the tissues of drought-tolerant and drought-sensitive wheat genotype seedlings. AoB Plants 10(2):ply016. https://doi.org/10.1093/aobpla/ply016

Gupta P, De B (2017) Metabolomics analysis of rice responses to salinity stress revealed elevation of serotonin, and gentisic acid levels in leaves of tolerant varieties. Plant Signal Behav 12:7. https://doi.org/10.1080/15592324.2017.1335845

Gupta PK, Rustgi S, Mir RR (2008) Array-based high-throughput DNA markers for crop improvement. Heredity 101(1):5

Hall RD (2006) Plant metabolomics: from holistic hope, to hype, to hot topic. New Phytol 169 (3):453–468

Hardison RC (2003) Comparative genomics. PLoS Biol 1(2):e58
Harris TW, Antoshechkin I, Bieri T, Blasiar D, Chan J, Chen WJ, Fernandes J (2009) WormBase: a comprehensive resource for nematode research. Nucleic Acids Res 38(suppl_1):D463–D467
Henry RJ (2014) Genomics strategies for germplasm characterization and the development of climate resilient crops. Front Plant Sci 5:68. https://doi.org/10.3389/fpls.2014.00068
Hodgkin J, Plasterk RH, Waterston RH (1995) The nematode *Caenorhabditis elegans* and its genome. Science 270(5235):410–414
Jegadeesan S, Chaturvedi P, Ghatak A, Pressman E, Meir S, Faigenboim A, Rutley N, Beery A, Harel A, Weckwerth W, Firon N (2018) Proteomics of heat-stress and ethylene-mediated thermotolerance mechanisms in tomato pollen grains. Front Plant Sci 9:1558. https://doi.org/10.3389/fpls.2018.01558
Jenks MA, Hasegawa PM (eds) (2005) Plant abiotic stress. Blackwell, London
Jex AR, Liu S, Li B, Young ND, Hall RS, Li Y, Chen F (2011) *Ascaris suum* draft genome. Nature 479(7374):529
Kaletta T, Hengartner MO (2006) Finding function in novel targets: *C. elegans* as a model organism. Nat Rev Drug Discov 5(5):387
Keller EF (2011) Genes, genomes, and genomics. Biol Theory 6(2):132–140
Kim ST, Kim SG, Hwang DH, Kang SY, Kim HJ, Lee BH, Lee JJ, Kang KY (2004) Proteomic analysis of pathogen-responsive proteins from rice leaves induced by rice blast fungus, *Magnaporthe grisea*. Proteomics 4:3569–3578
Knight H, Knight MR (2001) Abiotic stress signaling pathways: specificity and cross-talk. Trends Plant Sci 6(6):262–267
Kumaraswamya GK, Kushalappaa AC, Choob TM, Dionc Y, Riouxd S (2011) Differential metabolic response of barley genotypes, varying in resistance, to trichothecene-producing and -nonproducing (tri5) isolates of *Fusarium graminearum*. Plant Pathol 61:509–521
Le DT, Nishiyama R, Watanabe Y, Tanaka M, Seki M, Yamaguchi-Shinozaki K (2012) Differential gene expression in soybean leaf tissues at late developmental stages under drought stress revealed by genome-wide transcriptome analysis. PLoS One 7:e49522
Lee BJ, Kwon SJ, Kim SK, Kim KJ, Park CJ, Kim YJ, Park OK, Paek KH (2006) Functional study of hot pepper 26S proteasome subunit RPN7 induced by tobacco mosaic virus from nuclear proteome analysis. Biochem Biophys Res Commun 351:405–411
Lee Y-G, Jeong N, Kim J, Lee K, Kim KH, Pirani A, Ha B-K, Kang S-T, Park B-S, Moon J-K, Kim N, Jeong S-C (2014) Development, validation, and genetic analysis of a large soybean SNP genotyping array. Plant J 81:625–636. https://doi.org/10.1111/tpj.12755
Lewis EE, Campbell JF, Sukhdeo MV (eds) (2002) The behavioural ecology of parasites. CABI publication, Wallingford, New York
Lindon JC, Nicholson JK, Holmes E (2011) Cellular metabolomics: the quest for pathway structure. In: The handbook of metabonomics and metabolomics. Elsevier, Amsterdam, p 35
Lopez MM, Bertolini E, Olmos A, Caruso P, Gorris MT, Llop P, Penyalver R, Cambra M (2003) Innovative tools for the detection of plant pathogenic viruses and bacteria. Int Microbiol 6:233–243
Mashaki KM, Garg V, Nasrollahnezhad Ghomi AA, Kudapa H, Chitikineni A, Zaynali Nezhad K (2018) RNA-Seq analysis revealed genes associated with drought stress response in kabuli chickpea (*Cicer arietinum* L.). PLoS One 13(6):e0199774. https://doi.org/10.1371/journal.pone.0199774
Mahmood T, Jan A, Kakishima M, Komatsu S (2006) Proteomic analysis of bacterial-blight defense responsive proteins in rice leaf blades. Proteomics 6:6053–6065
Marks F, Klingmüller U, Müller-Decker (2008) Cellular signal processing: an introduction to the molecular mechanisms of signal transduction. Garland Science, New York
McCouch SR et al (2010) Development of genome-wide SNP assays for rice. Breed Sci 60:524–535
Mehta A, Brasileiro ACM, Souza DSL, Romano E, Campos MA, Grossi-de-Sa MF, Silva MS, Franco OL, Fragoso RR, Bevitori R, Rocha TL (2008) Plant–pathogen interactions: what is proteomics telling us? FEBS J 275:3731–3746

Michaletti A, Naghavi MR, Toorchi M, Zolla L, Rinalducci S (2018) Metabolomics and proteomics reveal drought-stress responses of leaf tissues from spring-wheat. Sci Rep 8(1):5710. https://doi.org/10.1038/s41598-018-24012-y
Mittler R (2006) Abiotic stress, the field environment and stress combination. Trends Plant Sci 11 (1):15–19
Mittler R, Blumwald E (2010) Genetic engineering for modern agriculture: challenges and perspectives. Annu Rev Plant Biol 61:443–462
Mosa KA, Ismail A, Helmy M (2017) Omics approaches to understand biotic stresses: a case study on plant parasitic nematodes. In: Plant stress tolerance. Springer, Cham, pp 35–54
Myles S, Peiffer J, Brown PJ, Ersoz ES, Zhang Z, Costich DE, Buckler ES (2009) Association mapping: critical considerations shift from genotyping to experimental design. Plant Cell 21 (8):2194–2202
Nandha AK, Mehta DR, Tulsani NJ, Umretiya N, Delvadiya N (2018) Proteomic analysis in wheat to study the effect of heat stress on flag leaf. Int J Curr Microbiol App Sci 7(2):3432–3439. https://doi.org/10.20546/ijcmas.2018.702.409
Naveed ZA, Ali GS (2018) Comparative Transcriptome analysis between a resistant and a susceptible wild tomato accession in response to *Phytophthora parasitica*. Int J Mol Sci 19:3735
Nicol JM, Turner SJ, Coyne DL, Den Nijs L, Hockland S, Maafi ZT (2011) Current nematode threats to world agriculture. In: Genomics and molecular genetics of plant-nematode interactions. Springer, Dordrecht, pp 21–43
Opperman CH, Bird DM, Williamson VM, Rokhsar DS, Burke M, Cohn J, Houfek TD (2008) Sequence and genetic map of *Meloidogyne hapla*: a compact nematode genome for plant parasitism. Proc Natl Acad Sci 105(39):14802–14807
Pang Q, Zhang T, Wang Y, Kong W, Guan Q, Yan X, Chen S (2018) Metabolomics of early stage plant cell–microbe interaction using stable isotope labeling. Front Plant Sci 9:760
Parkinson J, Whitton C, Schmid R, Thomson M, Blaxter M (2004) NEMBASE: a resource for parasitic nematode ESTs. Nucleic Acids Res 32(suppl_1):D427–D430
Passamani LZ, Barbosa RR, Reis RS, Heringer AS, Rangel PL, Santa-Catarina C (2017) Salt stress induces changes in the proteomic profile of micropropagated sugarcane shoots. PLoS One 12 (4):e0176076. https://doi.org/10.1371/journal.pone.0176076
Pérez-Bueno ML, Rahoutei J, Sajnani C, García-Luque I, Barón M (2004) Proteomic analysis of the oxygen-evolving complex of photosystem II under biotec stress: studies on Nicotiana benthamiana infected with tobamoviruses. Proteomics Syst Biol. https://doi.org/10.1002/pmic.200300655
Pombo MA, Ramos RN, Zheng Y, Fei Z, Martin GB, Rosli HG (2019) Transcriptome-based identification and validation of reference genes for plant-bacteria interaction studies using *Nicotiana benthamiana*. Sci Rep 9:1632
Putri SP, Yamamoto S, Tsugawa H, Fukusaki E (2013) Current metabolomics: technological advances. J Biosci Bioeng 116:9–16. https://doi.org/10.1016/j.jbiosc.2013.01.004
Rampitsch C, Bykova NV, McCallum B, Beimcik E, Ens W (2006) Analysis of the wheat and *Puccinia triticina* (leaf rust) proteomes during a susceptible host-pathogen interaction. Proteomics 6:1897–1907
Ramu VS, Paramanantham A, Ramegowda V, Mohan-Raju B, Udayakumar M, Senthil-Kumar M (2016) Transcriptome analysis of sunflower genotypes with contrasting oxidative stress tolerance reveals individual- and combined- biotic and abiotic stress tolerance mechanisms. PLoS One 11(6):e0157522
Rep M, Dekker HL, Vossen JH, Boer AD, Houterman PM, Speijer D, Back JW, Koster CG (2002) Mass spectrometric identification of isoforms of PR proteins in xylem sap of fungus infected tomato. Plant Physiol 130:904–917
Robinson JT, Thorvaldsdóttir H, Winckler W, Guttman M, Lander ES, Getz G, Mesirov JP (2011) Integrative genomics viewer. Nat Biotechnol 29(1):24
Rubin GM, Yandell MD, Wortman JR, Gabor GL, Nelson CR, Hariharan IK, Cherry JM (2000) Comparative genomics of the eukaryotes. Science 287(5461):2204–2215

Salt DE (2004) Update on plant ionomics. Plant Physiol 136(1):2451–2456
Seneviratne G, Weerasekara MLMAW, Kumaresan D, Zavahir JS (2017) Microbial signaling in plant—microbe interactions and its role on sustainability of agroecosystems. In: Agro-environmental sustainability. Springer, Cham, pp 1–17
Sharma N, Rahman MH, Strelkov S, Thiagarajah M, Bansal VK, Kav NNV (2007) Proteome-level changes in two *Brassica napus* lines exhibiting differential responses to the fungal pathogen *Alternaria brassicae*. Plant Sci 172:95–110
Shrivastava N, Jiang L, Li P, Sharma AK, Luo X, Wu S, Pandey R, Gao Q, Lou B (2018) Proteomic approach to understand the molecular physiology of symbiotic interaction between *Piriformospora indica* and *Brassica napus*. Sci Rep 8:5773
Singh UM, Sareen P, Sengar RS, Kumar A (2013) Plant ionomics: a newer approach to study mineral transport and its regulation. Acta Physiol Plant 35(9):2641–2653
Singh N, Jayaswal PK, Panda K et al (2015) Single-copy gene based 50 K SNP chip for genetic studies and molecular breeding in rice. Sci Rep 5:11600. https://doi.org/10.1038/srep11600
Suzuki N, Rivero RM, Shulaev V, Blumwald E, Mittler R (2014) Abiotic and biotic stress combinations. New Phytol 203(1):32–43
Taylor CE, Brown DJ (1997) Nematode vectors of plant viruses. CAB International, Wallingford
Tripathi P, Rabara CR, Rushton PJ (2012) Emergence of Omics-approaches for crop improvement during abiotic stress. Int J Envi Sci Tech 1(2):125–128
Tseng GC, Ghosh D, Feingold E (2012) Comprehensive literature review and statistical considerations for microarray meta-analysis. Nucleic Acids Res 40(9):3785–3799
Unterseer S, Bauer E, Haberer G et al (2014) A powerful tool for genome analysis in maize: development and evaluation of the high density 600 k SNP genotyping array. BMC Genomics 15:823. https://doi.org/10.1186/1471-2164-15-823
Varshney RK, Nayak SN, May GD, Jackson SA (2009) Next-generation sequencing technologies and their implications for crop genetics and breeding. Trends Biotechnol 27:522–530
Varshney RK, Bansal KC, Aggarwal PK, Datta SK, Craufurd PQ (2011) Agricultural biotechnology for crop improvement in a variable climate: hope or hype? Trends Plant Sci 16(7):363–371
Wang Z, Gerstein M, Snyder M (2009) RNA-Seq: a revolutionary tool for transcriptomics. Nat Rev Genet 10(1):57
Winfield MO, Allen AM, Burridge AJ, Barker GLA, Benbow HR, Wilkinson PA, Coghill J, Waterfall C, Davassi A, Scopes G, Pirani A, Webster T, Brew F, Bloor C, King J, West C, Griffiths S, King I, Bentley AR, Edwards KJ (2015) High-density SNP genotyping array for hexaploid wheat and its secondary and tertiary gene pool. Plant Biotechnol J. https://doi.org/10.1111/pbi.12485
Wu D, Shen Q, Cai S, Chen ZH, Dai F, Zhang G (2013) Ionomic responses and correlations between elements and metabolites under salt stress in wild and cultivated barley. Plant Cell Physiol 54:1976–1988. https://doi.org/10.1093/pcp/pct134
Ye G, Ma Y, Feng Z, Zhang X (2018) Transcriptomic analysis of drought stress responses of sea buckthorn (*Hippophae rhamnoides* subsp. sinensis) by RNA-Seq. PLoS One 13(8):e0202213. https://doi.org/10.1371/journal.pone.0202213
Zhang H, Ge Y (2011) Comprehensive analysis of protein modifications by top-down mass spectrometry. Circ Cardiovasc Genet 4(6):711
Zhang Y, Tian L, Yan DH, He W (2018) Genome-wide transcriptome analysis reveals the comprehensive response of two susceptible poplar sections to *Marssonina brunnea* infection. Genes 9:154
Zhao Q, Chen W, Bian J, Xie H, Li Y, Xu C, Ma J, Guo S, Chen J, Cai X, Wang X, Wang Q, She Y, Chen S, Zhou Z, Dai S (2018) Proteomics and Phosphoproteomics of heat stress-responsive mechanisms in spinach. Front Plant Sci 9:800. https://doi.org/10.3389/fpls.2018.00800
Zhou W, Kolb FL, Riechers DE (2005) Identification of proteins induced or upregulated by Fusarium head blight infection in the spikes of hexaploid wheat (*Triticum aestivum*). Genome 48:770–780
Zoeller M, Stingl N, Krischke M, Fekete A, Waller F, Berger S (2012) Lipid profiling of the Arabidopsis hypersensitive response reveals specific lipid peroxidation and fragmentation processes: biogenesis of pimelic and azelaic acid. Plant Physiol 160:365–378

Chapter 9
Role of Arbuscular Mycorrhizal Fungi in Amelioration of Drought Stress in Crop Plants

Pallavi and Anil Kumar Sharma

Abstract Present chapter addresses the importance of arbuscular mycorrhizal (AM) symbiosis in the mitigation of one of the most limiting environmental constraints in terms of global crop productivity loss, i.e. drought/water stress. Superior root morphology leading to enhanced water and nutrient uptake, better antioxidant machinery to fight the elevated reactive oxygen species (ROS), along with improved osmotic adjustment by accumulating high levels of compatible solutes like proline, glycine betaine, sugars, etc., are among the multifaceted factors which make AM association far more significant than any other symbiotic association in plants. So far little progress has been made on molecular level to understand the mechanisms of this miraculous organism which include the involvement of aquaporins, some binding protein genes (BiPs), MAPK pathway genes, and stress responsive genes like proline dehydrogenase, invertase, proline synthetase, etc. However, with the advancement of new age molecular techniques it seems possible that the unravelling of this mystery is not far away.

Abbreviations

AM	Arbuscular Mycorrhiza
ATP	Adenosine triphosphate
BiP	Binding proteins
CAT	Catalase
EEA	European Environment Agency
ERM	extraradical mycelium
GOGAT	Glutamine synthase
GS	Glutamine synthetase

Pallavi (✉)
Department of Botany, J.N.L. College, Patliputra University, Patna, Bihar, India

A. K. Sharma
Department of Biological Sciences, College of Basic Science and Humanities, G.B. Pant University of Agriculture and Technology, Pantnagar, Uttarakhand, India

A. Sharma (ed.), *Microbes and Signaling Biomolecules Against Plant Stress*, Rhizosphere Biology, https://doi.org/10.1007/978-981-15-7094-0_9

HNE	4-hydroxy-2-nonenal
IAA	Indole-3-acetic acid
IPCC	Intergovernmental Panel on Climate Change
LPO	Lipid peroxidation
MAPK	Mitogen-activated protein kinase
MDA	Malondialdehyde
NR	Nitrate reductase
ROS	Reactive oxygen species
RUBISCO	Ribulose-1,5-biphosphate carboxylase/oxygenase
RWC	Relative water content
SOD	Super oxide dismutase

9.1 Introduction

Plant–arbuscular mycorrhizal (AM) fungal interaction (symbiotic) is the most studied association, so far. Around 80% of land plants, especially agricultural and horticultural crop species establish symbiotic association with AM fungi belonging to the phylum Glomeromycota (Barea et al. 2002; Smith and Read 2008). Through this mutualistic beneficial association, AM fungi fulfil their carbon requirements in the form of photo-assimilates from the host plants and in return, fungi enhance plant's acquisition of water and mineral nutrients. Improved photosynthetic efficiency and antioxidant activity is detected in mycorrhized plants exposed to elongated water deficit conditions in comparison to non-mycorrhized plants. AM plants show changes in root architecture which is often accredited to the higher accumulation of indole-3-acetic acid (IAA), nitric oxide, methyl jasmonate, and calmodulin in roots (Zou et al. 2017), thus supporting the idea that AM fungal infection causes these morphological and biochemical changes in plants providing them a upper hand in facing drought conditions.

Amongst the yield deterring reasons, drought is the front runner among all the abiotic stresses. It is especially prominent in developing countries of tropical region where agricultural lands are primarily rain-fed and results in high production losses to the farmers. It can be defined as a stage of water deficiency resulting in stunted growth of plants. Drought or water stress happens when there is no significant rainfall or alternate irrigation facilities available. This leads to the deficiency of usable water to facilitate proper functioning of the plants in the soil which is further aggravated by atmospheric conditions causing continuous loss of water by transpiration or evaporation. According to the reports of IPCC (2007) and EEA (2015), climate change and growing water scarcity will result in further aggravation of drought conditions in coming years.

9.2 Plant Physiological Responses Towards Drought Stress

Basically, drought stress interferes with the normal metabolic activities of the plants by disrupting plant-water potential and turgor pressure (Rahdari and Hoseini 2012). Along the years, researchers have studied the impact of drought stress on all the major crops such as wheat (Farooq et al. 2014), rice (Korres et al. 2017), barley (Samarah 2005), and maize (Daryanto et al. 2016). Generally, intensity and time period of drought decide the effect it will have on the quantity and quality of plant growth by influencing the availability and uptake of soil nutrients to the shoot and carbon assimilating channels of the roots through water (Selvakumar et al. 2012). Drought stress changes chlorophyll contents and damages the photosynthetic machinery of plants (Ortiz et al. 2015). Root growth is severely hindered by water deficit. Similar to the other environmental stresses such as salinity, heavy metal toxicity, nutrient deficiency/excess, and biological stress like pathogen attack, drought also results in plant growth impairment by causing hormonal imbalance, susceptibility to diseases, nutrient deficiency, and metal toxicity (Nadeem et al. 2014).

All the developmental stages of plant are vulnerable to drought stress and this can be visualized on both morphological and molecular levels. Plants usually have some inherent drought tolerance but its extent varies in both interspecific and intraspecific cases. Drought causes loss in turgor pressure inducing stomatal closure which limits carbon fixation and reduces cell growth (Taiz and Zeiger 2006). In extended water deficient conditions, alteration in carbon and nitrogen metabolism results in reduced photosynthesis by inactivation of RUBISCO and retarded plant growth (Bota et al. 2004). Water deficit condition in the rhizosphere leads to an increased rate of root respiration, an imbalance is created inside the cell in the utilization of carbon resources, reduced ATP production and enhanced generation of reactive oxygen species (ROS) like hydroxyl radicals (OH-), superoxide anion radicals (O^{-2}), singlet oxygen (O^2), hydrogen peroxide (H_2O_2), and alkoxy radicals (RO) initiate the oxidative stress (Munné-Bosch and Peñuelas 2004; Nair et al. 2008). ROS may react with proteins, DNA, and lipids causing oxidative damage and destroy the normal functioning of the cells (Foyer and Fletcher 2001). Other parameters that are negatively influenced by drought stress are chlorophyll content (Anjum et al. 2011; Rahdari and Hoseini 2012) and tissue nutrient content (Garg 2003). Drought also increases ethylene biosynthesis, which inhibits normal functioning of the plants (Ali et al. 2014). Therefore, in order to achieve the designated goal of producing more food, amelioration of drought stress is extremely important.

9.3 Drought-Induced Oxidative Stress

Plant metabolic activities result in production of reactive oxygen species (ROS) and plant get rid of these by certain ROS scavenging activities, thus the balance is maintained. However, when plant is exposed to various abiotic environmental stresses (including drought), rate of production of these ROS becomes higher and the plant suffers from oxidative stress (Sies 1997). ROS are forms of oxygen that are more reactive than molecular oxygen (O_2). Plants use combination of signalling and survival tactics to deal with a variety of stressful conditions. Since, Earth's atmosphere consists of 21% oxygen, production of toxic reactive oxygen species as result of metabolism is inevitable (Finkel and Holbrook 2000). ROS are frequently produced as by-products of various metabolic pathways, for example, during photosynthesis oxygen is generated in the chloroplasts which after accepting electrons passing through the photosystems forms superoxide radical (O2•−) which has two unpaired electrons with same spin making it unstable, reactive, and harmful for the cells (Navrot et al. 2007). However, ROS are not harmful under normal circumstances, in fact, they act as key signalling molecules for cellular responses to developmental and environmental stimuli under non-stressed conditions (Pei et al. 2000; Carol and Dolan 2006), thus are also important in general metabolism of plants. Various stress conditions result in disruption of the balance between the production and the removal of ROS leading to loss in crop productivity worldwide (Shi et al. 2001).

9.4 Consequences of Oxidative Stress in Plants

Plants have evolved several mechanisms that permit them to adapt and survive the period of abiotic stress, particularly drought. As discussed in the previous section, one of the impending consequences of drought stress is enhanced ROS production in plant cell. ROS reactive nature makes them extremely dangerous, especially singlet oxygen and the hydroxyl radicals, unlike atmospheric oxygen, they are capable of oxidizing important building block of cellular machinery such as proteins, fatty acids/lipids, and nucleic acids. Uninhibited oxidation of these biomolecules can ultimately result in cell demise. Lipid peroxidation (LPO) is one of the first indications that the cell is under oxidative duress, wherein, a reactive radical initiates a chain reaction, by removal of a hydrogen atom from a polyunsaturated lipid, creating and rearranging the double bonds and is terminated by antioxidants. It is shown that LPO results in the release of small hydrocarbon fragments such as small aldehydes like malondialdehyde (MDA), and hydroxyalkenal like 4-hydroxy-2-nonenal (HNE), etc., as end products from the polyunsaturated precursors. (Garg and Manchanda 2009). Excessive LPO makes membrane porous thus increases its permeability and causes modification of membrane bound proteins, these changes disrupt the bilipid membrane arrangement (Møller et al. 2007). Another side effect of oxidative stress is protein oxidation or ROS mediated irreversible covalent modification of protein either by addition of nitrosyl moiety (nitrosylation), carbon monoxide molecule (carbonylation), disulphide bond formation, addition of glutathione (glutathionylation) or by reacting with breakdown products of poly unsaturated fatty acid peroxidation (Ghezzi and Bonetto 2003; Yamauchi et al. 2008). It advertently leads to cell death, which is characterized by disruption of the cell membrane and cellular organelles.

9.5 Physiological and Biological Responses of Plants towards Drought Stress Under AM Symbiosis

AMF facilitate alleviation of detrimental effect caused by drought stress in plants. Rapparini and Peñuelas (2014) in their review have compiled all the mechanisms that are speculated to play role in AMF mediated drought stress tolerance in plants. Among these mechanisms, the most important mechanism is, increase in water absorbing surface area of the roots by producing intensive intraradical and extraradical mycelium (ERM) for better utilization and effective transportation of nutrition and water to the plants. ERM has great ability to mobilize essential nutrients like phosphorus (P), zinc (Zn), nitrogen (N), or copper (Cu) which can be readily absorbed by host plants with the help of arbuscules and vesicles (Smith and Read 2010). One more benefit of fungal hyphal network is that they can enhance soil architecture by forming aggregates through the production of the glycoprotein glomalin (Miransari et al. 2007; Singh et al. 2013). These soil aggregates increase the water holding capacity of the soil (Augé 2001; Ruiz-Lozano 2003). Another parameter of drought tolerance is root hydraulic conductivity (L), it is found to be reduced under drought stress, but mycorrhized plants were partially able to avoid this diminution (El-Mesbahi et al. 2012; Bárzana et al. 2012).

Osmotic adjustment is the most vital biochemical adaptation by plants to resist drought stress. It is achieved by the accretion of osmolytes/compatible solutes like proline, glycine betaine, aspartic acid, protein, and soluble sugars which have the ability to decrease osmotic potential, while, maintaining a high relative water content (RWC), thereby optimal cellular environment required for plant survival (Farooq et al. 2009). Wu and Zou (2017) found that AMF were able to strongly alter plant tissue's sucrose and proline metabolism by regulating key enzymes important for osmotic adjustment of the host plant. Proline being an excellent osmoprotectant increases the plant tolerance by maintaining the osmotic balance in plant cells. Extensive studies reported that AMF-colonization-mediated proline accumulation on exposure to water deficit leads to subsequent drought tolerance in comparison to non-mycorrhizal counterparts (Yooyongwech et al. 2013; Doubková et al. 2013; Rapparini and Peñuelas 2014). In a study, drought-exposed *Poncirus trifoliata* when inoculated with *Funneliformis mosseae* resulted in decreased tissue proline accumulation and improved its growth performance along with biomass production (Zou et al. 2013). Researchers suggested that the inhibition of proline synthesis along with enhanced proline degradation might have resulted in lower proline concentration in the AM plants (Zou et al. 2013).

Oxidative stress occurs when plant's antioxidant defence system is overpowered and no longer able to maintain cellular redox balance (Rapparini and Peñuelas 2014). However, mycorrhiza-inoculated plants were reported to induce varied levels of major antioxidants/ROS scavengers and which helps in regaining the optimal cellular redox status (Rapparini and Peñuelas 2014). The amelioration of drought stress by AM symbiosis is often ascribed to the augmentation of antioxidant activities in plants (Ruiz-Sánchez et al. 2010; Baslam and Goicoechea 2012; Wu and Zou

2017). Drought-exposed mycorrhized plants were reported to have higher glutathione level which was correlated with a lowered cellular H_2O_2, decreased membrane lipid peroxidation, and enhanced photosynthetic performance. Also, although glutathione levels were increased, ascorbate levels decreased in mycorrhizal plants in comparison to control plants. It suggests differential up-regulation of various antioxidant systems, with preferential activation of the systems that are more effective in protecting plants against drought (Ruiz-Sánchez et al. 2010). This suggests that mycorrhizal infection offers protection against drought-induced oxidative stress (Ruiz-Lozano 2003). However, it has been contradicted by some studies. These antioxidant compounds can also be viewed as markers of drought stress. Less accumulation of glutathione and ascorbate in mycorrhizal plants of lavender under drought conditions was correlated with a high level of resistance in lavender plants against drought (Marulanda et al. 2007). Flavonoids are also found to play a defensive role against drought in mycorrhiza-inoculated plants (Abbaspour et al. 2012). Other compounds such as isoprenoids, specific isoprenoid-derived apocarotenoids, and strigolactones also contribute as a shield against abiotic stresses such as drought (Rapparini et al. 2008; Rapparini and Peñuelas 2014; Walter and Strack 2011; López Ráez et al. 2008).

It has been studied by various researchers that AM colonization alleviates ROS mediated oxidative burst by triggering the antioxidant machinery inside their host plant cell (Abbaspour et al. 2012; Baslam and Goicoechea 2012). Ascorbic acid is a powerful natural scavenger of OH that lowers the accumulations of malondialdehyde (MDA) in cell membrane, thus stabilizes the membrane integrity (Foyer and Noctor 2005). Increase in ascorbate content during salt stress in mycorrhized citrus plant was reported by Wu et al. (2010). Zou et al. (2015) observed significantly higher super oxide dismutase (SOD) or catalase (CAT) activity in leaves and roots in AM than in non-AM trifoliate orange seedlings, irrespective of water status, except for CAT activity that did not show any significant difference in AM and non-AM seedlings under non-stressed condition. These results suggest that the presence of AMF could induce higher antioxidant enzyme activities in the seedlings as a protective mechanism to prevent over accumulation of ROS under drought stress. Similar observations were reported by Bompadre et al. (2014) in pomegranate plants where *Rhizophagus intraradices* resulted in considerably higher SOD and CAT activity in shoots under irrigated and drought conditions.

Reports have also indicated that AM symbiosis under drought conditions enhances photochemical efficiency of photosystem II, which was assessed by chlorophyll fluorescence in rice, tomato, and lettuce (Ruiz-Sánchez et al. 2010; Ruiz Lozano et al. 2016), and in woody tree species (Yooyongwech et al. 2013). Such results indicate an improved performance of photosynthetic apparatus along with minimal drought-imposed damage in mycorrhized plants. These findings are consistent with another study investigating the effect of root inoculation of different tree species with a combination of both AM and ecto-mycorrhizal fungi (Fini et al. 2011).

Multiple studies have reported enhanced nutrient acquisition by plants after AMF inoculation which mitigates drought-induced deficiency of important nutrients especially with low mobility like P, Zn, and Fe. In addition, concentration of other nutrients like Ca, K, Mg in inoculated plants is also found to be improved (Bagheri et al. 2012; Gholamhoseini et al. 2013). However, the most significant role of AMF is observed in improving P concentration in plants which subsequently enhanced drought tolerance in plants (Gholamhoseini et al. 2013). Inoculation of *F. Mosseae* in sunflower improved P availability and minimized the drought-impact on seed oil percentage and oil yield (Gholamhoseini et al. 2013). AM symbiosis is considered as the most common strategy in enhancing P availability in the soil or capacity of P uptake (Smith et al. 2011). Recent findings have provided new evidences for the contribution of two well-recognized pathways (roots and fungal hyphae) by which P can be absorbed in mycorrhized plants. These results suggest a pivotal role of 'hidden P uptake' into plants via the AM fungal pathway (Smith et al. 2011), including when mycorrhizal plants experience conditions of drought stress (Smith et al. 2010). Wu et al. (2013) found diminished AMF colonization in roots of several citrus plants under drought. However, enhancement of plant growth by AM was still observed. In addition to P, in trifoliate orange (*Poncirus trifoliata* L. Raf.) seedlings, *G. versiforme* colonization significantly improved the concentration of Ca, K, and Fe in plants under well-watered and water stressed conditions (Wu and Zou 2011). Similarly, Pistachio plants, inoculated with *F. mosseae* and *R. intraradices* mediated significant augmentation in P, K, Zn, and Mn concentrations, irrespective of soil moisture conditions (Bagheri et al. 2012).

Mariotte et al. (2017) emphasized that drought induces changes in the nitrogen (N) and phosphorus (P) cycle making their availability to the plants sparse. Most plant species are unable to cope with limited nutrient supply causing an imbalance in optimal N:P ratio. They suggested that the plants which are able to sustain this imbalance under drought condition will be more successful in enduring drought. In a study, they observed that *Cynodon dactylon* with stronger symbiotic association was able to maintain the constant N:P, in comparison to *Paspalum dilatatum* with weaker association under drought condition.

AMF can also alter host water regulation through hormonal signalling modulation. Abscisic acid (ABA) reportedly play role in controlling stomatal conductance and AM symbiosis is found to enhance ABA production which result in stomatal closure and thus ensuring limited water loss in drought-exposed plants (Ludwig-Müller 2010; Doubková et al. 2013). Another method is by stimulating osmolytes or compatible solutes such as photosynthetic by-products and soluble sugars, enhanced levels of free polyamines and soluble nitrogenous compounds as compared to non-AM plants (Ruiz-Sánchez et al. 2010; Fan and Liu 2011; Rapparini and Peñuelas 2014).

According to Augé (2001), AMF inoculation improves the P status of the plant cells, thus the stomatal closure caused on exposure to the water deficit might be because of the elevated P concentration in the leaves possibly by either affecting the energetics involved in osmotic adjustments of the guard cell or by increased cell wall rigidity controlling the stomatal movements. This is further supported by the enhancement of soil acid phosphatase activity resulted drought stress alleviation in AMF inoculated plants (Wu et al. 2011). Additionally, plant tissues of mycorrhized maize cultivars exhibited higher nitrate reductase (NR), glutamine synthetase (GS), and glutamine synthase (GOGAT) activities than non-AM tissues under water stress conditions. Bernardo et al. (2019) observed metabolic responses of two AMF inoculated wheat cultivars exposed to water deficit conditions and reported some variations in the compounds like sugar and lipids in inoculated and un-inoculated wheat, metabolites involved in oxidative stress management along with modulation of phytohormones, specifically brassinosteroids biosynthetic pathway.

9.6 Molecular Mechanisms Involved in Stress Alleviation by Arbuscular Mycorrhiza

So far, many studies have confirmed the significance of aquaporins in modulation of root hydraulic properties under drought stress (Ruiz-Lozano and Aroca 2010; Barzana et al. 2014). Several studies confirmed that AM-inoculated plants can conserve more water under prolonged water deficit conditions. It was observed that hydraulic properties of the roots under AM symbiosis were regulated by aquaporin levels and state of phosphorylation. Authors argued that down-regulation of the activity of these proteins may provide a better explanation for the changes during water deficit. Additionally, the drought-induced reduction in the transpiration rate (observed in *Phaseolus vulgaris* mycorrhizal plants) occurred in conjunction with an increased free exuded sap flow and a higher water uptake in AM plants (Aroca et al. 2007; Rapparini and Peñuelas 2014). In another study, transformed carrot roots were cultured with *Glomus intraradices* and on exposure to drought, showed elevated expression of two aquaporin genes (GintAQPF1 and GintAQPF2) (Li et al. 2013). Studies on 'aquaporins and AMF' signify that simultaneous regulation of both expression and activity of aquaporins in host plants and fungi may represent an AM-dependent mechanism for enhancing plant tolerance to drought stress (Rapparini and Peñuelas 2014). Bárzana et al. (2012) reported activation of increased apoplastic water transport pathway in the mycorrhizal roots along with 'cell to cell' pathway during drought stress. Potential of AM plants to coordinate two transport pathways has been hypothesized as a mechanism that plays an important role in adding flexibility in drought responses of AM-plants in comparison to non-AM-plants (Rapparini and Peñuelas 2014). Aquaporins are also involved in nutrient transport during plant–mycorrhiza association (Maurel and Plassard 2011). Activation of genes encoding a crucial element of ER-luminal binding protein (BiP) is among the speculated mechanisms involved in drought stress alleviation by mycorrhized plants (Kapoor et al. 2013). From *G. intraradices*, a BiP gene has been discovered from the fungus growing in-vitro under drought stress (Porcel et al. 2007). Under in-vitro and in-vivo conditions, BiP expression was up-regulated during drought stress under natural symbiosis with plants. Liu et al., (2015) also reported a coordinated response in mitogen-activated protein kinase (MAPK) transcripts in mycorrhized soybean plant roots and fungi, suggesting a potential molecular mechanism responsible for mycorrhiza induced drought stress response. Table 9.1. summarizes the recent significant advances in the knowledge of molecular mechanism lying behind the AM-induced plant drought tolerance.

Table 9.1 Molecular and physiological basis of AM induced drought tolerance

S. No.	Crop	AMF	Effect/Mechanism	Study
1	Citrus (*Poncirus trifoliata*)	*F. Mosseae*	Mycorrhized roots were found to be having higher root unsaturation index which might have been achieved by the modulation of genes such as *FA desaturase 2 (PtFAD2)*, *PtFAD6*, and *Δ9 FA desaturase (Pt Δ9)* under irrigated condition, while under water stressed conditions *PtFAD2*, *PtFAD6*, and *Δ15 FA desaturase (PtΔ15)* were activated	Wu et al. (2019)
2	Soyabean (*Glycine max*)	*Glomus intraradices*	Upregulation of MAPK transcripts in mycorrhizal soybean roots, suggesting the possibility of its potential role in communication between the two symbiotic partners	Liu et al. (2015)
3	Citrus (*P. trifoliata*)	*F. Mosseae*	Alternation in the gene expression profile of various tonoplast intrinsic proteins (TIPs) in mycorrhized plants indicating involvement of AMF in improving water absorption under drought	Jia-Dong et al. (2019)
4	Wheat (*Triticum aestivum*)	*F. Mosseae*	Effect on expression of proteins constituting the plant defence system against drought like in sugar metabolism, cytoskeletal organization, and Sulphur-containing proteins, as well as cell wall rearrangement	Bernardo et al. (2017)
5	Maize (*Zea mays*)	*Rhizophagus intraradices*	Expression of D-*myo*-inositol-3phosphate synthase (IPS) gene in association with 14–3GF gene was found to be responsible for the AM mediated drought tolerance in maize plants	Li et al. (2016)
6	Carrot (*Daucus carota*)	*R. Intraradices*	GintAQPF1 and GintAQPF2 aquaporin genes were found to be upregulated, thus, in turn, improving plant water relations under drought	Li et al. (2013)
7	Maize (*Z. mays*)	*R. Intraradices*	A wide number of aquaporin subfamily genes were regulated by presence of AMF and different water level	Barzana et al. (2014)
8	Trifoliate orange (*P. trifoliata*)	*F. Mosseae*	AM-inoculation elevated the synthesis of calmodulin in orange leaves and upregulated antioxidant enzyme activity	Huang et al. (2014)
9	Tomato (*Solanum lycopersicum*)	*F. Mosseae and R. intraradices*	Expression of LeEPFL9-STOMAGEN, and genes encoding intercellular signalling factors EPF1 and EPF2, which play role in stomatal development were found to be promoted by AM inoculation	Chitarra et al. (2016)
10	Citrus (*P. trifoliata*)	*F. Mosseae*	RT-PCR analysis revealed increase in mRNA abundance of genes involved in antioxidant activities like flavonoid biosynthase, lactoyl glutathione lyase, and oxido-reductase under water stress	Fan and Liu (2011)

9.7 Conclusions

Importance of AMF in combating drought stress is undeniable and it has also been demonstrated in multiple studies by researchers across the world. Still, there is lacuna of information about the mechanism(s) behind AM mediated drought stress amelioration. Next generation molecular techniques like Single Molecule Real Time (SMRT) and nanopore sequencing approaches may generate new insight in stress alleviation by mycorrhizal symbiosis. They can help in identification of target genes for promoting growth under stress. This information can be used to transfer target genes into plants through biotechnology and developing stress tolerant plants.

Acknowledgement Dr. Pallavi gratefully acknowledges Department of Science and Technology, Govt. of India for providing Inspire fellowship (IF130963).

References

Abbaspour H, Saeidi-Sar S, Afshari H, Abdel-Wahhab MA (2012) Tolerance of mycorrhiza infected Pistachio (*Pistacia vera* L.) seedling to drought stress under glasshouse conditions. J Plant Physiol 169:704–709

Ali S, Charles TC, Glick BR (2014) Amelioration of high salinity stress damage by plant growth promoting bacterial endophytes that contain ACC deaminase. Plant Physiol Biochem 80:160–167

Anjum SA, Xie XY, Wang LC, Saleem MF, Man C, Lei W (2011) Morphological, physiological and biochemical responses of plants to drought stress. Afr J Agric Res 6:2026–2032

Aroca R, Porcel R, Ruiz-Lozano JM (2007) How does arbuscular mycorrhizal symbiosis regulate root hydraulic properties and plasma membrane aquaporins in Phaseolus vulgaris under drought, cold or salinity stresses? New Phytol 173(4):808–816

Augé RM (2001) Water relations, drought and vesicular-arbuscular mycorrhizal symbiosis. Mycorrhiza 11(1):3–42

Bagheri V, Shamshiri MH, Shirani H, Roosta HR (2012) Nutrient uptake and distribution in mycorrhizal pistachio seedlings under drought stress. J Agr Sci Tech 14:1591–1604

Barea JM, Azcón-Aguilar C, Azcón R (2002) Interactions between mycorrhizal fungi and rhizosphere micro-organisms within the context of sustainable soil-plant systems. In: Multitrophic interactions in terrestrial systems. Cambridge University Press, Cambridge, pp 65–68

Barzana G, Aroca R, Bienert GP, Chaumont F, Ruiz-Lozano JM (2014) New insights into the regulation of aquaporins by the arbuscular mycorrhizal symbiosis in maize plants under drought stress and possible implications for plant performance. Mol Plant-Microb Interact 27:349–363

Bárzana G, Aroca R, Paz JA, Chaumont F, Martinez-Ballesta MC, Carvajal M, Ruiz-Lozano JM (2012) Arbuscular mycorrhizal symbiosis increases relative apoplastic water flow in roots of the host plant under both well-watered and drought stress conditions. Ann Bot 109(5):1009–1017

Baslam M, Goicoechea N (2012) Water deficit improved the capacity of arbuscular mycorrhizal fungi (AMF) for inducing the accumulation of antioxidant compounds in lettuce leaves. Mycorrhiza 22(5):347–359

Bernardo L, Carletti P, Badeck FW, Rizza F, Morcia C, Ghizzoni R, Rouphael Y, Colla G, Terzi V, Lucini L (2019) Metabolomic responses triggered by arbuscular mycorrhiza enhance tolerance to water stress in wheat cultivars. Plant Physiol Biochem 137:203–212

Bernardo L, Morcia C, Carletti P, Ghizzoni R, Badeck FW, Rizza F, Lucini L, Terzi V (2017) Proteomic insight into the mitigation of wheat root drought stress by arbuscular mycorrhizae. J Prot 169:21–32

Bompadre MJ, Silvani VA, Bidondo LF, Ríos de Molina MDC, Colombo RP, Pardo AG, Godeas AM (2014) Arbuscular mycorrhizal fungi alleviate oxidative stress in pomegranate plants growing under different irrigation conditions. Botany 92(3):187–193

Bota J, Medrano H, Flexas J (2004) Is photosynthesis limited by decreased Rubisco activity and RuBP content under progressive water stress? New Phytol 162:671–681

Carol RJ, Dolan L (2006) The role of reactive oxygen species in cell growth: lessons from root hairs. J Exp Bot 57:1829–1834

Chitarra W, Pagliarani C, Maserti B, Lumini E, Siciliano I, Cascone P, Schubert A, Gambino G, Balestrini R, Guerrieri E (2016) Insights on the impact of arbuscular mycorrhizal symbiosis on tomato tolerance to water stress. Plant Physiol 171(2):1009–1023

Daryanto S, Wang L, Jacinthe PA (2016) Global synthesis of drought effects on maize and wheat production. PLoS One 11(5):e0156362

Doubková P, Vlasáková E, Sudová R (2013) Arbuscular mycorrhizal symbiosis alleviates drought stress imposed on *Knautia arvensis* plants in serpentine soil. Plant Soil 370(1–2):149–161

EEA (2015). https://www.eea.europa.eu/data-and-maps

El-Mesbahi MN, Azcón R, Ruiz-Lozano JM, Aroca R (2012) Plant potassium content modifies the effects of arbuscular mycorrhizal symbiosis on root hydraulic properties in maize plants. Mycorrhiza 22(7):555–564

Fan QJ, Liu JH (2011) Colonization with arbuscular mycorrhizal fungus affects growth, drought tolerance and expression of stress-responsive genes in *Poncirus trifoliata*. Acta Physiol Plant 33:1533–1531

Farooq M, Hussain M, Siddique KH (2014) Drought stress in wheat during flowering and grain-filling periods. Crit Rev Plant Sci 33(4):331–349

Farooq M, Wahid A, Kobayashi N, Fujita D, Basra SMA (2009) Plant drought stress: effects, mechanisms and management. In: Sustainable agriculture. Springer, Dordrecht, pp 153–188

Fini A, Frangi P, Amoroso G, Piatti R, Faoro M, Bellasio C, Ferrini F (2011) Effect of controlled inoculation with specific mycorrhizal fungi from the urban environment on growth and physiology of containerized shade tree species growing under different water regimes. Mycorrhiza 21(8):703–719

Finkel T, Holbrook NJ (2000) Oxidants, oxidative stress and the biology of ageing. Nature 408:239–247

Foyer CH, Fletcher JM (2001) Plant antioxidants: colour me healthy. Biologist (London, England) 48:115

Foyer CH, Noctor G (2005) Oxidant and antioxidant signalling in plants: a re-evaluation of the concept of oxidative stress in a physiological context. Plant, Cell and Environment 28(8):1056–1071

Garg BK (2003) Nutrient uptake and management under drought: nutrient-moisture interaction. Curr Agric 27:1–8

Garg N, Manchanda G (2009) ROS generation in plants: boon or bane? Plant Biosys 143:81–96

Ghezzi P, Bonetto V (2003) Redox proteomics: identification of oxidatively modified proteins. Proteomics 3:1145–1153

Gholamhoseini M, Ghalavand A, Dolatabadian A, Jamshidi E, Khodaei-Joghan A (2013) Effects of arbuscular mycorrhizal inoculation on growth, yield, nutrient uptake and irrigation water productivity of sunflowers grown under drought stress. Agric Water Manag 117:106–114

Huang YM, Srivastava AK, Zou YN, Ni QD, Han Y, Wu QS (2014) Mycorrhizal-induced calmodulin mediated changes in antioxidant enzymes and growth response of drought-stressed trifoliate orange. Front Microbiol 5:682

IPCC (2007) Climate change 2007: the physical science basis. In: Solomon SD (ed) Contribution of Working Group I to the fourth assessment report of the Intergovernmental Panel on Climate Change

Jia-Dong H, Tao D, Hui-Hui W, Zou YN, Wu QS, Kamil K (2019) Mycorrhizas induce diverse responses of root TIP aquaporin gene expression to drought stress in trifoliate orange. Sci Horti 243:64–69

Kapoor R, Evelin H, Mathur P, Giri B (2013) Arbuscular mycorrhiza: approaches for abiotic stress tolerance in crop plants for sustainable agriculture. In: Plant acclimation to environmental stress. Springer, New York, pp 359–401

Korres NE, Norsworthy JK, Burgos NR, Oosterhuis DM (2017) Temperature and drought impacts on rice production: an agronomic perspective regarding short-and long-term adaptation measures. Water Resour Rural Dev 9:12–27

Li T, Hu YJ, Hao ZP, Li H, Chen BD (2013) Aquaporin genes GintAQPF1 and GintAQPF2 from *Glomus intraradices* contribute to plant drought tolerance. Plant Signal Behav 8(5):e24030

Li T, Sun Y, Ruan Y, Xu L, Hu Y, Hao Z, Zhang X, Li H, Wang Y, Yang L, Chen B (2016) Potential role of D-myo-inositol-3-phosphate synthase and 14-3-3 genes in the crosstalk between *Zea mays* and *Rhizophagus intraradices* under drought stress. Mycorrhiza 26 (8):879–893

Liu Z, Li Y, Ma L, Wei H, Zhang J, He X, Tian C (2015) Coordinated regulation of arbuscular mycorrhizal fungi and soybean MAPK pathway genes improved mycorrhizal soybean drought tolerance. Mol Plant-Microbe Interact 28(4):408–419

López Ráez JA, Charnikhova T, Gómez Roldán V, Matusova R, Kohlen W, De Vos R, Verstappen F, Puech Pages V, Bécard G, Mulder P, Bouwmeester H (2008) Tomato strigolactones are derived from carotenoids and their biosynthesis is promoted by phosphate starvation. New Phytol 178(4):863–874

Ludwig-Müller J (2010) Hormonal responses in host plants triggered by arbuscular mycorrhizal fungi. In: Arbuscular mycorrhizas: physiology and function. Springer, Dordrecht, pp 169–190

Mariotte P, Canarini A, Dijkstra FA (2017) Stoichiometric N: P flexibility and mycorrhizal symbiosis favour plant resistance against drought. J Ecol 105(4):958–967

Marulanda A, Porcel R, Barea JM, Azcón R (2007) Drought tolerance and antioxidant activities in lavender plants colonized by native drought-tolerant or drought-sensitive Glomus species. Microb Ecol 54(3):543

Maurel C, Plassard C (2011) Aquaporins: for more than water at the plant-fungus interface? New Phytol 190(4):815–817

Miransari M, Bahrami HA, Rejali F, Malakouti MJ, Torabi H (2007) Using arbuscular mycorrhiza to reduce the stressful effects of soil compaction on corn (*Zea mays* L.) growth. Soil Biol Biochemist 39(8):2014–2026

Møller IM, Jensen PE, Hansson A (2007) Oxidative modifications to cellular components in plants. Annu Rev Plant Biol 58:459–481

Munné-Bosch S, Peñuelas J (2004) Drought-induced oxidative stress in strawberry tree (*Arbutus unedo* L.) growing in Mediterranean field conditions. Plant Sci 166(4):1105–1110

Nadeem SM, Ahmad M, Zahir ZA, Javaid A, Ashraf M (2014) The role of mycorrhizae and plant growth promoting rhizobacteria (PGPR) in improving crop productivity under stressful environments. Biotechnol Adv 32:429–448. https://doi.org/10.1016/j.biotechadv.2013.12.005

Nair AS, Abraham TK, Jaya DS (2008) Studies on the changes in lipid peroxidation and antioxidants in drought stress induced cowpea (*Vigna unguiculata* L.) varieties. J Environ Biol 29 (5):689–691

Navrot N, Rouhier N, Gelhaye E, Jacquot JP (2007) Reactive oxygen species generation and antioxidant systems in plant mitochondria. Physiol Plant 129:185–195

Ortiz N, Armada E, Duque E, Roldan A, Azcon R (2015) Contribution of arbuscular mycorrhizal fungi and/or bacteria to enhancing plant drought tolerance under natural soil conditions: effectiveness of autochthonous or allochthonous strains. J Plant Physiol 174:87–96

Pei ZM, Murata Y, Benning G, Thomine S, Klusener B, Allen GJ, Grill E, Schroeder JI (2000) Calcium channels activated by hydrogen peroxide mediate abscisic acid signalling in guard cells. Nature 406:731–734

Porcel R, Aroca R, Cano C, Bago A, Ruiz-Lozano JM (2007) A gene from the arbuscular mycorrhizal fungus *Glomus intraradices* encoding a binding protein is up-regulated by drought stress in some mycorrhizal plants. Environ Exp Bot 60(2):251–256

Rahdari P, Hoseini SM (2012) Drought stress: a review. Intl J Agron Plant Prod 3:443–446

Rapparini F, Peñuelas J (2014) Mycorrhizal fungi to alleviate drought stress on plant growth. In: Use of microbes for the alleviation of soil stresses, vol 1. Springer, New York, pp 21–42

Rapparini F, Llusia J, Penuelas J (2008) Effect of arbuscular mycorrhizal (AM) colonization on terpene emission and content of *Artemisia annua* L. Plant Biol 10:108–122

Ruiz-Lozano JM (2003) Arbuscular mycorrhizal symbiosis and alleviation of osmotic stress. New perspectives for molecular studies. Mycorrhiza 13(6):309–317

Ruiz-Lozano JM, Aroca R (2010) Modulation of aquaporin genes by the arbuscular mycorrhizal symbiosis in relation to osmotic stress tolerance. In: Symbioses and stress. Springer, Dordrecht, pp 357–374

Ruiz Lozano JM, Aroca R, Zamarreño ÁM, Molina S, Andreo Jiménez B, Porcel R, García Mina JM, Ruyter Spira C, López Ráez JA (2016) Arbuscular mycorrhizal symbiosis induces strigolactone biosynthesis under drought and improves drought tolerance in lettuce and tomato. Plant Cell Environ 39(2):441–452

Ruiz-Sánchez M, Aroca R, Muñoz Y, Polón R, Ruiz-Lozano JM (2010) The arbuscular mycorrhizal symbiosis enhances the photosynthetic efficiency and the antioxidative response of rice plants subjected to drought stress. J Plant Physiol 167(11):862–869

Samarah NH (2005) Effects of drought stress on growth and yield of barley. Agron Sustain Dev 25:145–149

Selvakumar G, Panneerselvam P, Ganeshamurthy AN (2012) Bacterial mediated alleviation of abiotic stress in crops. In: Bacteria in agrobiology: stress management. Springer, Berlin, pp 205–224

Shi WM, Muramoto Y, Ueda A, Takabe T (2001) Cloning of peroxisomal ascorbate peroxidase gene from barley and enhanced thermo-tolerance by overexpressing in *Arabidopsis thaliana*. Gene 273:23–27

Sies H (1997) Oxidative stress: oxidants and antioxidants. Exp Physiol 82:291–295

Singh PK, Singh M, Tripathi BN (2013) Glomalin: an arbuscular mycorrhizal fungal soil protein. Protoplasma 250(3):663–669

Smith SE, Read DJ (2010) Mycorrhizal symbiosis. Academic press, Amsterdam

Smith SE, Facelli E, Pope S, Smith FA (2010) Plant performance in stressful environments: interpreting new and established knowledge of the roles of arbuscular mycorrhizas. Plant Soil 326(1–2):3–20

Smith SE, Read DJ (2008) Mycorrhizal symbiosis, 3rd edn. Academic Press, New York. ISBN, 440026354, 605

Smith SE, Jakobsen I, Grønlund M, Smith FA (2011) Roles of arbuscular mycorrhizas in plant phosphorus (P) nutrition: interactions between pathways of P uptake in arbuscular mycorrhizal (AM) roots have important implications for understanding and manipulating plant P acquisition. Plant Physiol 156(3):1050–1057

Taiz L, Zeiger E (2006) Plant physiology, 4th edn. Massachusetts, Sinauer Associates Inc

Walter MH, Strack D (2011) Carotenoids and their cleavage products: biosynthesis and functions. Nat Prod Rep 28(4):663–692

Wu QS, Zou YN (2011) Citrus mycorrhizal responses to abiotic stresses and polyamines. Adv Plant Physiol 12:31–56

Wu QS, Zou YN (2017) Arbuscular mycorrhizal fungi and tolerance of drought stress in plants. In: Arbuscular Mycorrhizas and Stress Tolerance of Plants. Springer, Singapore, pp 25–41

Wu QS, He JD, Srivastava AK, Zou YN, Kuča K (2019) Mycorrhiza enhances drought tolerance of citrus by altering root fatty acid compositions and their saturation levels. Tree Physiol 39:1149–1158

Wu QS, Srivastava AK, Zou YN (2013) AMF-induced tolerance to drought stress in citrus: a review. Sci Horti 164:77–87

Wu QS, Zou YN, He XH (2011) Differences of hyphal and soil phosphatase activities in drought-stressed mycorrhizal trifoliate orange (*Poncirus trifoliata*) seedlings. Sci Horti 129(2):294–298

Wu QS, Zou YN, Liu W, Ye XF, Zai HF, Zhao LJ (2010) Alleviation of salt stress in citrus seedlings inoculated with mycorrhiza: changes in leaf antioxidant defense systems. Plant Soil Environ 56(10):470–475

Yamauchi Y, Furutera A, Seki K, Toyoda Y, Tanaka K, Sugimoto Y (2008) Malondialdehyde generated from peroxidized linolenic acid causes protein modification in heat-stressed plants. Plant Physiol Biochem 46:786–793

Yooyongwech S, Phaukinsang N, Cha-um S, Supaibulwatana K (2013) Arbuscular mycorrhiza improved growth performance in *Macadamia tetraphylla* L. grown under water deficit stress involves soluble sugar and proline accumulation. Plant Growth Regul 69(3):285–293

Zou YN, Huang YM, Wu QS, He XH (2015) Mycorrhiza-induced lower oxidative burst is related with higher antioxidant enzyme activities, net H_2O_2 effluxes, and Ca^{2+} influxes in trifoliate orange roots under drought stress. Mycorrhiza 25(2):143–152

Zou YN, Wang P, Liu CY, Ni QD, Zhang DJ, Wu QS (2017) Mycorrhizal trifoliate orange has greater root adaptation of morphology and phytohormones in response to drought stress. Sci Rep 7:41134

Zou YN, Wu QS, Huang YM, Ni QD, He XH (2013) Mycorrhizal-mediated lower proline accumulation in *Poncirus trifoliata* under water deficit derives from the integration of inhibition of proline synthesis with increase of proline degradation. PLoS One 8(11):e80568

Chapter 10
Drought Stress Tolerance: An Insight to Resistance Mechanism and Adaptation in Plants

Priyanka Khati, Pankaj Kumar Mishra, Manoj Parihar, Ashish Kumar Singh, Jaideep Kumar Bisht, and Arunava Pattanayak

Abstract Water is essentially required by all the living systems including plants to maintain normal function of growth and metabolism. Drought stress in plants may affect several activities like cell elongation, morphology, nutrient availability due to the production of ROS and photo oxidation. Present study discusses various aspects of drought stress in plants and possible solutions to improve the tolerance. Application of plant growth promoting bacteria to ameliorate drought stress is undeniable in addition to several other ways to fight drought stress. Endophytic fungi are also reported to improve the stress management in plants as they colonize the plant roots and impart several benefits by providing essential metabolites related to resistance and development. Genetic engineering has always been a solution to stress management through the insertion or deletion of required genes. However, nanotechnology also offers ways to tolerate drought in plants as it improves the water use efficiency for plants and associated microbes. Future basic research at molecular level is still needed to explore the exact mechanisms lying beneath the stress tolerance strategies in plants. This may help in developing proper understanding between drought and plant metabolites and its application in future research.

10.1 Introduction

Global climatic changes accelerate the occurrence of a variety of abiotic (drought, salinity, heavy metals, and extreme temperatures) and biotic stresses (phytopathogens), which considerably affect agricultural productivity and bioremediation efficiency of plants even in forest ecosystems. Water is the basic requirement of plants which helps in maintaining normal function and turgor pressure to perform growth and metabolism. Drought, among all the other environmental stresses, is one of the major limiting factor for plant health and productivity worldwide. Drought stress

P. Khati · P. K. Mishra (✉) · M. Parihar · A. K. Singh · J. K. Bisht · A. Pattanayak
ICAR-Vivekananda Parvatiya Krishi Anusandhan Sansthan, Almora, Uttarakhand, India
e-mail: pankaj.mishra@icar.gov.in

A. Sharma (ed.), *Microbes and Signaling Biomolecules Against Plant Stress*, Rhizosphere Biology, https://doi.org/10.1007/978-981-15-7094-0_10

arises in plants due to inadequate water supply during agricultural practices (Jaleel et al. 2007) and it is expected to cause serious problems in crop productivity in more than 50% of the arable lands by 2050 (Ashraf 1994; Kasim et al. 2013).

Negative Impact of Drought Stress on Plants

(a) Water stress may affect the plant growth as the turgor pressure is responsible for cell elongation.
(b) Drought may also affect the nutrient availability to plants which depends upon the availability of water.
(c) Increase in reactive oxygen species during drought stress may damage several cell functions and structures.
(d) The photo oxidation caused by drought may decrease chlorophyll content (Rahdari et al. 2012).
(e) Reduced growth due to obstructed cell elongation and impaired cell mitosis.
(f) Affects photosynthesis due to reduced leaf area and decrease in chlorophyll content.

Crop production in the rainfed region (~60%) of North West Himalaya is still a challenge due to insufficient utilization of water for irrigation (Sharma et al. 2017). Despite of adequate rainfall, this area shows water shortage for 3–4 confined months because of runoff water due to steep slopes. According to Gupta (2006), water availability in India is likely to be reduced to 1500 m^3/annum/person by 2025. Root development is affected by the availability of water (soil water deficit), however, root growth is less affected in comparison to aerial parts, so increase in overall root to shoot ratio is mostly observed. Root texture is mainly affected during drought as reduction of root weight in wheat, without an appreciable decline in their number was observed. Roots were fine and fibrous under dry conditions in comparison to favourable moisture conditions (Ashraf 1994). Seedling growth is affected by water scarcity as the turgor pressure which helps in cell elongation is absent. However, the root size and ratio of root/shoot improved under dry conditions. Coleoptiles growth is inhibited due to the reduction of water supply from the root to coleoptile under drought stress (Monayeri et al. 1984). Eziz et al. (2017) stated that drought significantly enhanced the root mass while creating negative impact on the leaf, stem, and reproductive mass of the plant.

10.1.1 Signalling Pathways During Drought Stress

(a) *Abscisic acid (ABA) dependent pathway*: In plants, drought response is observed in the operation of ABA pathway which synthesizes ABA with the help of enzyme 9-*cis*-epoxycarotenoid dioxygenase (NCED). Once ABA is synthesized it activates ABA-responsive *cis*-element binding protein/ABA-responsive *cis*-element binding factor (AREB/ABF) which in turns activates ABA-responsive *cis*-element (ABER). ABER gets accumulated in the promoter region of drought responsive genes and upregulates their transcription.

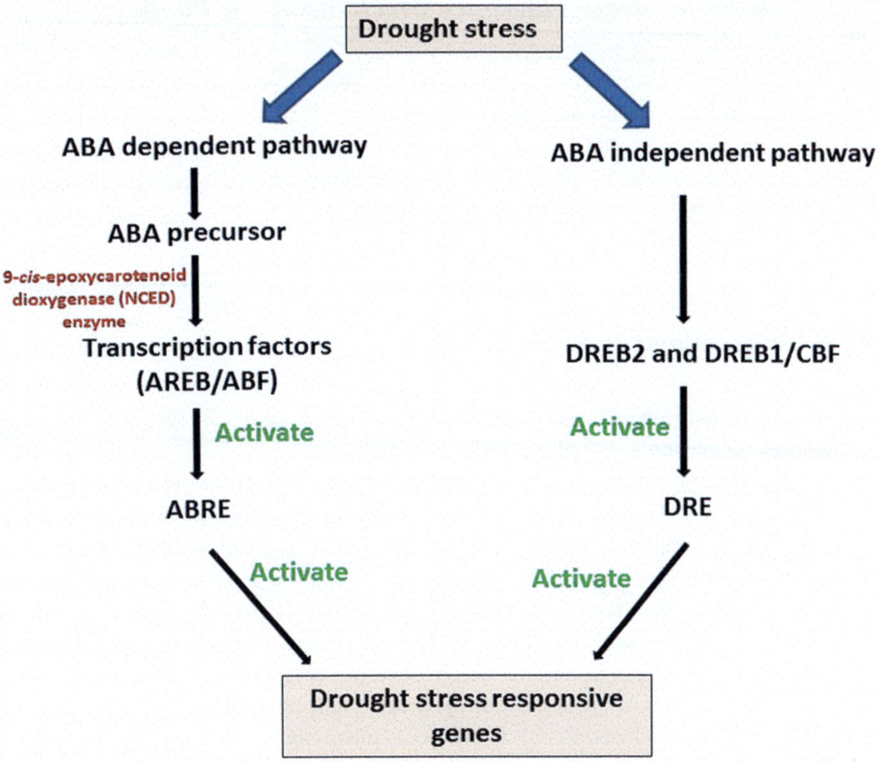

Fig. 10.1 Signalling pathway involved in drought stress tolerance

(b) *ABA independent*: In the absence of ABA, drought stress is controlled by cold-binding factor/dehydration responsive (DREB2 and DEB1/CBF) transcription factors. These transcription factors through dehydration responsive element (DRE) regulate drought stress-responsive genes (Todaka et al. 2015). The signalling mechanism is shown in Fig. 10.1.

Plant tolerance to water stress may occur either by drought avoidance or drought tolerance mechanisms (Blum 2005, 2009). Drought avoidance is the maintenance of high water status or water conservation under water deficit conditions which promotes water use efficiency in plants (Karaba et al. 2007). Drought tolerance is the ability of one crop genotype to yield better than the other in a drought state (Blum 2005). According to Blum (2005) it depends on the following factors:

(a) Enhanced capture of soil moisture.
(b) Reduced water use.
(c) Osmotic adjustment and conservation of cellular water content.

10.1.2 Drought Stress Tolerance Mechanisms in Plants

Plants have several adaptation mechanisms to tolerate drought stress by secreting different metabolites, which may help in management of drought stress at different stages. Number of molecular mechanisms act for physical action of plants during drought stress. These mechanisms help in maintaining normal growth and metabolism of plants. Some of these are discussed below.

10.1.3 Physiological Traits

(a) Several metabolites like polyols, xylitol, and sorbitol, tyrosine, 2-hydroxypalmitate, flavonoid, vannilate, ϒ-glutamylglutamate, ϒ-glutamylphenylalanine, and ϒ-glutamylvaline from glutathione metabolism, erythritol, arabitol, glycerol, inositol 1-phosphate, and myo-inositol are affected during drought stress (Yobi et al. 2013; Pampurov and Dijck 2014).
(b) Signalling pathway induced by jasmonate (JA) is also found responsible for stress tolerance in plants (Ahmad et al. 2016). They showed role of JA (a signalling molecule) in plants under stress and non-stress condition. JA mediated stress response is caused by several cross talk molecules like nitric oxide, calcium, ABA, ethylene, salicylic acid, and reactive oxygen species.
(c) Curling of leaves fronds and folding of plasmalema to prevent water loss (Pampurov and Dijck 2014).
(d) Arrangement of organelle packaging (Pampurov and Dijck 2014).
(e) Accumulation of osmoprotectants like proline is very crucial step for water stress tolerance. These protectants help plants to maintain normal function of life even in extreme stress. Increase in proline content decreases the water potential of plant and reduces the water loss (Wang et al. 2019).
(f) Controlling stomatal conductance for water loss through transpiration (Taiz and Zeiger 1991; Pietragalla and Alistair 2012).
(g) Accumulation of antioxidants {superoxide dismutase (SOD), peroxidase (POD) e.g., guaiacol peroxidase and ascorbate peroxidase and catalase (CAT)} (Yuan et al. 2016) and reactive oxygen species {superoxide anion ($O_2.-$), singlet oxygen ($^-O_2$)}, and hydroxyl anion (OH^-), while non-radical forms include hydrogen peroxide (H_2O_2) (Sharma et al. 2012).
(h) Genes for antioxidant enzymes are induced by nitrous oxide (Siddiqui et al. 2011). Zhao et al. (2001) observed an increase in NOS-like activities in wheat seedlings under drought stress, and ABA accumulation was inhibited by NOS inhibitor. ROS and NO may be signal molecules by which the plants sense the drought stress, and may participate in the leaf water maintenance by stimulating ABA synthesis.
(i) Accumulation of potassium plays a central role as it acts as an osmolyte, aiding osmotic adjustment in plants (Kant and Kafkafi 2002; Marschner 2011).

(j) 2R, 3R-butanediol, a bacterial volatile induces systemic tolerance to drought stress by reducing the stomata aperture (Cho et al. 2008).

10.1.4 Molecular Mechanisms

The response shown by the plants during stress management is mainly regulated and controlled at molecular level. Stress factors cause the alteration of different stress specific genes and enzyme activities to cope up with stress conditions. A wide array of genes and gene products associated with stress response have been identified using experimental approaches.

(a) N-terminus Glycine Rich Ribosomal binding protein (Nt-GR-RBP 1) affects m RNA binding protein function (Jabeen et al. 2017). Ribosome inactivating proteins which remove adenine from rRNA control the drought stress at translation level (Jiang et al. 2008).
(b) Translationally controlled tumour protein (TCTP) and proliferating cells nuclear antigen (PCNA) induced systemic resistance in plants interact with DNA polymerase and enhance its fidelity and efficiency (Ghabooli et al. 2013).
(c) Late embryogenesis abundant proteins (LEA proteins) are one of the major groups of proteins upregulated during stress tolerance to help in survival of plants during extreme water loss (Battaglia et al. 2008). LEA proteins are responsible for protection of cellular and molecular structures including dehydration, buffering, ion sequestration, ROS scavenging, and metal ion-binding properties (Goyal et al. 2005; Wu et al. 2014).
(d) Wheat boiling stable stress-responsive protein (wBsSRP) was isolated from drought tolerant cultivar of wheat (PBW 175) under drought conditions (Rakhra et al. 2017).
(e) AtMYC2/JIN1, encoding the basic helix-loop-helix, leucine zipper transcription factor, is not only involved in ABA-mediated drought stress signalling pathway, but also in other wound response and pathogen defence responses. Over expression of AtMYC2 encodes ABA-mediated drought stress signalling (Wu et al. 2009).

10.2 Exogenous Solutions to Drought Stress

10.2.1 Plant Growth Promoting Rhizobacteria (PGPR)

Plant growth promoting rhizobacteria (PGPR) are well known bacterial species which help in plant growth promotion in addition to elicit drought stress response in plants through alteration in biochemical parameters (Grover et al. 2011). These PGPR colonize the roots of plants and modulate the secretion of various hormones which help them tolerate different stress. Jaleel et al. (2007) investigated the role of plant growth promoting bacteria (PGPB) in combating drought stress of plants and suggested the role of *Pseudomonas fluorescens* which induced the production of ajmalicine in *Catharanthus roseus* under drought stress. They observed enhanced plant biomass and alkaloid content in *Pseudomonas* treated seedlings. Similarly, Garcíaa et al. (2017) also studied the effect of *Azospirillum* on the growth of maize under drought stress and observed that inoculation of seeds with the PGP strain resulted in better response to water deficit conditions. Polyamines are known to impart abiotic stress tolerance to plants. Increase in the level of 3 major polyamines, i.e. putrescine, spermidine, and spermine in *Arabidopsis thaliana* was observed after the influence of plant growth promoting *Pseudomonas putida* strain by Sen et al. (2018). These PGPB, known to modify phytohormonal activities in plants (Egamberdieva 2013), reduce the production of volatiles (Timmusk et al. 2014) and thus improve photosynthesis and produce antioxidants, osmolytes and exo-polysaccharides which help plants to combat stress like drought. PGPR have been reported to improve drought tolerance in *Arabidopsis* (Timmusk and Wagner 1999), pepper (Mayak et al. 2004), wheat (Yandigeri et al. 2012), tomato (Mayak et al. 2004), and pea (Belimov et al. 2009). PGPRs under stress trigger several mechanisms in plants like phytohormone synthesis, inhibition of plant pathogens and induction of systemic resistance (Sathya et al. 2017). Yang et al. (2009) proposed the term "induced systemic tolerance" (IST) to refer defence-related physical and chemical changes in plants induced by microbes under abiotic stresses. Certain phytohormones, such as indole-3-acetic acid (IAA), cytokinin (CTK), and gibberellin can be synthesized by PGPRs as well as by plants. Filamentous bacteria (Actinobacteria) might help improve the resilience of plants under water (Kohler et al. 2008; Yandigeri et al. 2012). *Streptomyces* species reportedly can produce 1-aminocyclopropane-1-carboxylate (ACC) deaminase, IAA, and siderophores, which can improve plant growth under stress conditions (Palaniyandi et al. 2014). The mechanisms underlying the drought tolerance-promoting effects of *Streptomyces* sp. in plants have been documented which include increasing the osmotic pressure in plant cells, callose accumulation, and lignification of the cell walls (Hasegawa et al. 2007, 2008).

Tolerance to drought was correlated with reduced water loss in *P. chlororaphis* O6-colonized plants and with stomatal closure, indicated by size of stomatal aperture and percentage of closed stomata. Stomatal closure and drought resistance were mediated by production of 2R, 3R-butanediol, and production of salicylic acid (SA)-, ethylene-, and jasmonic acid-signalling pathways. Increase in SA during drought stress also suggests the primary role of SA in management of drought stress (Cho et al. 2008). The mechanism of drought stress tolerance by PGPRs can be categorized by the following mechanisms:

(a) The PGPR secrete plant growth promotory substances (IAA and siderophores) that help plant to continue normal metabolism during stress (Egamberdieva 2013).
(b) Secretion of polyamines like putrescine, spermidine, and spermine is another mechanism to control drought stress (Sen et al. 2018).

(c) Induced systemic resistance (ISR), a combat mechanism of plant is induced by the microbial colonizers of plants.
(d) The phytohormone, ACC is secreted by PGPR that inhibits the conversion of ACC to ethylene, which is responsible to shut down the normal mechanism of plants during stress.
(e) The PGP are also known to secrete osmolytes and exopolysaccharides which protect the plant from environmental stress (Zhang et al. 2018a, b).
(f) 2R, 3R butanediol are the chemical compounds, which in addition to plants are also secreted by PGPRs and found responsible for stomata closure and hence reduced water loss.
(g) The microbes are also responsible for increasing the osmotic pressure in plant cells through callose accumulation and lignifications of the cell walls (Hasegawa et al. 2008). This mechanism thickens the cell wall and thus reduces the water loss.
(h) Hairpin proteins produced by several gram-negative bacteria induce defence response against pathogens and drought (Liu et al. 2016).

Extensive studies are still needed to explore more potential microbes which can address the drought stress management in plants.

10.2.2 Fungal Inoculants in the Form of Endosymbiont or Arbuscular Mycorrhiza (AM)

Ghabooli et al. (2013) studied the role of endophytic fungus *Piriformospora indica,* which imparts several plant health benefits and maintains productivity during drought conditions of Sebacinales. This endophytic fungus promotes plant growth and resistance to abiotic stress, including drought. Zhang et al. (2018a, b) studied the effect of *Piriformospora indica* inoculation on maize growth under drought stress. The maize after colonization showed improved leaf size, root length, and number of tap roots. Gene ontology reveals the role of carbon and sulphur metabolic pathways as major target. Higher transcripts for microtubule associated processes improved oxidative potential of roots and stimulated genes for hormone functions, including those which respond to abscisic acid, auxin, salicylic acid, and cytokinins.

Fungal symbionts have been found to be associated with every plant studied in the natural ecosystem, where they colonize and reside entirely or partially in the internal tissues of their host plants. Fungal endophytes can form a range of different relationships with different hosts including mutualistic, symbiotic, commensalistic, and parasitic in response to host genotype and environmental factors (Singh et al. 2011). In mutualistic association fungal endophyte can enhance growth, increase reproductive success, and confer biotic and abiotic stress tolerance to its host plant. Since abiotic stress such as, drought, high soil salinity, cold, heat, oxidative stress, and heavy metal toxicity are some common adverse environmental conditions that affect and limit crop productivity worldwide. It may be a promising alternative strategy to exploit fungal endophytes to overcome the limitations to crop production brought by abiotic stress. There is an increasing interest in developing the potential biotechnological applications of fungal endophytes for improving plant stress tolerance and sustainable production of food crops (Singh et al. 2011).

10.2.3 Nanotechnology

According to Prasad et al. (2017), nanotechnology may revolutionize agriculture by innovative new techniques. Research in the area of application of nanotechnology in agriculture to reduce the water requirement for irrigation can be carried out to address drought stress in rainfed area of hills (Fig.10.2).

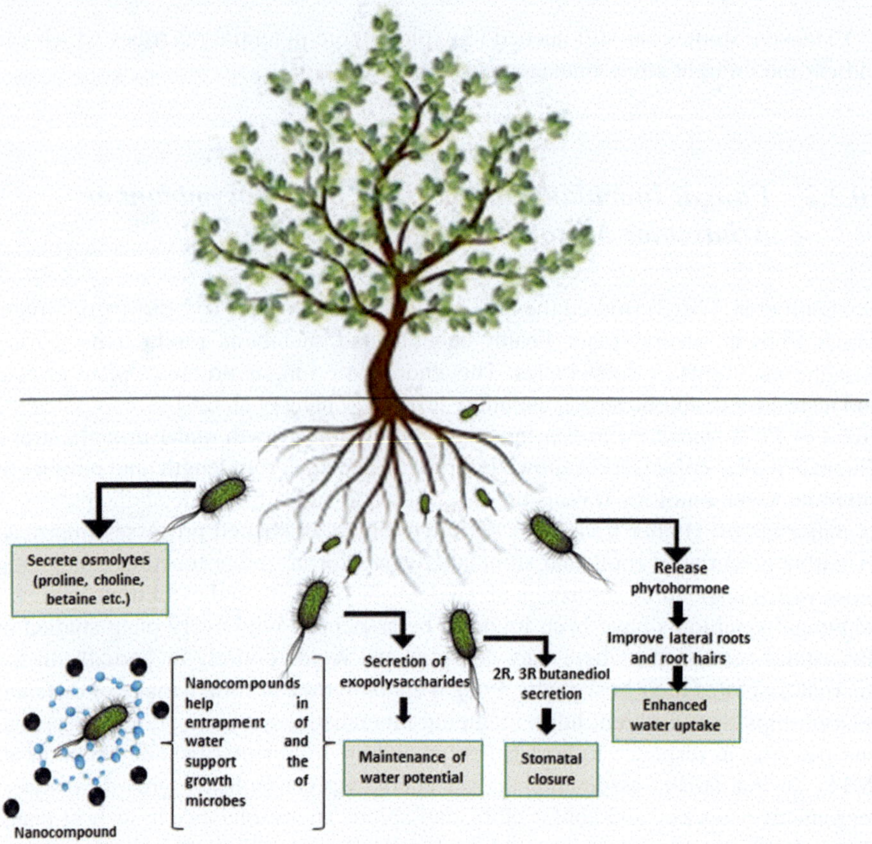

Fig. 10.2 Impact of nanocompounds and microbial inoculants on drought stress tolerance mechanisms of plants

Dimkpa et al. (2017) assessed the micronutrient nanoparticles (ZnO, B_2O_3, and CuO) for their role in improving drought stress and observed positive results. They suggested the application of nanoparticles boosted the crop performance in drought stress. Bioformulations are already known for their potential to improve plant productivity, but due to the shelf life issue they still lack behind the chemical fertilizers. Meagre information is available regarding the shelf life of microbial population (especially vegetative cells of *Pseudomonas, Azospirillum,* etc.) after 6 months in nano-bioformulation, which brings the necessity for the research in this area. The soil based nanocompounds (nanoclay, nanozeolite) improve the soil structure and microbiome due to their good water and nutrient holding capacity (Khati et al. 2017).

The nanocompounds can support the life of microbial inoculants in formulation due to controlled release of nutrients to them. The poor survival of microbial inoculants in field condition is due to several factors, the major one is insufficiency of nutrients and moisture. A chelating system that can entrap the resources and allow their slow release to sustain the microbial population may solve the problem of nutrient and moisture scarcity in any formulation. The nanocompounds (nanozeolite and nanochitosan) due to their physical and chemical nature are reported to enhance the survival efficiency of bacteria in broth (Khati et al. 2019). Further research is needed to be carried to bring a better insight in this field and utilize it for the improvement of agriculture production. The role of nanocompounds and microbial inoculants to mitigate drought stress is elucidated in Fig. 10.2.

10.3 Genetic Engineering of Plants and Bacteria

The present technology has allowed us to engineer the genome of plants for the selected traits and impart any character of interest. The stress tolerance can also be improved by altering the stress-responsive traits to solve the management issue. Molecular chaperons like Glycine Rich RNA binding protein (GR-RBPs) were also reported to have important role in stress response management. The role of mRNA of GR-RBPs in drought stress response was reported by Khan et al. (2013). A Ribosomal inactivation protein (RIP), N-glycosidase inhibits protein synthesis by depurinating the rRNA. Jiang et al. (2008) studied the RIPs in *Oryza sativa* and reported their expression under biotic (pathogen) and abiotic stress (drought, salinity, and temperature).

Hairpin proteins are elicitors and produced by several gram-negative plant pathogenic bacteria, trigger multiple beneficial responses in plants, such as growth promotion, induction of defence against diverse pathogens and insects, and drought tolerance. The hairpin-encoding gene (popW) derived from *Ralstonia solanacearum* ZJ3721 was observed to impart tolerance to tobacco plant towards drought stress. These transgenic lines of tobacco displayed a significant increase in different enzymes like superoxide dismutase, peroxidase, catalase, and ascorbic acid content as compared to control plants under drought stress. Overexpression of popW in

Table 10.1 Genes potentially responsible for drought stress tolerance

Sl. No	Gene	Role	Reference
1	*PopW*	Hairpin-encoding gene that enhanced the relative transcript levels of oxidative stress-responsive genes (NtAPX, NtCAT1, NtGST, and NtCu/Zn-SOD) under drought stress	Liu et al. (2016)
2	*2b*	Encoded by cucumber mosaic virus and cause suppression of RNA silencing and improve drought stress tolerance	Westwood et al. (2013)
3	*AGO1*	Controls a microRNA-mediated drought tolerance mechanism	Westwood et al. (2013)
4	*LEA gene*	Improvement of growth characteristics under water deficient conditions, more biomass and more efficient water use	Wu et al. (2014)
5	*Sac B gene*	Encodes for levansucrase, which takes part in fructan synthesis. Fructan promotes the process of root branching, thus increasing root surface and water uptake	Pilon-Smits et al. (1995)
6	*P5CS gene*	Responsible for proline accumulation which is associated with abiotic stress	Delauney and Verma (1993)
7	*ADC*	In plants, polyamine, accumulate under several abiotic stress stimuli, including drought and salt	Kumar and Minocha (1998)
8	*DRE*	The dehydration responsive element (DRE) was identified as a cis-acting element regulating *gene expression* in response to dehydration	Van Rensburg and Kruger (1994)
9	*Fe-SOD*	Within a cell, the SODs constitute the first line of the defence against ROS. Transgenic tobacco plants containing oxidative stress-related genes showed elevated levels of glutathione reductase, superoxide dismutase and ascorbate peroxidase, resulting in enhance drought tolerance	Van Rensburg and Kruger (1994)
10	*WXP1*	The gene designated WXP1 is able to activate wax production and confer drought tolerance in alfalfa (*Medicago sativa*)	Zhang et al. (2005)
11	*MAPK*	Expression of mitogen-activated protein kinase gene (MAPK) genes activates an oxidative signal cascade and lead to the tolerance of freezing, drought, heat and salinity stressing	Shou et al. (2004)
12	*ERA1*	The gene identified in *Arabidopsis* encodes β-subunit of a farnesyltransferase and involves in ABA signalling	Wang et al. (2005)
13	*bZIP*	Act as ABA dependent transcription factor for drought responsive genes	Xiang et al. (2008)
14	*OsSIK1*	Enhanced tolerance to drought and salinity stresses	Ouyang et al. (2010)
15	*OsCPK4*	A calcium-dependent protein kinase showed enhanced tolerance to drought and salinity stresses	Campo et al. (2014)
16	*EDT1/ HDG11*	Encodes a homeodomain—leucine zipper transcription factor, which is likely involved in reproductive development during stress	Yu et al. (2008)
17	*SNAC1*	Responsible for reduced water loss due to increased stomatal closure and enhanced expression of a large number of stress-related genes	Hu et al. (2006)
18	*AP2/ ERF*	Enhanced photosynthesis and reduced transpiration under drought stress	Karaba et al. (2007)

tobacco also significantly enhanced the relative transcript levels of oxidative stress-responsive genes under drought stress. (Liu et al. 2016).

Suárez et al. (2008) showed a new strategy to increase drought tolerance and yield in legumes by overexpressing trehalose-6-phosphate synthase activity, if inoculated with *R. etli* over expressing trehalose-6-phosphate synthase gene. *Phaseolus vulgaris* plants had more nodules with increased nitrogenase activity and higher biomass compared with plants inoculated with wild-type *R. etli*. They suggested the role of trehalose metabolism in rhizobia for signalling plant growth, yield, and adaptation to abiotic stress. Manipulation in trehalose metabolism using symbiotic rhizobia may show major agronomical impact on leguminous plants under drought stress.

Westwood et al. (2013) proved the role of RNA silencing in drought stress tolerance. Cucumber mosaic virus (CMV) encodes 2b protein, which plays a role in local and systemic virus movement, symptom induction, and suppression of RNA silencing. It also disrupts signalling regulated by salicylic acid and jasmonic acid. CMV induced tolerance to drought in *Arabidopsis thaliana*. The silencing effector ARGONAUTE1 (AGO1) controls a microRNA-mediated drought tolerance mechanism. Similarly, Wu et al. (2014) cloned an LEA gene (SmLEA), from *Salvia miltiorrhiza* Bunge. SmLEA-transformed *S. miltiorrhiza* plants showed better root elongation and a lower level of malondialdehyde. SmLEA-overexpressing transgenic plants showed a less rapid rate of water loss and had greater superoxide dismutase activity and a higher glutathione concentration (Table 10.1).

10.4 Conclusion

Under the influence of global warming all over the world, changes in the climatic condition are creating unusual weather phenomena often in the form of water stress. Moisture limitation significantly affects most of the metabolic processes in plants, including membrane conformation, chloroplast organization, and enzyme activity at cellular level, which affects whole plant growth and yield. Drought stress is a severe environmental constraint to agricultural productivity. Drought stress induces stomatal closure, decreases transpiration and photosynthetic rates, and leads to earlier crop maturity and poor productivity. To minimize the negative effects of water stress, plants have various signalling pathways. Plants respond immediately for any stress by changing their growth pattern, upregulation of antioxidants, accumulation of compatible solutes and by producing stress proteins and chaperones. Microbial bioinoculants (PGPR and Fungal endosymbiont) play important role in conferring resistance/tolerance and adaptation of plants to drought stresses and have the potential role in solving issues for future food security. The interaction between plants and microbes under drought conditions affects not only the plant but also changes the soil properties. Application of microbial inoculants is a potential and natural way to combat drought stress in plants. This technique still faces challenges due to its slow impact in actual condition. Development of drought tolerant crop varieties through genetic engineering and plant breeding is another important way to fight drought stress. The study also focuses on application of nanocompounds to mitigate drought stress, which may help in improvement of water use efficiency by plants and microbes. More investigations into the mechanisms by which microbes elicit tolerance to specific stress factors are needed. Studies are needed to elucidate the signal transduction pathways that result from treatment of plants with PGPR under stress conditions.

Acknowledgement The authors are thankful to the Director, ICAR-VPKAS, Almora-263601, Uttarakhand, India for providing the opportunity and guidance for writing this chapter.

References

Ahmad P, Rasool S, Gul A, Sheikh SA, Akram NA, Ashraf M, Kazi AM, Gucel S (2016) Jasmonates: multifunctional roles in stress tolerance. Front Plant Sci 7:813

Ashraf M (1994) Breeding for salinity tolerance in plants. Crit Rev Plant Sci 13:17–42

Battaglia M, Olvera-Carrillo Y, Garciarrubio A, Campos F, Covarrubias AA (2008) The enigmatic LEA proteins and other hydrophilins. Plant Physiol 148:6e24

Belimov AA, Dodd IC, Hontzeas N, Theobald JC, Safronova VI, Davies WJ (2009) Rhizosphere bacteria containing 1-aminocyclopropane-1-carboxylatedeaminase increase yield of plants grown in drying soil via both local and systemic hormone signaling. New Phytol 181:413–423

Blum A (2005) Drought resistance, water-use efficiency, and yield potential–are they compatible, dissonant, or mutually exclusive? Aust J Agric Res 56:1159–1168

Blum A (2009) Effective use of water (EUW) and not water-use efficiency (WUE) is the target of crop yield improvement under drought stress. Field Crops Res 112:119–123

Campo S, Baldrich P, Messeguer J, Lalanne E, Coca M, San Segundo B (2014) Overexpression of a calcium-dependent protein kinase confers salt and drought tolerance in rice by preventing membrane lipid peroxidation. Plant Physiol 165:688–704

Cho SM, Kang BR, Han SH, Anderson AJ, Park JY, Lee YH, Cho BH, Yang KY, Ryu CM, Kim YC (2008) 2R, 3R-butanediol, a bacterial volatile produced by *Pseudomonas chlororaphis* O6, is involved in induction of systemic tolerance to drought in *Arabidopsis thaliana*. MPMI 21 (8):1067–1075

Delauney AJ, Verma DPS (1993) Proline biosynthesis and osmoregulation in plants. Plant J 4:215–223

Dimkpa C, Bindraban PS, Fugice J, Hellums D (2017) Composite micronutrient nanoparticles and salts decrease drought stress in soybean. Agron Sustain Dev 37(1). https://doi.org/10.1007/s13593-016-0412-8

Egamberdieva D (2013) The role of phytohormone producing bacteria in alleviating salt stress in crop plants. In: Miransari M (ed) Biotechnological techniques of stress tolerance in plants. Stadium Press LLC, Houston, TX, pp 21–39

Eziz A, Yan Z, Tian D, Han W, Tang Z, Fang J (2017) Drought effect on plant biomass allocation: a meta- analysis. Ecol Evol 7:11002–11010

Garcíaa JE, Maronicheb G, Creus C, Suárez-Rodríguezd R, Ramirez-Trujillod JA, Gropp MD (2017) In vitro PGPR properties and osmotic tolerance of different *Azospirillum* native strains and their effects on growth of maize under drought stress. Microbiol Res 202:21–29

Ghabooli M, Khatabi B, Shahriary F, Ahmadi Sepehri M, Mirzaei M, Amirkhani A, Jorrín-Novo JV, Salekdeh GH (2013) Proteomics study reveals the molecular mechanisms underlying water stress tolerance induced by *Piriformospora indica* in barley. J Proteome 94:289–301

Goyal K, Walton LJ, Tunnacliffe A (2005) LEA proteins prevent protein aggregation due to water stress. Biochem J 388:151e157

Grover M, Ali SZ, Sandhya V, Venkateswarlu B (2011) Role of microorganisms in adaptation of agricultural crops to abiotic stresses. World J Microbiol Biotechnol 27:1231–1240

Gupta (2006) Hill agriculture: status of research and production potential. In: Gupta HS, Srivastava AK, Bhatt JC (eds) Sustainable production from agricultural watershed in North West Himalaya. Vivekananda Parvatiya Krishi Anusandhan Sansthan, Almora, 615p

Hasegawa S, Meguro A, Toyoda K, Nishimura T, Kunoh H (2007) Drought tolerance of tissue-cultured seedlings of mountain Laurel (*Kalmia latifolia* L.) induced by an endophytic actinomycete: II. Acceleration of callose accumulation and lignification. Actinomycetologica 19:13–17

Hasegawa S, Meguro A, Nishimura T, Kunoh H (2008) Drought tolerance of tissue-cultured seedlings of Mountain Laurel (*Kalmia latifolia* L.) induced by an endophytic actinomycete: I. Enhancement of osmotic pressure in leaf cells. Actinomycetologica 18:43–47

Hu H, Dai M, Yao J, Xiao B, Xianghua L, Zhang Q, Xiong L (2006) Over expressing a NAM, ATAF, and CUC (NAC) transcription factor enhances drought resistance and salt tolerance in rice. Proc Natl Acad Sci U S A 103:12987–12992

Jabeen B, Naqvi SMS, Mahmood T, Sultana T, Arif M, Khan F (2017) Ectopic expression of plant RNA chaperone offering multiple stress tolerance in *E. coli*. Mol Biotechnol 59:66–72

Jaleel AC, Manivannan P, Sankar B, Kishorekumar A, Gopi R, Somasundaram R, Panneerselvam R (2007) *Pseudomonas fluorescens* enhances biomass yield and ajmalicine production in *Catharanthus roseus* under water deficit stress. Colloids Surf B: Biointerfaces 60:7–11

Jiang SY, Ramamoorthy R, Bhalla R, Luan HF, Nori P, Venkatesh Cai M, Ramachandran S (2008) Genome-wide survey of the RIP domain family in *Oryza sativa* and their expression profiles under various abiotic and biotic stresses. Plant Mol Biol 67:603–614

Kant S, Kafkafi U (2002) Potassium and abiotic stresses in plants. In: Pasricha NS, Bansal SK (eds) Potassium for sustainable crop production. Potash Institute of India, Gurgaon, pp 233–251

Karaba A, Dixit S, Greco R, Aharoni A, Trijatmiko KR, Marsch-Martinez N, Krishnan A, Nataraja KN, Udayakumar M, Pereira A (2007) Improvement of water use efficiency in rice by expression of HARDY, an *Arabidopsis* drought and salt tolerance gene. Proc Natl Acad Sci U S A 104:15270–15275

Kasim WA, Osman ME, Omar MN, Abd El-Daim IA, Bejai S, Meijer J (2013) Control of drought stress in wheat using plant growth promoting bacteria. J Plant Growth Regul 32:122–130

Khan F, Sultana T, Deeba F, Saqlan Naqvi SM (2013) Dynamics of mRNA of glycine-rich RNA-binding protein during wounding, cold and salt stresses in *Nicotiana tabacum*. Pak J Bot 45:297–300

Khati P, Chaudhary P, Gangola S, Bhatt P, Sharma A (2017) Nanochitosan supports growth of *Zea mays* and also maintains soil health following growth. Biotech 7:81

Khati P, Chaudhary P, Gangola S, Sharma A (2019) Influence of nanozeolite on plant growth promotory bacterial isolates recovered from nanocompound infested agriculture field. Environ Ecol 37(2):521–527

Kohler J, Hernandez JA, Caravaca F, Roldán A (2008) Plant-growth promoting rhizobacteria and arbuscular mycorrhizal fungi modify alleviation biochemical mechanisms in water stressed plants. Funct Plant Biol 35:141–151

Kumar A, Minocha SC (1998) Transgenic manipulation of polyamine metabolism. In: Lindsey K (ed) Transgenic plant research. Harwood Academic, London, pp 187–199

Liu H, Wang Y, Zhou X, Wang C, Wang C, Fu J, Wei T (2016) Overexpression of a harpin-encoding gene popW from *Ralstonia solanacearum* primed antioxidant defenses with enhanced drought tolerance in tobacco plants. Plant Cell Rep 35:1333–1344

Marschner H (2011) Marschner's mineral nutrition of higher plants. In: Marschner P (ed) . Elsevier/ Academic Press, Amsterdam, p 684

Mayak S, Tirosh T, Glick BR (2004) Plant growth-promoting bacteria that confer resistance to water stress in tomatoes and peppers. Plant Sci 166:525–530

Monayeri MO, Higazi AN, Ezzat NH, Saleem HM, Tohoun SN (1984) Growth and yield of some wheat and barley varieties grown under different moisture levels. Ann Agric Sci Mushtohor 20:231–243

Ouyang SQ, Liu YF, Liu P, Lei G, He SJ, Ma B, Zhang WK, Zhang JS, Chen SY (2010) Receptor-like kinase OsSIK1 improves drought and salt stress tolerance in rice (*Oryza sativa*) plants. Plant J 62(2):316–329

Palaniyandi SA, Damodharan K, Yang SH, Suh JW (2014) *Streptomyces* sp. strain PGPA39 alleviates salt stress and promotes growth of "Micro Tomato plants". J Appl Microbiol 117 (3):766–773

Pampurov S, Dijck PV (2014) The desiccation tolerant secrets of *Selaginella lepidophylla*: what we have learned so far? Plant Physiol Biochem 80:285–290

Pietragalla J, Alistair P (2012) Canopy temperature, stomatal conductance and water relation traits-chapter 2: stomatal conductance. In: Pask A, Pietragalla J, Mullan D, Reynolds M (eds) Physiological breeding II: a field guide to wheat phenotyping. CIIMMYT, Mexico

Pilon-Smits EAH, Ebskamp MJM, Paul MJ, Jeuken MJW, Weibeck PJ, Smeekens SCM (1995) Improved performance of transgenic fructan-accumulating tobacco under drought stress. Plant Physiol 107:125–130

Prasad R, Bhattacharyya A, Nguyen QD (2017) Nanotechnology in sustainable agriculture: recent developments, challenges, and perspectives. Front Microbiol 8:1014

Rahdari P, Hoseini SM, Tavakoli S (2012) The studying effect of drought stress on germination, proline, sugar, lipid, protein and chlorophyll content in purslane (*Portulaca oleracea* L.) leaves. J Med Plants Res 6:1539–1547

Rakhra G, Kaur T, Vyas D, Sharma AD, Singh J, Ram G (2017) Molecular cloning, characterization, heterologous expression and in silico analysis of disordered boiling soluble stress-responsive wBsSRP protein from drought tolerant wheat cv.PBW 175. Plant Physiol Biochem 112:29e44

Sathya A, Vijayabharathi R, Gopalakrishnan S (2017) Plant growth-promoting actinobacteria: a new strategy for enhancing sustainable production and protection of grain legumes. 3. Biotech 7:102

Sen S, Ghosh D, Mohapatra S (2018) Modulation of polyamine biosynthesis in *Arabidopsis thaliana* by a drought mitigating *Pseudomonas putida* strain. Plant Physiol Biochem 129:180–188

Sharma NK, Singh RJ, Mandal D, Kumar A, Alam NM (2017) Increasing farmer's income and reducing soil erosion using intercropping in rainfed maize – wheat rotation in Himalaya, India. Agric Ecosyst Environ 247:43–53

Sharma P, Jha AB, Dubey RS, Pessarakli M (2012) Reactive oxygen species, oxidative damage, and antioxidative defense mechanism in plants under stressful conditions. J Bot 2012:217037

Shou HX, Bordallo P, Wang K (2004) Expression of *Nicotiana* protein kinase (NPK1) enhances drought tolerance in transgenic maize. J Exp Bot 55:1013–1019

Siddiqui MH, Al-Whaibi MH, Basalah MO (2011) Role of nitric oxide in tolerance of plants to abiotic stress. Protoplasma 248:447–455

Singh LP, Gill SS, Tuteja N (2011) Unraveling the role of fungal symbionts in plant abiotic stress tolerance. Plant Signal Behav 6(2):175–191

Suárez R, Wong A, Ramírez M, Barraza A, Orozco MDC, Cevallos MA, Lara M, Hernández G, Iturriaga G (2008) Improvement of drought tolerance and grain yield in common bean by overexpressing trehalose-6-phosphate synthase in rhizobia. MPMI 21(7):958–966

Taiz L, Zeiger E (1991) Plant physiology. Benjamin/Cummings Publishing Company, Redwood City, CA

Timmusk S, Wagner EG (1999) The plant-growth-promoting rhizobacterium *Paenibacillus polymyxa* induces changes in *Arabidopsis thaliana* gene expression: a possible connection between biotic and abiotic stress responses. Mol Plant Microb Interact 12:951

Timmusk S, Islam A, El-Daim SA, Lucian C, Tanilas T, Kannaste A, Behers L, Nevo E, Seisenbaeva G, Stenström E, Niinemets U (2014) Drought-tolerance of wheat improved by rhizosphere bacteria from harsh environments: enhanced biomass production and reduced emissions of stress volatiles. PLoS One 9:1–13

Todaka D, Shinozaki K, Shinozaki KY (2015) Recent advances in the dissection of drought stress regulatory network and strategies for development of drought tolerant transgenic rice plants. Front Plant Sci 6(84):00084

Van Rensburg L, Kruger GHJ (1994) Applicability of abscisic acid and (or) proline accumulation as selection criteria for drought tolerance in *Nicotiana tabacum*. Can J Bot 72:1535–1540

Wang Z, Solanki MK, Yu ZX, Yang LT, An QL, Dong DF, Li YR (2019) Draft genome analysis offers insights into the mechanism by which *Streptomyces chartreusis* WZS021 increases drought tolerance in sugarcane. Front Microbiol 9:3262

Wang Y, Ying J, Kuzma M, Chalifoux M, Sample A, McArthur C, Uchacz T, Sarvas C, Wan J, Dennis DT, McCourt P, Huang Y (2005) Molecular tailoring of farnesylation for plant drought tolerance and yield protection. Plant J 43:413–424

Westwood JH, Mccann L, Naish M, Dixon H, Murphy AM, Stancombe MA, Bennett MH, Powell G, Webb AAR, Carr JP (2013) A viral RNA silencing suppressor interferes with abscisicacid-mediated signaling and induces drought tolerance in *Arabidopsis thaliana*. Mol Plant Pathol 14(2):158–170

Wu Y, Deng Z, Lai J, Zhang Y, Yang C, Yin B, Zhao Q, Zhang L, Li Y, Yang C, Xie Q (2009) Dual function of *Arabidopsis* ATAF1 in abiotic and biotic stress responses. Cell Res 19 (11):1279–1290

Wu Y, Liu C, Kuang J, Ge Q, Zhang Y, Wang Z (2014) Overexpression of SmLEA enhances salt and drought tolerance in *Escherichia coli* and *Salvia miltiorrhiza*. Protoplasma 251:1191–1199

Xiang Y, Tang N, Du H, Ye H, Xiong L (2008) Characterization of OsbZIP23 as a key player of the basic leucine zipper transcription factor family for conferring abscisic acid sensitivity and salinity and drought tolerance in rice. Plant Physiol 148:1938–1952

Yandigeri MS, Meena KK, Singh D, Malviya N, Singh DP, Solanki MK, Yadav AK, Arora DK (2012) Drought-tolerant endophytic actinobacteria promote growth of wheat (*Triticum aestivum*) under water stress conditions. Plant Growth Regul 68:411–420

Yang J, Kloepper JW, Ryu CM (2009) Rhizosphere bacteria help plants tolerate abiotic stress. Trends Plant Sci 14:1–4

Yobi A, Wone BW, Xu W, Alexander DC, Guo L, Ryals JA, Oliver MJ, Cushman JC (2013) Metabolomic profiling in *Selaginella lepidophylla* at various hydration states provides new insights into the mechanistic basis of desiccation tolerance. Mol Plant 6:369e385

Yuan XK, Yang ZQ, Li YX, Liu Q, Han W (2016) Effects of different levels of water stress on leaf photosynthetic characteristics and antioxidant enzyme activities of greenhouse tomato. Photosynthetica 54:28–39

Yu H, Chen X, Hong YY, Wang Y, Xu P, Ke SD, Liu HY, Zhu JK, Oliver DJ, Xiang CB (2008) Activated expression of an *Arabidopsis* HD-START protein confers drought tolerance with improved root system and reduced stomatal density. Plant Cell 20:1134–1151

Zhang SZ, Yang BP, Feng CL, Tang HL (2005) Genetic transformation of tobacco with the trehalose synthase gene from *Grifola frondosa* Fr. enhances the resistance to drought and salt in tobacco. J. Intensive Plant Biol 47:579–587

Zhang L, Li M, Li Q, Chen C, Qu M, Li M, Wang Y, Shen X (2018a) The catabolite repressor/ activator *Cra* is a bridge connecting carbon metabolism and host colonization in the plant drought resistance-promoting bacterium *Pantoea alhagi* LTYR-11Z. Appl Environ Microbiol 84:e00054–e00018

Zhang W, Wang J, Xu L, Wang A, Huang L, Du H, Qiu L, Oelmfuller R (2018b) Drought stress responses in maize are diminished by *Piriformospora indica*. Plant Signal Behav 13(1): e1414121

Zhao Z, Chen G, Zhang C (2001) Interaction between reactive oxygen species and nitric oxide in drought-induced abscisic acid synthesis in root tips of wheat seedlings. Australian J Plant Physiol 28:1055–1061

Chapter 11
Phytoremediation: A Synergistic Interaction Between Plants and Microbes for Removal of Unwanted Chemicals/ Contaminants

Sheel Ratna, Swati Rastogi, and Rajesh Kumar

Abstract Environmental pollution with obnoxious contaminants is detrimental to plant growth and poses health hazards to humans and other life forms. Thus, remediation of such antagonistic environment has become a key issue for environmentalists all around the world. Phytoremediation, a cooperative association between plants and microbes, is an emerging in situ cost-effective technology and provides a viable option in the treatment of such contaminated environments. Present chapter emphasizes on plant–microbes interactions during phytoremediation and how such beneficial interactions lead to improved plant growth and contamination free environment.

11.1 Introduction

Environmental contamination is one of the most intractable concerns worldwide to ensure the safest and healthiest environment. The source of pollution is natural and anthropogenic activities. Key anthropogenic sources of pollution are related to the burning of fossil fuels, mining and untreated or partially treated disposal of municipal solid wastes and wastewater discharges or use for irrigation and excessive utilization of fertilizers and pesticides (Pinto et al. 2016, 2018). Continuous and consistent increase in a variety of organic and inorganic pollutants has been reported to cause environmental pollution which results in severe health hazards in living beings. Inorganic contaminants are salts of nitrate, ammonia, sulfate, phosphate, cyanide, heavy metals (HMs), while alkanes, antibiotics, dioxins, phenols, polychlorinated biphenyls (PCBs), polycyclic aromatic hydrocarbons (PAHs), Persistent Organic Pollutants (POPs), pesticides, synthetic azo dyes, polyaromatic, chlorinated, and nitro-aromatic compounds constitute organic contaminants.

S. Ratna · S. Rastogi · R. Kumar (✉)
Rhizosphere Biology Laboratory, Department of Microbiology, Babasaheb Bhimrao Ambedkar University (A Central University), Lucknow, India

A. Sharma (ed.), *Microbes and Signaling Biomolecules Against Plant Stress*, Rhizosphere Biology, https://doi.org/10.1007/978-981-15-7094-0_11

Among all the inorganic and organic contaminants, HMs and POPs are most problematic in the atmosphere due to their toxicity, bioaccumulation factor, resistance to degradation in an environment through natural processes. They are highly obnoxious and dangerous to animals, including humans (Ahmadpour et al. 2012; Saxena and Bharagava 2017; Arslan et al. 2017; Shayler et al. 2017). Most of the HMs are naturally occurring elements and have density higher than 5 g cm^{-3}. Heavy metals are not easily degradable as can lose their electron(s) to form cations and in addition to these, other organic toxic contaminants such as POPs exist persistently in the environment for several years before getting totally mineralized (Adrees et al. 2015; UNEP 2006). Carson (1962) reported the harmful effect of DDT first time on bird's eggs, which were unable to hatch due to the impact of DDT. Due to this phenomenal incidence, DDT was banned in 1973 (Carson 1962; Staniforth 2013). Recently, Stockholm Convention (COP9) on POPs held at Zeneva has issued a list of hazardous chemicals like dicofol, perfluorooctanoic acid (PFOA), its salts and PFOA-related compounds in Annex A(elimination category) (the Stockholm Convention (COP9), https://www.meti.go.jp/english/press/2019/0514_001.html). Jamieson et al. (2017) have reported the presence of some POPs in fauna from the deepest ocean trenches (>10 km). Recently a research group has reported that consumption of leafy vegetables contaminated with PCBs and DDTs is not suitable for human health and causes death (Olatunji 2019). Soil acts as the hindmost acceptor of contaminants and these contaminants limit the development of plants through bioaccumulation and biomagnification along the whole food chain (He et al. 2014).

Phytoremediation concept was foremost postulated by Chaney. This term was used for the first time by Raskin and Salt to describe the use of plants for removal of toxic HMs (Raskin et al. 1994; Salt et al. 1995). Researchers used this promising approach for the remediation of soil and water, contaminated with organic and inorganic pollutants (Chaney 1983; Khan et al. 2014). Plants used in the process of phytoremediation of contaminants from the soil or water may be terrestrial and macrophytes or microphytes like free-floating, emerged or submerged (da Conceicao Gomes et al. 2017; Farid et al. 2017; Gu et al. 2017; Wani et al. 2017). Contamination of HMs and POPs in soil and water affects the phytoremediation efficiency of the plants (Ibanez et al. 2012; Khan et al. 2014; Gerhardt et al. 2017). Mutual relationships of the plants and microbes have been projected to enhance the efficiency of phytoremediation of HMs and POPs from contaminated soil and water (Becerra-Castro et al. 2013; Haslmayr et al. 2014; Kong and Glick 2017; Balazs et al. 2018; Boudh and Singh 2019; Gupta et al. 2019). In this chapter, we will discuss the toxic effects of HMs and POPs and their phytoremediation through synergistic interaction between plants and microbes in current scenario.

11.1.1 Toxicity of Heavy Metal and POPs

HMs and POPs show their obnoxious effect on biotic and abiotic components of the ecosystem due to their less solubility and high carcinogenicity and mutagenicity (Yadav et al. 2017; Lal et al. 2018). Some metals such as As, Pb, Cd, and Hg may be toxic even at low concentrations (Ma et al. 2016a; Rastogi et al. 2019). They are absorbed by biotic/abiotic components of the environment and then transferred to the top consumers via food web from where they continue to accumulate in their vital organs like brain, liver, bones, and kidneys for years and ultimately lead to health hazards in human beings (Lal et al. 2018). POPs show toxic effects on both targeted and non-target populations and may lead to several physiological, biochemical, developmental, and behavioral stress responses at different trophic levels of the food web (Inostroza et al. 2016). Researchers have measured the extent of adversity of HMs, POPs, etc. in an ecosystem on the basis of their toxicity, concentration, and exposure period. Different studies have been conducted to elucidate the acute and chronic effects of a variety of POPs on living beings (Ashraf et al. 2017; Rajkumar et al. 2013). Recently a study reported that insulin resistance, obesity, and endocrine disruption are caused by acute exposure of POPs in humans (Gregoraszczuk and Ptak 2013). Effects of PCBs and PBDEs on life-history traits like somatic growth, spawning, and larval survival of Zebrafish have been analyzed by Horri et al. (2018). Numerous studies have demonstrated the adverse effect of different metal (loid)s on the microbial diversity of polluted soils (Kerfahi et al. 2019; Srut et al. 2019; Zhao et al. 2019). Excessive concentration of metal(loid)s in soils can affect the growth, morphology, and metabolism of microorganisms mainly through destruction of cellular membranes and organelles, enzyme denaturation, and functional or conformational disturbance (Rajkumar et al. 2013; Ashraf et al. 2017).

11.2 Phytoremediation

A variety of physical, chemical, and biological approaches for the remediation of persistent (organic) and non-degradable inorganic pollutants (heavy metals and metalloids) from water and soil have been in practice but because of some demerits, phytoremediation has gained a momentum and used as a good alternative for decontamination (Mahar et al. 2016; Coninx et al. 2017; Saxena et al. 2019). Phytoremediation is an efficient, inexpensive, and eco-friendly rehabilitation strategy that uses plants to absorb, accumulate, immobilize or biodegrade organic and inorganic pollutants from different environmental matrices (air, soil, and water) through physical, chemical, and biological processes (Vangronsveld et al. 2009; Cristaldi et al. 2017; Wang et al. 2017; Kumar 2019). Terrestrial and aquatic plants such as submerged/emerged and free-floating play an important role in phytoremediation of contaminants from the aquatic and terrestrial environment (da Conceicao Gomes et al. 2017; Farid et al. 2017; Wani et al. 2017).

Phytoremediation is therefore a suitable option to remediate soil and water bodies contaminated with HMs and POPs (Khan et al. 2018; Mahar et al. 2016; Mishra et al. 2020). Phytoremediation encompasses distinct mechanisms namely phytostabilization, phytodegradation, phytovolatilization, and rhizodegradation (Kong and Glick 2017) Fig. 11.1.

11.2.1 Phytovolatilization

Phytovolatilization or phytoevaporation is the uptake of pollutants from the soil and water by plants and their conversion into volatile form and succeeding release into the atmosphere. It can be applied for organic pollutants and some volatile heavy metals, i.e. Selenium (Se) and Mercury (Hg) (Ali et al. 2013; Coninx et al. 2017). Complete removal of the pollutant from the site in gaseous form, without further need for plant harvesting and disposal holds a great promise as an attractive technology to control pollution (Pilon-Smits 2005; Lim et al. 2016). Phytoevaporation is the most controversial technology of the phytoremediation because this technique does not fix the contaminants completely but transfers pollutants from soil to atmosphere from where it can be redeposited (Padmavathiamma and Li 2007).

11.2.2 Phytostabilization

Phytostabilization includes immobilization of the contaminants in soil either simply by preventing erosion, leaching, or dispersion or by transforming them through precipitation in the rhizosphere to less bioavailable forms (Pulford and Watson 2003). Phytostabilization is the best option for the remediation of large contaminated areas but excessive soil pollution may inhibit the survival of plants used for phytostabilization (Vangronsveld et al. 2009). Some hyperaccumulator (HA) plants accumulate HMs in the root and confine the HMs transport to aerial parts of the plants. Therefore, these HAPs are used for phytostabilization as they can help in reducing soil pollution thus survival of the plants (Vangronsveld et al. 2009; Coninx et al. 2017; Sheoran et al. 2010). Phytoremediation efficiency of *Sarcocornia fruticosa* is increased after bioaugmentation using autochthonous metal resistant *Saccharomyces cerevisiae* isolated from HMs (Cd, Pb, Ni, Cr, and Cu) contaminated /sedimented site (Said et al. 2018). Microbial assisted phytostabilization should be a promising technique which could be used for the removal of organic and inorganic contaminants. Inoculation of *Bradyrhizobium* and *Azospirillum* with *Glycine max* reduced the translocation and improved phytostabilization of As (Armendariz et al. 2019). *Achromobacter spanium* and *Serratia plymuthica* isolated from *Medicago sativa* nodules are some suggested

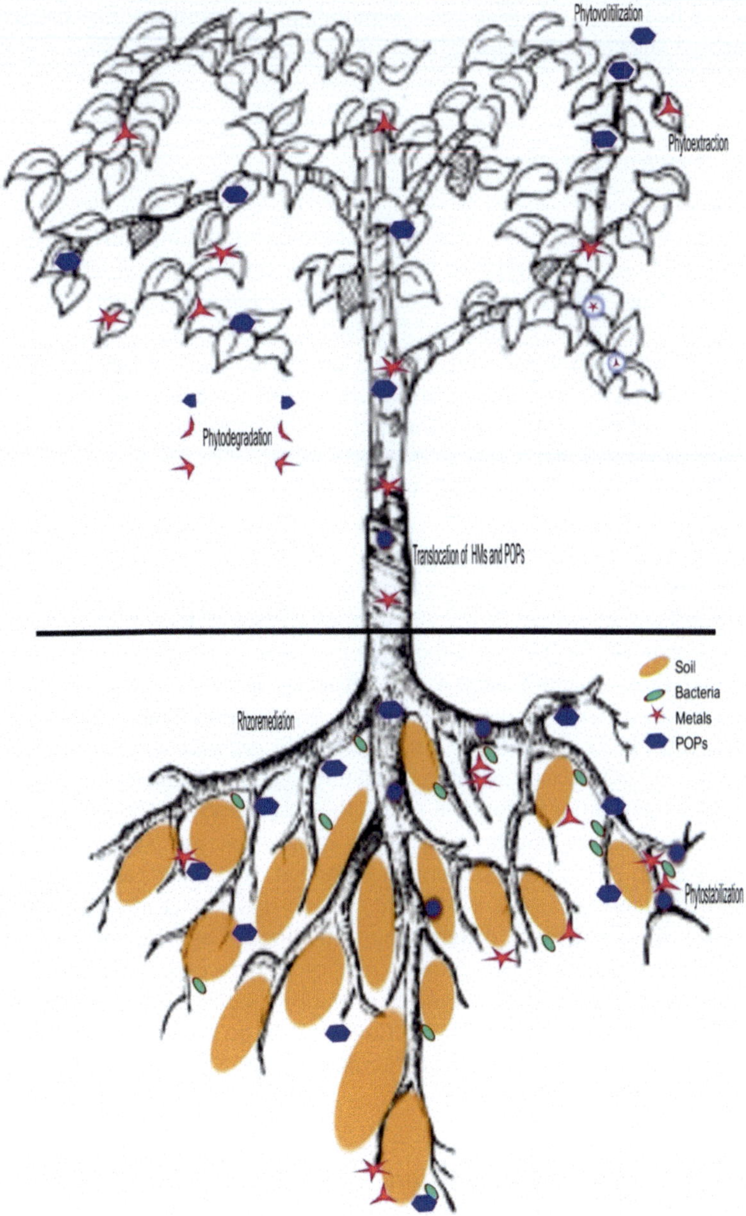

Fig. 11.1 A proposed model of plant and microbe interaction during phytoremediation

microbes which can be used as potential phytostabilizer for the pesticide-contaminated sites (Aroua et al. 2019).

11.2.3 Phytoextraction

Phytoextraction is popularly known as phytoaccumulation, phytoabsorption or phytosequestration. It is a type of phytoremediation through which uptake of the contaminants from soil, water, or sediments takes place by plant roots which are further translocated and accumulated in aerial parts, i.e., stem or other harvestable plant parts (Glick 2010; Pinto et al. 2015; Mahar et al. 2016). Phytoextraction is the most important and most useful phytoremediation technique for the removal of heavy metals and metalloids from polluted soils (Pinto et al. 2016) but it limits its use for non-food crop plants due to their accumulation in the aerial parts hence the plant biomass produced (Sharma and Pandey 2014; Burges et al. 2017). Some plants belonging to a particular family have the competence to grow and accumulate extremely high concentration of HMs and POPs without suffering from phytotoxic effect via enhancing uptake, translocation, detoxification, and sequestration in their airy parts and are known as Hyper Accumulator Plants (HAPs) (Samardjieva et al. 2015; Pinto et al. 2016). In environment, contaminants are present in complex mixed forms but HAPs can accumulate one metal from complex metal-polluted soils which often contains a mixture of metals and other organic pollutants (Pinto et al. 2018; Cristaldi et al. 2017). Some HAP (e.g. *Thlaspica erulescens*, *Alyssum bertolonii*, *Arabidopsis halleri*) species are able to accumulate contaminants but produce little biomass, and therefore it is possible to use those species that accumulate as well as produces more biomass like *Brassica* sp., *Arundo donax,* and *Typha* sp. (Pinto et al. 2016; Shiri et al. 2015) but the storage, treatment, placement, and disposal of the contaminated plant biomass are of great concern (Mahar et al. 2016; Ghosh and Singh 2005). In recent years, different biotechnological approaches have also been used to increase biomass and growth rate of hyperaccumulator plants through genetic engineering tools, i.e., synthesis of various metal-binding peptides (Mahar et al. 2016). The success of phytoextraction depends on high plant biomass production and ability to accumulate metals (Nayak et al. 2018).

11.2.4 Phyto/Rhizodegradation

In an integrated approach, phyto and rhizodegradation can be addressed as a mutually beneficial form of phytoremediation, where both plants and microorganisms mediate the breakdown of the contaminants via the use of their enzymatic machinery. Sampaio et al. (2019) reported the interaction between *Rhizophora mangle* and bacterial strains (*Pseudomonas aeruginosa* and *Bacillus*) as a potential association for the degradation of eight PAHs from diesel oil-contaminated site.

11.3 Phytoremediation Using Microbes (Fungi, Bacteria, Algae, and Mycorrhizae)

Plants and microbes reside together and interact with each other. A plant provides nutrients and space to the microbes and in return, microbes improve the bioavailability, mineralization, and detoxification of various organic and inorganic pollutants (Balseiro-Romero et al. 2017; Liao et al. 2016; Chirakkara et al. 2016; Wang et al. 2017). The interface between the microbes and plant roots in the rhizosphere is believed to vastly influence the growth and survival of the plants. Rhizospheric microbes have the potential to biodegrade organic pollutants through rhizodegradation, biotransformation, and volatilization (Ashraf et al. 2017). Optimization of the phytoremediation process is analyzed by the dynamics between the plants, microbes, and pollutants (Rajkumar et al. 2010, 2013; Ashraf et al. 2017). Recently, inoculation of plants with selected and acclimatized microbes (bioaugmentation) has gained attention for phytoremediation of metal(loid)s polluted soils and technologies equipped with phytoremediation techniques are gaining momentum to decontaminate polluted soils on large scales (Sarwar et al. 2017; Kong and Glick 2017; Rehman et al. 2018). Agnello et al. (2016) reported highest degree of bioaugmentation-assisted phytoremediation of heavy metals and petroleum hydrocarbons in alfalfa plants when inoculated with *Pseudomonas aeuroginosa.* PGP yeast *Candida* VITJzN04 inoculated with *Saccharum officinarum* showed the degradation of lindane via biostimulation thus helped in phytoremediation (Salam et al. 2017).

11.4 Mechanisms Between the Plant and Microbes During Phytoremediation

Plant-associated microbes (endophytic, phyllospheric, and rhizospheric) have a variety of interactions with host plants which are generally mutual beneficial associations for both the systems (Newman and Reynolds 2004; Weyens et al. 2010). Some species of bacteria that can resist the high concentration of metals interact with hyperaccumulators (HA) to promote their growth are of prime importance in phytoremediation (Wang et al. 2017). These bacterial species either directly or indirectly assist the HA plants to remediate the soil while growing well in a polluted environment (Kumari et al. 2016).

11.4.1 Plant Growth-Promoting Bacteria (PGPB)

Plant growth-promoting bacteria (PGPB) are so named due to their potential to enhance plant growth. These bacteria may develop symbiotic/mutualistic associations with plants and may be found as free-living rhizospheric or endophytic bacteria. Genera including *Gluconacetobacter*, *Flavobacterium*, *Beijerinckia*, *Klebsiella*, *Erwinia*, *Enterobacter*, *Burkholderia*, *Pseudomonas*, *Serratia*, and *Bacillus* are among the beneficial PGP bacteria that control/promote plant growth (Dardanelli et al. 2010; Ashraf et al. 2017). Some species of microorganisms including bacteria (pollutant degrading and plant growth-promoting, etc.) and filamentous fungi alter the rhizospheric environment through bioavailability or stabilization and thus diminish the toxicity of HMs and POPs and ultimately appear to be beneficial in phytoremediation by increasing plant biomass (Wang et al. 2017; Cao and Liu 2015; Compant et al. 2010). PGPB help plant in the process of phytoremediation directly and indirectly (Glick 1995). Indirectly like acting as a biocontrol agent where they prevent or decrease the deleterious effect of pathogens on plants via several mechanisms such as competition for the binding sites on roots, production of cell wall-lysing enzymes against pathogenic fungi, iron depletion in the rhizosphere, antibiotic production, and providing systemic resistance in plants (Khasheii et al. 2016). The direct approach is through the production of substances that promotes plant growth and also increases mineral absorption. These functions are accomplished by the production of various metabolites such as siderophores, biosurfactants, indole acetic acid, ACC (1-aminocyclopropane-1-carboxylate) deaminases, and organic acids (Kong and Glick 2017). Secretion of different organic acids helps in solubilization and mobilization of HMs (Ma et al. 2016b). The role of siderophores in phytoremediation is to increase the availability of iron to the plant and simultaneously to decrease the iron for pathogens, besides this, they also act as scavenger of Reactive Oxygen Species (ROSs) generated by HMs (Chen et al. 2017; Ma et al. 2016b; Khasheii et al. 2016). On the other hand, ACC deaminase reduces the inhibitory effect of ethylene produced as a result of HM stress. It also increases the length and width of roots so as to increase the effectiveness of phytoremediation (Kumari et al. 2016). Indole acetic acid (IAA) induces the physiological changes and promotes root growth and metal absorption and finally upgrades the phytoremediation process. Nutrient deficiency is also an important factor that can limit phytoremediation. Many microorganisms play an important role in solubilizing minerals such as P, K, releasing nutrients and supplying them to plants and help in upgradation of phytoremediation (Schmalenberger et al. 2013; Grobelak et al. 2018).

Recently a research team reported that metal tolerant PGPB (*Ralstonia eutropha* 1C2 and *Chryseobacterium humi* ECP37) increases the Zn bioavailability in the soil along with enhancement in the bioaccumulation and biomass of Maize plants (Moreira et al. 2019). In addition to this, enzyme ACC deaminase synthesized by microorganisms are essential for plant growth since it facilitates plant growth by lowering plant's ethylene levels (Glick et al. 2007; Shahid et al. 2019). Usually microorganisms facilitate plant growth by using one or more PGP properties (Valentin-Vargas et al. 2014; Nayak et al. 2019). Fungi show a distinct advantage in phytoremediation due to their high tolerance for extreme pH, climate, nutrients, and heavy metals. Concomitant use of arbuscular mycorrhizae (AM) and *Aspergillus niger* is reported to increase the bioavailability and reducing the toxicity of HMs in plants (Gaur and Adholeya 2004; Shafiq and Jamil 2012). Ma and his research team inoculated bacteria (*Pseudomonas libanensis* TR1) and arbuscular mycorrhizal fungus (AMF) *Claroideoglomus claroideum* BEG210 with *Helianthus annuus* and found that this combination improves Ni exclusion, detoxification, phytostabilization and finally remediates nickel from contaminated saline soils (Ma et al. 2019). During plant–microbe interactions, POPs could be degraded by rhizobacteria (Glick 2010; Weyens et al. 2009). The interactions between plants and bacteria having catabolic genes have led to the evolution of a diverse variety of catabolic enzymes that can metabolize and detoxify the xenobiotics (Hong et al. 2015). The combined use of plants and bacteria has been exploited to enhance the phytoremediation of soil and water contaminated with different organic pollutants (Afzal et al. 2014a, b). Similarly, several studies were performed to explore the potential of plant-bacteria partnership for the remediation of POPs contaminated soil and water (Aken et al. 2009; Becerra-Castro et al. 2013; Jha and Jha 2015). Rhizobacteria capable of degrading different POPs have been isolated from rhizospheric soil of different plants and studied for POP degradation pathways and genes involved in POP degradation (Brazil et al. 1995; Fatima et al. 2015; Nicoara et al. 2014). A rhizobacterial strain *Bacillus* sp. PRB101 enhanced 92% endosulfan degradation after 120 days of inoculation (Rani et al. 2019). During POPs degradation, co-metabolism, i.e., similar metabolite secretions from the host plant supports bacterial population in the remediation processes (Bedard et al. 1986). Endophytic bacteria resides in plant tissues beneath the epidermal cell layers, from where they can colonize the internal tissues and form a range of different interaction thus help in the phytoremediation (Rajkumar et al. 2009). Endophytes also have PGPR properties like other rhizospheric microbes and they help in the phytoremediation process (Rajkumar et al. 2009; Ma et al. 2016b). The synergistic associations of different plant and microorganisms in the phytoremediation of POPs and HMs have been summarized in Table 11.1.

11.4.2 Biosurfactant

Biosurfactants are low-molecular-weight amphiphilic molecules consisting of a hydrophilic and a hydrophobic moiety, comprising a wide range of chemical structures, such as mycolic acid, glycolipids, lipopeptides, polysaccharide-protein complexes, phospholipids, fatty acids, etc. (Ma et al. 2016b; Banat et al. 2010). Biosurfactants produced by endophytic and rhizobacteria seemed to increase the bioavailability of poorly soluble HMs and POPs and hence upgrade phytoremediation (Aslund and Zeeb 2010; Federici et al. 2012; Ma et al. 2016b; Rastogi and Kumar 2020). Mehetre et al. (2019) studied the role of individual and mixed cultures of thermophilic and thermo-tolerant biosurfactant-producing bacteria in biodegradation of polycyclic aromatic hydrocarbons (PAHs). They reported relatively higher degradation of four PAHs namely ANT (92–96%), FLU (83–86%), PHE (16–54%), and PYR (51–71%) at 50 °C by *Aeribacillus* sp. (UCPS2) and mixed cultures. Biosurfactants bind with POPs as soluble

Table 11.1 Phytoremediation of POPs and HMs via synergistic interaction between plant and microbes

S. No.	Plants	Microbe(s)	Mechanisms	Results	Targeted HMs and POPs	Source of sample	Research area	Study
1	*Silene vulgaris*	*Proteus vulgaris* H7, *Pseudomonas* sp. H15, *Pseudomonas helmanticensis* H16	Phytoextraction with PGPB	*P. helmanticensis* H16 enhanced highest Zn and Cd accumulation in shoot by 43.8% and 112%, respectively	Cd, Zn	Soil	Katowice, Poland	Płociniczak et al. (2019)
2	*Glycine max*	*Mucor circinelloides*	Combined plant, microbe and through amendment material assisted phytoremediation	Bioavailability decreased by 4.7–20.50%, 4.3–11.9%, 21.9–183.9%, 13.6–27.3%, and 1–16.4% for Cu, Zn, Pb, Cd, and Mn, respectively	Cu, Zn, Pb, Cd, Mn	Mine tailings	Liaoning province, eastern China	Li et al. (2019a)
3	*Saxifraga stellaris* var. *alpigena*	*Mortierella* sp., *Stenotrophomonas maltophilia*	Endophytic fungus and bacterium influenced Cu tolerance of *S. stellaris*	Growth and survival rate enhanced by 70% and 130%, respectively, after microbial inoculation	Cu	Copper mine wastes	Vienna, Austria	Ryszka et al. (2019)
4	*Lactuca sativa*	*Rhodotorula mucilaginosa* CAM4	Multiple PGP properties assisted phytoremediation	Removal efficiency up to 80–88%	Al	Culture medium	La Araucania Region, Southern Chile	Silambarasan et al. (2019)
5	*Sedum plumbizincicola*	*Rhodococcus* sp. NSX2	Bacterial assisted phytoremediation	NSX2 promoted Cd accumulation in plant	Cd	Metal-contaminated soil	Guangdong Province, South China	Hou et al. (2019)
6	*Linum usitatissium*	*Serratia* sp. CP-13	PGPB assisted phytoremediation	CP-13 inhibited the Cd uptake and sequestered it in rhizosphere zone and reduced lipid peroxidation and maintained the optimum level of nutrients in plant	Cd	Wastewater	Faisalabad, Pakistan	Shahid et al. (2019)

7	*Brassica mapus*	*Cupriavidus necator* GX_ 5, *Sphingomonas* sp. GX _ 15 and *Curtobacterium* sp. GX_ 31	PGPR associated phytoremediation	Increased dry weight of above ground tissues and root biomass by 40.90–80.55% and 1.099–103.13%, respectively, Cd uptake in aerial parts and roots by 7.38–11.98% and 48.09–79.73%, respectively	Cd	Pot experiment	Guangxi, China	Li et al. (2019b)
8	*Solanum torvum* and *Cymbopogon citratus*	*Bacillus cereus* T1B3	PGPB assisted phytoremediation	Increased biomass, photosynthetic pigment content and act as Phytostabilizer of HMs in roots	Cr (VI), Fe, Mn, Cu, Cd, Ni, and Zn	Mine tailings contaminated soil	Bhubaneshwar, India	Nayak et al. (2019)
9	*Acacia mangium* and *Eucalyptus camaldulensis*	*Pseudomonas psychrophila*	Endophytic bacterial promoted phytoremediation	Increased Pb concentration in *A. mangium* root approximately 9.5% but not in *E. camalduensis*	Pb	Hydroponic	Bangkok, Thailand	Yongpisanphop et al. (2019)
10	*Lens culinaris*	*Bacillus* sp. KUJM2	PGP assisted bioremediation	Bacterial inoculation lead to higher shoot length (7.1–27.61%), shoot dry weight (18.22–36.3%), seed production (19.23–29.17%), and uptake was reduced in shoot by 67.02–79.85% and 65.94–78.08% and in seed up to 72.82–86.62 and 68.68–85.94%, respectively, under single and combined PTEs	As, Cd, Cu, Ni	PTEs (potentially toxic elements) contaminated state	West Bengal, India	Mondal et al. (2019)
11	*Glycine max*	*Kocuria rhizophila*	PGPB and citric acid (CA) assisted phytoextraction	Co-application of *K. rhizophila and* CA enhanced plant biomass about 38.73% and accumulation of Cd, Cr, Cu, and Ni about 40.63%, 56.39%,59.1%, and 39.76%, respectively	Cd, Cr, Cu, and Ni	Multi-metal-contaminated soil	Islamabad, Pakistan	Hussain et al. (2019)

(continued)

Table 11.1 (continued)

S. No.	Plants	Microbe(s)	Mechanisms	Results	Targeted HMs and POPs	Source of sample	Research area	Study
12	*Helianthus annus*	*Pseudomonas libanensis* TR1 and *Claroideoglomus claroideum* BEG210	Salt resistant plant beneficial Bacteria (PBB) and Arbuscular mycorrhizal fungus (AMF) assisted phytoremediation	Enhanced plant growth changed physiological status as well as Ni and Na^+ accumulation potential of *H. annus*	Ni	Metal-polluted saline soils	Coimbra, Portugal	Ma et al. (2019)
13	*Medicago sativa, Festuca arundinacea, Lolium perenne*	*Pseudomonas putida* UW4	PGPB enhanced phytoremediation (PEP)	UW4 significantly enhanced dissipation of phenanthrene and fluoranthene, and plant biomass dissipation depended on PGPB inoculum biomass, plant species, plant microbe specificity, and type of contaminants	PAH, Pb	PAH and HMs spiked soils	Birmingham, UK	Afegbua and Batty (2019)
14	*Helianthus annus*	*Paenibacillus* sp. IITOSM08, *Bacillus* sp. PRB77 and *Bacillus* sp. PRB101	PGPB boosted phytoremediation	Plant biomass was increased and maximum degradation (92%) was found after inoculation with PRB10	Endosulfan	Contaminated soil	Dhanbad, India	Rani et al. (2019)
15	*Arabidopsis thaliana*	*Mucor circinelloides* (MC) and *Trichoderma asperellum* (TA)	Endogenous fungi assisted phytoremediation	MC and TA strengthen *A. thaliana* tolerance to Cd and Pb with higher root length (40.19–117.50%) and shoot fresh weight (58.31–154.14%) of plant compared to non-inoculation, MC enhanced microbial diversity, and soil enzyme activity than TA	Cd and Pb	Contaminated soil	Jinan, China	Zhang et al. (2018)

16	*Rosmarinus officinalis*	*Glomus mosseae* and *Glomus interadices*	Bioaugmentation-assisted phytoremediation	Accumulation in root and shoot had a sequence Cu > Zn > Mn > Cd > Pb > Fe	Cd, Pb, Fe, Zn, and Cu	Ore mine soil contaminated with HMs	Kerman, Iran	Abbaslou et al. (2018)
17	*Pennisetum purpureum* and *Lumbricus terrestris*	VITMVCJ1*Klebsiella variicola*	Bioaugmentation-assisted bioremediation	Bioaugmentation process enhanced root, shoot length, chlorophyll content of plant and biofilm forming ability of microbes, reduction in the toxic effects of Pb on earth worm and plant, Pb accumulated mainly in the root of the plant and body moss of the earthworm	Pb	Industrial contaminated site	Vellore, India	Das and Osborne (2018)
18	*Solanum nigrum*	*Enterobacter* sp. LC1, *Kocuria* sp. LC2, *Kocuria* sp. LC3, *Enterobacter* sp. LC4, *Kocuria* sp. LC5, *Enterobacter* sp. LC6, and *Kosakonia* sp. LC7	Microbe assisted phytoremediation	Microbial consortium most effectively increase plant growth and strong upregulation of an MRP (multidrug resistance-associated protein) transporter in the root by As + endophytes which mediated As (V) to As (III) for detoxification/transportation	As	As contaminated habitats	Kolkata, India	Mukherjee et al. (2018)
19	*Leptochloa fusca* and *Brachiaria mutica*	*Enterobacter* sp. HU38, *Microbacterium arborescens* HU33, and *Pantoea stewartii* ASI11	Endophyte-assisted phytoremediation	Augmented endophytes increased root length (24–45%), shoot height (39–64%), chlorophyll content (20–55%), and overall biomass (32–61%) of plant	Cr	Contaminated soil	Faisalabad, Pakistan	Ahsan et al. (2018)
20	*Sarcocornia fruticosa*	*Saccharomyces cerevisiae*	Bioaugmentation with autochthonous yeast assisted phytoremediation	Yeast increased phytoremediation capacity of *S. fruticosa*	Cd, Pb, Ni, Cr, and Cu	Metal-contaminated sediments	Ria Formosa lagoon, Portugal	Said et al. (2018)
20	*Vetiveria zizanioides*	*Bacillus cereus* (T1B1)	PGPB assisted phytoremediation	Removal efficiency of *V. zizanioides* withT1B3 for Cr^{6+},		HM amended medium	Bhubaneshwar, India	Nayak et al. (2018)

(continued)

Table 11.1 (continued)

S. No.	Plants	Microbe(s)	Mechanisms	Results	Targeted HMs and POPs	Source of sample	Research area	Study
				Fe, Mn, Zn, Cd, Cu and Ni is 82%, 92%, 67%, 36%, 31%, 25%, and 43%, respectively	Cr^{6+}, Fe, Mn, Zn, Cd, Cu, Ni			
21	*Catharanthus roseus*	*Burkholderia cepacian* CS8	PGPB assisted phytoremediation	CS8 treated plant show higher root-shoot length, fresh and dry weight, chlorophyll content, reduced levels of malondialdehyde (MDA) and H_2O_2 under Cd stress	Cd	Cd contaminated soil	Lahore, Pakistan	Khan et al. (2018)
22	*Saccharum officinarum*	*Candida* VitJzN04	Bioaugmentation assisted phytoremediation	Removal rates in terms of half-life period were recorded as 7.1 days, 9.8 days, 43.3 days, and 13.3 days for immobilized yeast-plant, free yeast-plant, *Saccharum* and yeast, respectively, due to biostimulation of soil by PGP activity of yeast	Lindane	Doped garden soil	Cochin, India	Salam et al. (2017)
23	*Festuca arundinacea* and *Lolium perenne*	*Pseudomonas* sp. SB	Biosurfactant-producing bacterial-assisted phytoremediation	*Pseudomonas* sp. SB + fertilizer + tall fescue, fertilizer + perennial ryegrass, fertilizer + perennial ryegrass + Pseudomonas treatment effectively helped to increase the bioavailability of DDT and increased the removal rate about 65.6% than fertilizer control rate by 40.3%	DDT	DDT- contaminated soil	Nanjing, China	Wang et al. (2017)

24	*Populus* sp.	*Enterobacter* sp. PDN_3	Endophyte-assisted phytoremediation	Inoculated popular tree exhibited increased growth and reduced TCE toxicity effects with increase in trunk diameter, excreted 50% more chloride ion into the rhizosphere and inactivation of increased TCE metabolism than non-inoculated tree	TCE	Trichloroethylene (TCE) contaminated ground water	Washington, United States	Doty et al. (2017)
25	*Brassica napus*	*Fusarium* sp. $CBRF_{44}$, *Penicillium* sp. $CBRF_{65}$ and *Alternaria* sp. $CBSF_{68}$	Endophytic fungi assisted phytoremediation	$CBRF_{44}$ and $CBRF_{65}$ significantly increased the rape biomass and promoted the extraction efficacy of Pb and Cd, while $CBSF_{68}$ did not show similar results as above	Cd and Pb	Multi-metal-contaminated soils	Guangdong, Peoples of China	Shi et al. (2017)
26	*Catharanthus roseus*	*Pseudomonas fluorescens* RB4 and *Bacillus subtilis* 189	Microbe assisted phytoremediation	Co-cultivation of *P. fluorescens* RB4 and *B. subtilis* 189 exhibited 121%, 102%, and 117% higher fresh weight in Cu, Pb and both elements together also showed higher accumulation of Cu and Pb in *C. roseus* shoot	Cu and Pb	Cu and Pb contaminated soils	Lahore, Pakistan	Khan et al. (2017)

aggregates which ultimately release the pollutant from soil particles. However, the release of surfactants in the root exudates seems more promising as it may provide easy solubilization of POPs in plant rhizosphere (Passatore et al. 2014). These studies reveal that the combined use of plants and biosurfactant-producing bacteria can improve the bioavailability of organic pollutants through biosurfactant exudation and/or production and consequently the remediation of POPs contaminated environment. Rhizo-engineering has also gained attention for the removal of POPs from the environment (Thijs et al. 2016). A research team reported that biosurfactant amended phytoremediation may be a useful biotechnological approach for the remediation of petroleum hydrocarbon polluted soil (Liao et al. 2016). Biosurfactants work by decreasing the bonding between the metal and soil particles thus increasing the metal and POPs availability (Wang et al. 2017; Lal et al. 2018).

11.4.3 Exopolysaccharides (EPS)

Extracellular polymeric substance (EPS) consists mainly of polysaccharides, proteins, lipids, and nucleic acids (Rajkumar et al. 2013; Joshi and Juwarkar 2009). EPS has the ability to reduce the accessibility and bioavailability of HMs in the soil by forming metal complexes. They also provide essential micronutrient based environment to the plant thus help in indirect phytoremediation of HMs (Ma et al. 2016b; Lal et al. 2018). Wani et al. (2019) explored the role of *Bacillus subtilis* PAW_3 in the reduction of Cr (VI) and cowpea growth through its various PGP substances such as EPS along with ACC deaminase, siderophore, etc. that reduced Cr (VI) to Cr (III) and also showed the presence of chromium reductase gene (ChR) in PAW_3.

11.5 Conclusion

A consequence of anthropogenic activities is observed as an increase of contaminated areas, which should be remediated to prevent or mitigate the transfer of contaminants into terrestrial, atmospheric, or aquatic environments. Point and diffuse contamination by organic and inorganic pollutants cause wide concerns, and intentional or accidental introduction of these substances in the environment may represent serious impacts on public health. Therefore, urgent action is needed to address the world about elimination, remediation, cleanup, and safe disposal of such chemicals/ contaminants. The use of transgenic plants and biotechnological potential of microorganisms to resist and/or remove metals directly from polluted media and their beneficial effects on plant growth may lead to environmental-friendly and cost-effective strategies toward the reclamation of polluted soils.

Acknowledgement One of the author (Sheel Ratna) is highly thankful to research grant F.15-6 (NOV. 2017)/2018(NET) University Grant Commission, New Delhi, India. Facilities provided by the University (BBAU) and citation of research work of all the researchers are duly acknowledged.

References

Abbaslou H, Bakhtiari S, Hashemi SS (2018) Rehabilitation of iron ore mine soil contaminated with heavy metals using rosemary phytoremediation-assisted mycorrhizal arbuscular fungi bioaugmentation and fibrous clay mineral immobilization. Iran J Sci Technol, Trans A: Sci 42 (2):431–441

Adrees M, Ali S, Rizwan M, Ibrahim M, Abbas F, Farid M, Zia-ur-Rehman M, Irshad MK, Bharwana SA (2015) The effect of excess copper on growth and physiology of important food crops: a review. Environ Sci Pollut Res 22(11):8148–8162

Afegbua SL, Batty LC (2019) Effect of plant growth promoting bacterium; *Pseudomonas putida* UW4 inoculation on phytoremediation efficacy of monoculture and mixed culture of selected plant species for PAH and lead spiked soils. Int J Phytoremediation 21(3):200–208

Afzal M, Khan QM, Sessitsch A (2014a) Endophytic bacteria: prospects and applications for the phytoremediation of organic pollutants. Chemosphere 117:232–242

Afzal M, Shabir G, Tahseen R, Islam EU, Iqbal S, Khan QM, Khalid ZM (2014b) Endophytic *Burkholderia sp.* strain Ps JN improves plant growth and phytoremediation of soil irrigated with textile effluent. Clean: Soil, Air, Water 42(9):1304–1310

Agnello AC, Bagard M, Van Hullebusch ED, Esposito G, Huguenot D (2016) Comparative bioremediation of heavy metals and petroleum hydrocarbons co-contaminated soil by natural attenuation, phytoremediation, bioaugmentation and bioaugmentation-assisted phytoremediation. Sci Total Environ 563:693–670

Ahmadpour P, Ahmadpour F, Mahmud TMM, Abdu A, Soleimani M, Tayefeh FH (2012) Phytoremediation of heavy metals: a green technology. Afr J Biotechnol 11(76):14036–14043

Ahsan MT, Saeed A, Mustafa T, Afzal M (2018) Augmentation with potential endophytes enhances phytostabilization of Cr in contaminated soil. Environ Sci Pollut Res 25(7):7021–7032

Aken BV, Correa PA, Schnoor JL (2009) Phytoremediation of polychlorinated biphenyls: new trends and promises. Environ Sci Technol 44(8):2767–2776

Ali H, Khan E, Sajad MA (2013) Phytoremediation of heavy metals-concepts and applications. Chemosphere 91(7):869–881

Das A, Osborne JW (2018) Enhanced lead uptake by an association of plant and earthworm bioaugmented with bacteria. Pedosphere 28(2):311–322

Armendariz AL, Talano MA, Nicotra MFO, Escudero L, Breser ML, Porporatto C, Agostini E (2019) Impact of double inoculation with *Bradyrhizobium japonicum* E109 and *Azospirillum brasilense* Az39 on soybean plants grown under arsenic stress. Plant Physiol Biochem 138:26–35

Aroua I, Abid G, Souissi F, Mannai K, Nebli H, Hattab S, Borgi Z, Jebara M (2019) Identification of two pesticide-tolerant bacteria isolated from *Medicago sativa* nodule useful for organic soil phytostabilization. Int Microbiol 22(1):111–120

Arslan M, Imran A, Khan QM, Afzal M (2017) Plant–bacteria partnerships for the remediation of persistent organic pollutants. Environ Sci Pollut Res 24(5):4322–4336

Ashraf MA, Hussain I, Rasheed R, Iqbal M, Riaz M, Arif MS (2017) Advances in microbe-assisted reclamation of heavy metal contaminated soils over the last decade: a review. J Environ Manag 198:132–143

Aslund MW, Zeeb BA (2010) A review of recent research developments into the potential for phytoextraction of persistent organic pollutants (POPs) from weathered, contaminated soil. In:

Application of phytotechnologies for cleanup of industrial, agricultural, and wastewater contamination. Springer, Dordrecht, pp 35–59

Balazs HE, Schmid CA, Feher I, Podar D, Szatmari PM, Marincaş O, Zoltan R, Balazs ZR, Schroder P (2018) HCH phytoremediation potential of native plant species from a contaminated urban site in Turda, Romania. J Environ Manag 223:286–296

Balseiro-Romero M, Gkorezis P, Kidd PS, Van Hamme J, Weyens N, Monterroso C, Vangronsveld J (2017) Use of plant growth promoting bacterial strains to improve *Cytisus striatus* and *Lupinus luteus* development for potential application in phytoremediation. Sci Total Environ 581:676–688

Banat IM, Franzetti A, Gandolfi I, Bestetti G, Martinotti MG, Fracchia L, Smyth TJ, Marchant R (2010) Microbial biosurfactants production, applications and future potential. Appl Microbiol Biotechnol 87(2):427–444

Becerra-Castro C, Prieto-Fernandez A, Kidd PS, Weyens N, Rodriguez-Garrido B, Touceda-Gonzalez M, Acra MJ, Vangronsveld J (2013) Improving performance of *Cytisus striatus* on substrates contaminated with hexachlorocyclohexane (HCH) isomers using bacterial inoculants: developing a phytoremediation strategy. Plant Soil 362(1–2):247–260

Bedard DL, Unterman R, Bopp LH, Brennan MJ, Haberl ML, Johnson C (1986) Rapid assay for screening and characterizing microorganisms for the ability to degrade polychlorinated biphenyls. Appl Environ Microbiol 51(4):761–768

Boudh S, Singh JS (2019) Pesticide contamination: environmental problems and remediation strategies. In: Emerging and eco-friendly approaches for waste management. Springer, Singapore, pp 245–269

Brazil GM, Kenefick L, Callanan M, Haro A, De Lorenzo V, Dowling DN, O'gara F (1995) Construction of a rhizosphere pseudomonad with potential to degrade polychlorinated biphenyls and detection of bph gene expression in the rhizosphere. Appl Environ Microbiol 61 (5):1946–1952

Burges A, Epelde L, Blanco F, Becerril JM, Garbisu C (2017) Ecosystem services and plant physiological status during endophyte-assisted phytoremediation of metal contaminated soil. Sci Total Environ 584:329–338

Cao XF, Liu LP (2015) Using microorganisms to facilitate phytoremediation in mine tailings with multi heavy metals. In: Advanced materials research, vol 1094. Trans Tech Publications, Zurich, pp 437–440

Carson RL (1962) Silent spring. Riverside Press, Cambridge

Chaney RL (1983) Plant uptake of inorganic waste. In: Parr JF (ed) Land treatment of hazardous waters. OSTI GOV US Department of Energy N.P, 1983

Chen Y, Yang W, Chao Y, Wang S, Tang YT, Qiu RL (2017) Metal-tolerant *Enterobacter sp.* strain EG16 enhanced phytoremediation using *Hibiscus cannabinus* via siderophore-mediated plant growth promotion under metal contamination. Plant Soil 413(1–2):203–216

Chirakkara RA, Cameselle C, Reddy KR (2016) Assessing the applicability of phytoremediation of soils with mixed organic and heavy metal contaminants. Rev Environ Sci Biotechnol 15 (2):299–326

Compant S, Clement C, Sessitsch A (2010) Plant growth-promoting bacteria in the rhizo-and endosphere of plants: their role, colonization, mechanisms involved and prospects for utilization. Soil Biol Biochem 42(5):669–678

Coninx L, Martinova V, Rineau F (2017) Mycorrhiza-assisted phytoremediation. In: Advances in botanical research, vol 83. Academic Press, San Diego, pp 127–188

Cristaldi A, Conti GO, Jho EH, Zuccarello P, Grasso A, Copat C, Ferrante M (2017) Phytoremediation of contaminated soils by heavy metals and PAHs. A brief review. Environ Technol Innov 8:309–326

da Conceicao Gomes MA, Hauser-Davis RA, Suzuki MS, Vitoria AP (2017) Plant chromium uptake and transport, physiological effects and recent advances in molecular investigations. Ecotoxicol Environ Saf 140:55–64

Dardanelli MS, Manyani H, Gonzalez-Barroso S, Rodriguez-Carvajal MA, Gil-Serrano AM, Espuny MR, Lopez-Baena FJ, Bellogin RA, Megias M, Ollero FJ (2010) Effect of the presence of the plant growth promoting rhizobacterium (PGPR) *Chryseobacterium balustinum* Aur9 and salt stress in the pattern of flavonoids exuded by soybean roots. Plant Soil 328(1–2):483–493

Doty SL, Freeman JL, Cohu CM, Burken JG, Firrincieli A, Simon A, Khan Z, Iseberands JG, Lukas J, Blaylock MJ (2017) Enhanced degradation of TCE on a superfund site using endophyte-assisted poplar tree phytoremediation. Environ Sci Technol 51(17):10050–10058

Farid M, Ali S, Akram NA, Rizwan M, Abbas F, Bukhari SAH, Saeed R (2017) Phyto-management of Cr-contaminated soils by sunflower hybrids: physiological and biochemical response and metal extractability under Cr stress. Environ Sci Pollut Res 24(20):16845–16859

Fatima K, Afzal M, Imran A, Khan QM (2015) Bacterial rhizosphere and endosphere populations associated with grasses and trees to be used for phytoremediation of crude oil contaminated soil. Bull Environ Contam Toxicol 94(3):314–320

Federici E, Giubilei MA, Covino S, Zanaroli G, Fava F, D'Annibale A, Petruccioli M (2012) Addition of maize stalks and soybean oil to a historically PCB-contaminated soil: effect on degradation performance and indigenous microbiota. New Biotechnol 30(1):69–79

Gaur A, Adholeya A (2004) Prospects of arbuscular mycorrhizal fungi in phytoremediation of heavy metal contaminated soils. Curr Sci 86(4):528–534

Gerhardt KE, Gerwing PD, Greenberg BM (2017) Opinion: taking phytoremediation from proven technology to accepted practice. Plant Sci 256:170–185

Ghosh M, Singh SP (2005) A review on phytoremediation of heavy metals and utilization of it's by products. Asian J Energy Environ 6(4):214–231

Glick BR (1995) The enhancement of plant growth by free-living bacteria. Can J Microbiol 41 (2):109–117

Glick BR (2010) Using soil bacteria to facilitate phytoremediation. Biotechnol Adv 28(3):367–374

Glick BR, Todorovic B, Czarny J, Cheng Z, Duan J, McConkey B (2007) Promotion of plant growth by bacterial ACC deaminase. Crit Rev Plant Sci 26(5–6):227–242

Gregoraszczuk EL, Ptak A (2013) Endocrine-disrupting chemicals: some actions of POPs on female reproduction. Int J Endocrinol 2013:828532

Grobelak A, Kokot P, Hutchison D, Grosser A, Kacprzak M (2018) Plant growth-promoting rhizobacteria as an alternative to mineral fertilizers in assisted bioremediation-sustainable land and waste management. J Environ Manag 227:1–9

Gu CS, Liu LQ, Deng YM, Zhang YX, Wang ZQ, Yuan HY, Huang SZ (2017) De novo characterization of the *Iris lactea* var. *chinensis* transcriptome and an analysis of genes under cadmium or lead exposure. Ecotoxicol Environ Saf 144:507–513

Gupta P, Rani R, Usmani Z, Chandra A, Kumar V (2019) The role of plant-associated bacteria in phytoremediation of trace metals in contaminated soils. In: New and future developments in microbial biotechnology and bioengineering. Elsevier, Amsterdam, pp 69–76

Haslmayr HP, Meißner S, Langella F, Baumgarten A, Geletneky J (2014) Establishing best practice for microbially aided phytoremediation. Environ Sci Pollut Res 21(11):6765–6774

He L, Yang H, Yu Z, Tang J, Xu L, Chen X (2014) Arbuscular mycorrhizal fungal phylogenetic groups differ in affecting host plants along heavy metal levels. J Environ Sci 26(10):2034–2040

Hong Y, Liao D, Chen J, Khan S, Su J, Li H (2015) A comprehensive study of the impact of polycyclic aromatic hydrocarbons (PAHs) contamination on salt marsh plants Spartina alterniflora: implication for plant-microbe interactions in phytoremediation. Environ Sci Pollut Res 22(9):7071–7081

Horri K, Alfonso S, Cousin X, Munschy C, Loizeau V, Aroua S, Begout ML, Ernande B (2018) Fish life-history traits are affected after chronic dietary exposure to an environmentally realistic marine mixture of PCBs and PBDEs. Sci Total Environ 610:531–545

Hou J, Liu W, Wu L, Ge Y, Hu P, Li Z, Christie P (2019) *Rhodococcus sp.* NSX2 modulates the phytoremediation efficiency of a trace metal-contaminated soil by reshaping the rhizosphere microbiome. Appl Soil Ecol 133:62–69

Hussain A, Kamran MA, Javed MT, Hayat K, Farooq MA, Ali N, Ali M, Manghwar H, Jan F, Chaudhary HJ (2019) Individual and combinatorial application of *Kocuria rhizophila* and citric acid on phytoextraction of multi-metal contaminated soils by *Glycine max L.* Environ Exp Bot 159:23–33

Ibanez SG, Alderete LGS, Medina MI, Agostini E (2012) Phytoremediation of phenol using *Vicia sativa* L. plants and its antioxidative response. Environ Sci Pollut Res 19(5):1555–1562

Inostroza PA, Wicht AJ, Huber T, Nagy C, Brack W, Krauss M (2016) Body burden of pesticides and wastewater-derived pollutants on freshwater invertebrates: method development and application in the Danube River. Environ Pollut 214:77–85

Jamieson AJ, Malkocs T, Piertney SB, Fujii T, Zhang Z (2017) Bioaccumulation of persistent organic pollutants in the deepest ocean fauna. Nat Ecol Evol 1(3):1

Jha P, Jha PN (2015) Plant-microbe partnerships for enhanced biodegradation of polychlorinated biphenyls. In: Plant microbes symbiosis: applied facets. Springer, New Delhi, pp 95–110

Joshi PM, Juwarkar AA (2009) In vivo studies to elucidate the role of extracellular polymeric substances from *Azotobacter* in immobilization of heavy metals. Environ Sci Technol 43 (15):5884–5889

Kerfahi D, Ogwu MC, Ariunzaya D, Balt A, Davaasuren D, Enkhmandal O, Purevsuren T, Batbaatar A, Tibbett M, Undrakhbold S, Boldgiv B (2019) Metal-tolerant fungal communities are delineated by high zinc, lead, and copper concentrations in Metalliferous Gobi Desert Soils. Microb Ecol 79:420. https://doi.org/10.1007/s00248-019-01405-8

Khan A, Sharif M, Ali A, Shah SNM, Mian IA, Wahid F, Jan B, Adnan M, Nawaz S, Ali N (2014) Potential of AM fungi in phytoremediation of heavy metals and effect on yield of wheat crop. Am J Plant Sci 5(11):1578–1586

Khan WU, Ahmad SR, Yasin NA, Ali A, Ahmad A (2017) Effect of *Pseudomonas fluorescens* RB4 and *Bacillus subtilis* 189 on the phytoremediation potential of *Catharanthus roseus* (L.) in Cu and Pb-contaminated soils. Int J Phytoremediation 19(6):514–521

Khan WU, Yasin NA, Ahmad SR, Ali A, Ahmad A, Akram W, Faisal M (2018) Role of *Burkholderia cepacia* CS8 in Cd-stress alleviation and phytoremediation by *Catharanthus roseus*. Int J Phytoremediation 20(6):581–592

Khasheii B, Anvari S, Jamalli A (2016) Frequency evaluation of genes encoding siderophores and the effects of different concentrations of Fe ions on growth rate of uropathogenic *Escherichia coli*. Iran J Microbiol 8(6):359–365

Kong Z, Glick BR (2017) The role of plant growth-promoting bacteria in metal phytoremediation. In: Advances in microbial physiology, vol 71. Academic Press, San Diego, pp 97–132

Kumar V (2019) Synergism between microbes and plants for soil contaminants mitigation. In: Amelioration technology for soil sustainability. IGI Global, Hershey, pp 101–134

Kumari S, Varma A, Tuteja N, Choudhary DK (2016) Bacterial ACC-deaminase: an eco-friendly strategy to cope abiotic stresses for sustainable agriculture. In: Plant-microbe interaction: an approach to sustainable agriculture. Springer, Singapore, pp 165–185

Lal S, Ratna S, Said OB, Kumar R (2018) Biosurfactant and exopolysaccharide-assisted rhizobacterial technique for the remediation of heavy metal contaminated soil: an advancement in metal phytoremediation technology. Environ Technol Innov 10:243–263

Li X, Wang X, Chen Y, Yang X, Cui Z (2019a) Optimization of combined phytoremediation for heavy metal contaminated mine tailings by a field-scale orthogonal experiment. Ecotoxicol Environ Saf 168:1–8

Li X, Yan Z, Gu D, Li D, Tao Y, Zhang D, Su L, Ao Y (2019b) Characterization of cadmium resistant rhizobacteria and their promotion effects on *Brassica napus* growth and cadmium uptake. J Basic Microbiol 59(6):579–590

Liao C, Xu W, Lu G, Deng F, Liang X, Guo C, Dang Z (2016) Biosurfactant-enhanced phytoremediation of soils contaminated by crude oil using maize (*Zea mays.* L). Ecol Eng 92:10–17

Lim MW, Von Lau E, Poh PE (2016) A comprehensive guide of remediation technologies for oil contaminated soil-present works and future directions. Mar Pollut Bull 109(1):14–45

Ma Y, Zhang C, Oliveira RS, Freitas H, Luo Y (2016a) Bioaugmentation with endophytic bacterium E6S homologous to *Achromobacter piechaudii* enhances metal rhizoaccumulation in host *Sedum plumbizincicola*. Front Plant Sci 7:75. https://doi.org/10.3389/fpls.2016.00075

Ma Y, Rajkumar M, Zhang C, Freitas H (2016b) Beneficial role of bacterial endophytes in heavy metal phytoremediation. J Environ Manag 174:14–25

Ma Y, Rajkumar M, Oliveira RS, Zhang C, Freitas H (2019) Potential of plant beneficial bacteria and arbuscular mycorrhizal fungi in phytoremediation of metal-contaminated saline soils. J Hazard Mater 379:120813

Mahar A, Wang P, Ali A, Awasthi MK, Lahori AH, Wang Q, Li R, Zhang Z (2016) Challenges and opportunities in the phytoremediation of heavy metals contaminated soils: a review. Ecotoxicol Environ Saf 126:111–121

Mehetre GT, Dastager SG, Dharne MS (2019) Biodegradation of mixed polycyclic aromatic hydrocarbons by pure and mixed cultures of biosurfactant producing thermophilic and thermo-tolerant bacteria. Sci Total Environ 679:52–60

Mishra A, Mishra SP, Arshi A, Agarwal A, Dwivedi SKS (2020) Plant-microbe interactions for bioremediation and phytoremediation of environmental pollutants and agro-ecosystem development. In: Bioremediation of industrial waste for environmental safety. Springer, Singapore, pp 415–436

Mondal M, Biswas JK, Tsang YF, Sarkar B, Sarkar D, Rai M, Sarkar SK, Hooda PS (2019) A wastewater bacterium *Bacillus sp.* KUJM2 acts as an agent for remediation of potentially toxic elements and promoter of plant (*Lens culinaris*) growth. Chemosphere 232:439–452

Moreira H, Pereira SI, Marques AP, Rangel AO, Castro PM (2019) Effects of soil sterilization and metal spiking in plant growth promoting rhizobacteria selection for phytotechnology purposes. Geoderma 334:72–81

Mukherjee G, Saha C, Naskar N, Mukherjee A, Mukherjee A, Lahiri S, Majumadar AL, Seal A (2018) An endophytic bacterial consortium modulates multiple strategies to improve arsenic phytoremediation efficacy in *Solanum nigrum*. Sci Rep 8(1):6979

Nayak AK, Panda SS, Basu A, Dhal NK (2018) Enhancement of toxic Cr (VI), Fe, and other heavy metals phytoremediation by the synergistic combination of native *Bacillus cereus* strain and *Vetiveria zizanioides* L. Int J Phytoremediation 20(7):682–691

Nayak AK, Basu A, Panda SS, Dhal NK, Lal RK (2019) Phytoremediation of heavy metal-contaminated tailings soil by symbiotic interaction of *Cymbopogon citratus* and *Solanum torvum* with *Bacillus Cereus* T1B3. Soil Sediment Contam Int J 28(6):547–568

Newman LA, Reynolds CM (2004) Phytodegradation of organic compounds. Curr Opin Biotechnol 15(3):225–230

Nicoara A, Neagoe A, Stancu P, de Giudici G, Langella F, Sprocati AR, Lordache V, Kothe E (2014) Coupled pot and lysimeter experiments assessing plant performance in microbially assisted phytoremediation. Environ Sci Pollut Res 21(11):6905–6920

Olatunji OS (2019) Evaluation of selected polychlorinated biphenyls (PCBs) congeners and dichlorodiphenyltrichloroethane (DDT) in fresh root and leafy vegetables using GC-MS. Sci Rep 9(1):538

Padmavathiamma PK, Li LY (2007) Phytoremediation technology: hyper-accumulation metals in plants. Water Air Soil Pollut 184(1–4):105–126

Passatore L, Rossetti S, Juwarkar AA, Massacci A (2014) Phytoremediation and bioremediation of polychlorinated biphenyls (PCBs): state of knowledge and research perspectives. J Hazard Mater 278:189–202

Pilon-Smits E (2005) Phytoremediation. Annu Rev Plant Biol 56:15–39

Pinto AP, De Varennes A, Fonseca R, Teixeira DM (2015) Phytoremediation of soils contaminated with heavy metals: techniques and strategies. In: Phytoremediation. Springer, Cham, pp 133–155

Pinto AP, de Varennes A, Lopes ME, Teixeira DM (2016) Biological approaches for remediation of metal-contaminated sites. In: Phytoremediation. Springer, Cham, pp 65–112

Pinto AP, de Varennes A, Dias CMB, Lopes ME (2018) Microbial-assisted phytoremediation: a convenient use of plant and microbes to clean up soils. In: Phytoremediation. Springer, Cham, pp 21–87

Płociniczak T, Chodor M, Pacwa-Płociniczak M, Piotrowska-Seget Z (2019) Metal-tolerant endophytic bacteria associated with *Silene vulgaris* support the Cd and Zn phytoextraction in non-host plants. Chemosphere 219:250–260

Pulford ID, Watson C (2003) Phytoremediation of heavy metal-contaminated land by trees—a review. Environ Int 29(4):529–540

Rajkumar M, Ae N, Freitas H (2009) Endophytic bacteria and their potential to enhance heavy metal phytoextraction. Chemosphere 77(2):153–160

Rajkumar M, Ae N, Prasad MNV, Freitas H (2010) Potential of siderophore-producing bacteria for improving heavy metal phytoextraction. Trends Biotechnol 28(3):142–149

Rajkumar M, Prasad MNV, Swaminathan S, Freitas H (2013) Climate change driven plant-metal-microbe interactions. Environ Int 53:74–86

Rani R, Kumar V, Usmani Z, Gupta P, Chandra A (2019) Influence of plant growth promoting rhizobacterial strains *Paenibacillus sp.* IITISM08, *Bacillus sp.* PRB77 and *Bacillus sp.* PRB101 using *Helianthus annuus* on degradation of endosulfan from contaminated soil. Chemosphere 225:479–489

Raskin I, Kumar PN, Dushenkov S, Salt DE (1994) Bioconcentration of heavy metals by plants. Curr Opin Biotechnol 5(3):285–229

Rastogi S, Kumar J, Kumar R (2019) An investigation into the efficacy of fungal biomass as a low cost bio-adsorbent for the removal of lead from aqueous solutions. Int Res J Eng Technol 6 (3):7144–7149

Rastogi S, Kumar R (2020) Remediation of heavy metals using non-conventional adsorbents and biosurfactant-producing bacteria. In: Kumar V, Singh J, Kumar P (Eds) Environmental Degradation: Causes and Remediation Strategies. pp 133–153

Rehman K, Imran A, Amin I, Afzal M (2018) Inoculation with bacteria in floating treatment wetlands positively modulates the phytoremediation of oil field wastewater. J Hazard Mater 349:242–251

Ryszka P, Lichtscheidl I, Tylko G, Turnau K (2019) Symbiotic microbes of *Saxifraga stellaris spp. alpigena* from the copper creek of Schwarzwand (Austrian Alps) enhance plant tolerance to copper. Chemosphere 228:183–194

Said OB, da Silva MM, Hannier F, Beyrem H, Chicharo L (2018) Using *Sarcocornia fruticose* and *Saccharomyces cerevisiae* to remediate metal contaminated sediments of the Ria Formosa lagoon (SE Portugal). Ecohydrol Hydrobiol 19:588. https://doi.org/10.1016/j.ecohyd.2018.10.002

Salam JA, Hatha MA, Das N (2017) Microbial-enhanced lindane removal by sugarcane (*Saccharum officinarum*) in doped soil-applications in phytoremediation and bioaugmentation. J Environ Manag 193:394–399

Salt DE, Blaylock M, Kumar NP, Dushenkov V, Ensley BD, Chet I, Raskin I (1995) Phytoremediation: a novel strategy for the removal of toxic metals from the environment using plants. Biotechnology 13(5):468–475

Samardjieva KA, Tavares F, Pissarra J (2015) Histological and ultrastructural evidence for zinc sequestration in *Solanum nigrum* L. Protoplasma 252(1):345–357

Sampaio CJ, de Souza JR, Damiao AO, Bahiense TC, Roque MR (2019) Biodegradation of polycyclic aromatic hydrocarbons (PAHs) in a diesel oil-contaminated mangrove by plant growth-promoting rhizobacteria. 3 Biotech 9(4):155. https://doi.org/10.1007/s13205-019-1686-8

Sarwar N, Imran M, Shaheen MR, Ishaque W, Kamran MA, Matloob A, Rehim A, Hussain S (2017) Phytoremediation strategies for soils contaminated with heavy metals: modifications and future perspectives. Chemosphere 171:710–721

Saxena G, Bharagava RN (2017) Organic and inorganic pollutants in industrial wastes: ecotoxicological effects, health hazards, and bioremediation approaches. In: Environmental pollutants and their bioremediation approaches. CRC Press, Milton, pp 23–56

Saxena G, Purchase D, Mulla SI, Saratale GD, Bharagava RN (2019) Phytoremediation of heavy metal-contaminated sites: eco-environmental concerns, field studies, sustainability issues, and future prospects. Rev Environ Contam Toxicol 249:71–131

Schmalenberger A, O'Sullivan O, Gahan J, Cotter PD, Courtney R (2013) Bacterial communities established in bauxite residues with different restoration histories. Environ Sci Technol 47 (13):7110–7119

Shafiq M, Jamil S (2012) Role of plant growth regulators and a saprobic fungus in enhancement of metal phytoextraction potential and stress alleviation in pearl millet. J Hazard Mater 237:186–193

Shahid M, Javed MT, Masood S, Akram MS, Azeem M, Ali Q, Gilani R, Basit F, Abid A, Lindberg S (2019) *Serratia sp.* CP13 augments the growth of cadmium (Cd) stressed *Linum usitatissimum* L. by limited Cd uptake, enhanced nutrient acquisition and antioxidative potential. J Appl Microbiol 126(6):1708–1721

Sharma P, Pandey S (2014) Status of phytoremediation in world scenario. Int J Environ Bioremed Biodegrad 2(4):178–191

Shayler H, McBride M, Harrison E (2017) Cornell Waste Management Institute. Retrieved from Department of Crop & Soil Sciences, http://cwmi.css.cornell.edu

Sheoran V, Sheoran AS, Poonia P (2010) Role of hyperaccumulators in phytoextraction of metals from contaminated mining sites: a review. Crit Rev Environ Sci Technol 41(2):168–214

Shi Y, Xie H, Cao L, Zhang R, Xu Z, Wang Z, Deng Z (2017) Effects of Cd-and Pb-resistant endophytic fungi on growth and phytoextraction of *Brassica napus* in metal-contaminated soils. Environ Sci Pollut Res 24(1):417–426

Shiri M, Rabhi M, Abdelly C, Amrani AEI (2015) The halophytic model plant *Thellungiella salsuginea* exhibited increased tolerance to phenanthrene-induced stress in comparison with the glycophitic one *Arabidopsis thaliana*: application for phytoremediation. Ecol Eng 74:125–134

Silambarasan S, Logeswari P, Cornejo P, Abraham J, Valentine A (2019) Simultaneous mitigation of aluminum, salinity and drought stress in *Lactuca sativa* growth via formulated plant growth promoting *Rhodotorula mucilaginosa* CAM4. Ecotoxicol Environ Saf 180:63–72

Srut M, Menke S, Hockner M, Sommer S (2019) Earthworms and cadmium-heavy metal resistant gut bacteria as indicators for heavy metal pollution in soils? Ecotoxicol Environ Saf 171:843–853

Staniforth S (ed) (2013) Historical perspectives on preventive conservation. Getty Publications, Los Angeles, p 6

Stockholm convention (COP9) on POPs report online (n.d.) https://www.meti.go.jp/english/press/2019/0514_001.html. Accessed 05 Jul 2019

Thijs S, Sillen W, Rineau F, Weyens N, Vangronsveld J (2016) Towards an enhanced understanding of plant–microbiome interactions to improve phytoremediation: engineering the metaorganism. Front Microbiol 7:341. https://doi.org/10.3389/fmicb.2016.00341

UNEP (2006) https://www.epa.gov/international-cooperation/persistent-organic-pollutants-global-issue-global-response. Accessed 2019

Valentin-Vargas A, Root RA, Neilson JW, Chorover J, Maier RM (2014) Environmental factors influencing the structural dynamics of soil microbial communities during assisted phytostabilization of acid-generating mine tailings: a mesocosm experiment. Sci Total Environ 500:314–324

Vangronsveld J, Herzig R, Weyens N, Boulet J, Adriaensen K, Ruttens A, Thewys T, Vassilev A, Meers E, Nehnevajova E, van der Lelie D (2009) Phytoremediation of contaminated soils and groundwater: lessons from the field. Environ Sci Pollut Res 16(7):765–794

Wang B, Wang Q, Liu W, Liu X, Hou J, Teng Y, Luo Y, Christie P (2017) Biosurfactant-producing microorganism Pseudomonas sp. SB assists the phytoremediation of DDT-contaminated soil by two grass species. Chemosphere 182:137–142

Wani RA, Ganai BA, Shah MA, Uqab B (2017) Heavy metal uptake potential of aquatic plants through phytoremediation technique—a review. J Bioremed Biodegr 8(404):2. https://doi.org/10.4172/2155-6199.1000404

Wani PA, Garba SH, Wahid S, Hussaini NA, Mashood KA (2019) Prevention of oxidative damage and phytoremediation of Cr (VI) by chromium (VI) reducing *Bacillus subtilus* PAW3 in cowpea plants. Bull Environ Contam Toxicol 103(3):476–483

Weyens N, van der Lelie D, Taghavi S, Newman L, Vangronsveld J (2009) Exploiting plant-microbe partnerships to improve biomass production and remediation. Trends Biotechnol 27 (10):591–598

Weyens N, Croes S, Dupae J, Newman L, van der Lelie D, Carleer R, Vangronsveld J (2010) Endophytic bacteria improve phytoremediation of Ni and TCE co-contamination. Environ Pollut 158(7):2422–2427

Yadav A, Chowdhary P, Kaithwas G, Bharagava RN (2017) Toxic metals in environment, threats on ecosystem and bioremediation approaches. In: Handbook of metal microbe interactions and bioremediation. CRC Press, Boca Raton, p 813

Yongpisanphop J, Babel S, Kurisu F, Kruatrachue M, Pokethitiyook P (2019) Isolation and characterization of Pb-resistant plant growth promoting endophytic bacteria and their role in Pb accumulation by fast-growing trees. Environ Technol:1–9. https://doi.org/10.1080/09593330.2019.1615993

Zhang X, Li X, Yang H, Cui Z (2018) Biochemical mechanism of phytoremediation process of lead and cadmium pollution with *Mucor circinelloides* and *Trichoderma asperellum*. Ecotoxicol Environ Saf 157:21–28

Zhao X, Huang J, Lu J, Sun Y (2019) Study on the influence of soil microbial community on the long-term heavy metal pollution of different land use types and depth layers in mine. Ecotoxicol Environ Saf 170:218–226

Chapter 12
Involvement of Synergistic Interactions Between Plant and Rhizospheric Microbes for the Removal of Toxic/Hazardous Contaminants

Sandhya Mishra, Sikandar I. Mulla, Sudipta Saha, Arun S. Kharat, Nandkishor More, and Ram Naresh Bharagava

Abstract The abundance of toxic heavy metals, organic and inorganic pollutants, xenobiotic chemicals and contaminants pose a major threat to the environment which is linked to the health of the living systems including human beings and needs an eco-friendly remediation technology. This technology may not only reduce the pollution level by degradation or detoxification of such hazardous pollutants but also does not generate any secondary or intermediate pollutants. Phytoremediation is an advanced and efficient plant-based treatment technology that takes advantage of the remarkable ability of the plants to tolerate and grow at higher concentration of pollutants/contaminants. The mechanisms used by the plants for the removal of toxic components from the surroundings include accumulation, absorption or transformation of the toxic compounds in their tissues, leaf, stem and roots. In addition to this, plant-associated microbes also play an important role to enhance the efficiency of

---

S. Mishra
Key Laboratory of Microbial Signals and Disease Control, Integrative Microbiology Research Centre, South China Agricultural University, Guangzhou, P.R. China

S. I. Mulla
Department of Biochemistry, School of Applied Sciences, REVA University, Bangalore, India

S. Saha
Department of Pharmaceutical Sciences, B B Ambedkar University, Lucknow, Uttar Pradesh, India

A. S. Kharat
Laboratory of Applied Microbiology, School of Life Sciences, Jawaharlal Nehru University, New Delhi, India

N. More
Department of Environmental Science, B B Ambedkar University, Lucknow, Uttar Pradesh, India

R. N. Bharagava (✉)
Laboratory of Bioremediation and Metagenomics Research, Department of Microbiology, B B Ambedkar University, Lucknow, Uttar Pradesh, India

A. Sharma (ed.), *Microbes and Signaling Biomolecules Against Plant Stress*, Rhizosphere Biology, https://doi.org/10.1007/978-981-15-7094-0_12

phytoremediation process. In recent years, understanding the knowledge of the synergistic relationship between plants and microbes for the removal of unwanted chemicals becomes an interesting area of research. Understanding the physiological and molecular mechanisms of plant–microbe interaction helps the environmentalists to get an insight of this technology and to improve and expand new horizons of phytoremediation.

12.1 Introduction

Widespread pollution of aquatic and terrestrial ecosystems with organic and inorganic pollutants becomes a major problem for the entire earth. These pollutants not only affect living beings but also disturb the ecological balance, which results in ozone depletion and problems of global warming. The number of toxic and restricted pollutants, chemicals and metals are used regularly in industrial and agricultural practices. For instance, long-term applications of pesticides, herbicides, chlorinated compounds, urea, phosphates, nitrates and chemical fertilizers for crop production and pest control are the principal sources of soil contamination. Wastewaters from textiles, tanneries, distilleries, electroplating and printing industries, pharmaceutical, food, paper and metallurgical industries contain tons of hazardous toxic pollutants, endocrine disrupting chemicals, phthalates, recalcitrant organic pollutants (ROPs), polycyclic aromatic hydrocarbons (PAHs), polychlorinated bis-phenyls (PCBs) hydrocarbons, absorbable organic halides (AOX), radioactive elements, organic solvents and corrosive infectious metals like cadmium, lead, chromium, arsenic and mercury (Dixit et al. 2015; Bharagava and Mishra 2018; Mishra et al. 2019).

These pollutants are highly dangerous, persistent, carcinogenic and mostly non-biodegradable in nature due to their chemical composition and toxicity. They get accumulated in aquatic and terrestrial ecosystems and cause severe soil and water pollution by deteriorating their natural and physicochemical qualities (Mishra and Bharagava 2016). Gradual increase in soil and water pollution leads to severe health hazards, poor soil fertility, unsafe food and water quality. Therefore, cleaning of the contaminated environment is a subject of utmost concern to human beings and needs an effective and advanced remedial technology that can treat all kinds of pollutants and chemicals in an effective and eco-friendly manner (Lee 2013). However, microbial degradation is an effective approach, many of these pollutants can be metabolized by bacterial, fugal and algal species but this process is very slow and inefficient because of the relatively low number of degradative microorganisms and after some time, the degradative ability decreases due to their stationary phase and tolerance capacity (Glick 2003). Besides, microbial remediation, "phytoremediation" could be the most promising and potential approach for the speedy degradation and mitigation of contaminated sites.

Phytoremediation is a relatively new concept for the removal of unwanted contaminants from the environment. It may be considered as advanced, solar-driven, environment friendly, cost-effective and economically feasible technology for the betterment of our environment (Glick 2012; Lee 2013). Phytoremediation, a plant-based technology directly utilizes green plants in situ or ex situ for the removal or transformation of toxic environmental pollutants into less toxic state (Ma et al. 2016). Generally, phytoremediation is carried out by using hyperaccumulators plants (Glick 2010; Lotfy and Mostafa 2014). Hyperaccumulators plants are capable to absorb high concentration of pollutants through roots. The root region of these plants is highly rich in microbial diversity including endophytes, mycorrhiza and plant growth promoting rhizospheric bacteria (PGPR). These root microorganisms are very helpful in phytoremediation process as they enhance plant growth and tolerance level for toxic chemicals and uptake of nutrients and supplements (Bhaduri et al. 2015).

Thus, the interaction between the plants and their root microorganism is a very interesting and extensive topic for the current research trend. Hence, this chapter describes about phytoremediation technology, which deals with not only plants but also root microorganism, plant–microbe interaction and their role in decontamination or degradation of environmental pollutants.

12.2 Classification of Environmental Pollutants

Based on the chemical nature, biodegradability and toxicity, environmental pollutants can be classified into two major classes: biodegradable and non-biodegradable (Table 12.1).

12.2.1 Biodegradable Pollutants

Those pollutants, which are converted, degraded or transformed from their highly toxic state to non-toxic or less toxic states by microbial, plants or natural processing, are called biodegradable pollutants. These pollutants mostly originated from plants, animals or other organic sources and can be commonly found in municipal solid wastes like food waste, paper waste, manure, slaughterhouse waste, etc. Biodegradable pollutants can be of two types based on their degradability.

a. *Degradable non-persistent pollutants*: These pollutants are rapidly decomposed by natural processes.
b. *Degradable persistent pollutants:* These pollutants are degraded by natural processes, but their rate of decomposition is very slow and they persist in the environment for long time. Examples of this class are: pesticides, hydrocarbons, biodegradable plastics, etc.

Table 12.1 Comparison between biodegradable and non-biodegradable waste

S. No.	Properties	Biodegradable pollutants	Non-biodegradable pollutants
1.	Origin	These pollutants are mostly of plant, animal or organic origin	These pollutants are mostly of inorganic origin
2.	Nature	These are degraded and decomposed through microbial or natural processing	Non-degradable pollutants cannot be degraded by microbial actions
3.	Persistence	They can be or cannot be persisted for a long time into the environment	They are persisted for a long time into the environment
4.	Accumulation	Cannot be accumulated and if accumulated can be used up in short time as being a part of bio-geochemical cycle and give back rapid turnover	Rapidly accumulated into the environment, highly toxic and cannot enter into the biogeo-chemical cycle
5.	Degradation rate	The degradation process is rapid	The degradation process is very slow and takes a long time to process
6.	Toxicity	The toxicity is reduced by the degradation process	These are highly toxic and hazardous
7.	Uses	After degradation process their degraded products are used to produce biogas, manure and organic compost	These can be recycled but the process is highly expansive and inefficient
	Examples	Municipal waste, pesticides, hydrocarbons, sewage, etc.	Plastics and radio-nuclear waste

12.2.2 *Non-biodegradable*

The pollutants that cannot be degraded/converted or transformed by plants or microbial actions are called non-biodegradable wastes. These pollutants are highly toxic and hazardous to human health and the environment. Non-biodegradable pollutants persist in the environment for a long time and these are originated from inorganic substances.

12.3 Sources, Courses and Toxicity of Environmental Pollutants

12.3.1 *Toxic Heavy Metals*

Heavy metals are highly toxic and belong to a heterogeneous group of elements having higher density (<5 g/cm^3) and atomic weight. Examples of heavy metals are arsenic (As) cadmium (Cd), chromium (Cr), lead (Pb), nickel (Ni), mercury (Hg) and zinc (Zn). These are non-degradable in nature because of their complex chemical alignment and thus can persist in the air, soil and water for years (Nagajyoti et al. 2010). Industrial activities are mainly responsible for the entrance of these metals into the environment beyond their acceptable limits. Industrial processes like smelting, tanning and mining, pulping, dyeing, food processing, electroplating, chemical manufacturing release tons of metals in wastewater or in the solid waste form (Gadd 2010). Prolonged exposure of such toxic metals may result in their easy accumulation inside the living organisms which causes carcinogenic, genotoxic, mutagenic, teratogenic and neurogenic disabilities (Wuana and Okieimen 2011).

The excess concentration of Cd affects the metabolic rate of Ca^{+2} ions resulting in low muscle power, bone fractures, cartilage disorders and kidney damage (ATSDR 2008). It also shows enzyme inhibiting reactions and causes nutritional deficiency in plants (Irfan et al. 2013). Lead causes infertility and memory dysfunction. It prominently pays for the severe Pb-poisoning in infants and young children, which may even lead to death by altering cell signalling, protein folding and inhibitory effects on di- and monovalent cations (Flora et al. 2007). Arsenic causes neurotoxicity, gastrointestinal toxicity and nephrotoxicity. Chromium toxicity may lead to severe health implications like nasal irritation, lung carcinoma and hearing impairment. It causes mutations and genotoxic abnormalities (IARC 1990; Tchounwou et al. 2003).

12.3.2 Polycyclic Aromatic Hydrocarbons (PAHs)

The level of polycyclic aromatic hydrocarbons (PAHs), a group of predominantly recalcitrant organic compounds having two rings is found apparently increased in the environment because of several anthropogenic activities. PAHs are widely distributed in the atmosphere, volcanic eruptions and forest fire (Lee and Vu 2010). PAHs are well-known "*persistent organic pollutants*" that are continuously concentrated in the environment by incomplete combustion of fossil fuels, vehicular traffic, waste incineration and industrial processing (Zhang and Tao 2009). Ingestion and inhalation of smoked grilled vegetables and meats, tobacco smoking, polluted air and water are some common sources of PAHs exposure to humans and the environment (Lundstedt et al. 2007; Bansal and Kim 2015; Ramos and Moorthy 2005).

PAHs compounds are regarded as cancer and mutation causing agents and long-term exposure to these compounds results in potential health risks. These are acutely toxic even at very low concentrations to cause necrotic toxicity. Environmental factors such as solar radiation can considerably increase the toxicity level of some PAHS as these compounds absorb ultraviolet and visible light and generate reactive oxygen species (ROS), which can significantly harm living systems (Landrum et al. 1987; Marzooghi and DI Toro 2017).

Long-term exposure of hydrocarbons can severely affect central nervous systems in humans and animals and may cause skin, lung and kidney cancer, respiratory dysfunction, inhibits enzymatic activities and disrupts endocrine system (Singh et al. 2004; Gkorezis et al. 2016). Anthracene, benzopyrene and naphthalene are well-known skin irritant and sensitizer and cause allergic and inflammatory skin problems

(Unwin et al. 2006). Increased pollution of air, soils and water with excessive concentration of PAHs is a serious worldwide environmental problem, which drastically affects the normal functioning of human, animals and plants.

12.3.3 *Pesticides and their Residues*

Extensive and widespread use of chemical pesticides in agricultural practices and households to kill pests, insects and weeds to protect plant crops is also contributing towards environmental pollution and contamination of soil and water ecosystems. Pesticides are highly toxic, recalcitrant and persistent organic pollutants, with low-biodegradability due to their high solubility, mobility, long-term stability and chemical complexity (Olette et al. 2008; Bolognesi and Merlo 2011). Owing to their extreme toxicity, chemical composition and mode of action, pesticides can be classified into four main classes like organochlorine, organophosphorus, carbamates and pyrethroids compounds (Barlow 1985). All these pesticides pose a dangerous health risk to target and non-target organisms and can cause endocrine disruption by interfering with the production, release and metabolism of hormones and enzymes.

Organophosphate and carbamate pesticides are reported for their neurotoxicity as they act as an inhibitor of acetylcholinesterase, which hydrolyses the neurotransmitter acetylcholine at the neuromuscular junction (Galloway and Handy 2003; Lushchaka et al. 2018), whereas organochlorines are endocrine disrupting chemicals and interfere with the functioning of molecular circuitry and cause dysfunction of the endocrine system (Sohail et al. 2004). Higher concentrations of pesticides get accumulated in human and animal bodies directly through intake of food items, dairy products, meats and vegetables and cause severe health hazards such as neuritis, psychiatric manifestations, immunological disorders resulting in genotoxic and carcinogenic diseases (Ascarrunz et al. 2006; Rusiecki et al. 2008). The excess use of pesticides in agricultural fields for high crop productivity has also led to increase the problem of soil, air and water pollution. It drastically affects the biological and physical properties of these environmental compartments, which adversely affect the life span of aquatic and terrestrial flora and fauna (Jayaraj et al. 2016).

12.4 Phytoremediation: A Synergistic Interaction Between Plant and Microbes

The use of plants for the treatment of contaminated soil and water has given a new way to the thought of clean, green, sustainable and safe environment. Phytoremediation has gained increasing attention since the last decade, as an emerging cheaper, environmentally friendly, aesthetically pleasing and cost-effective

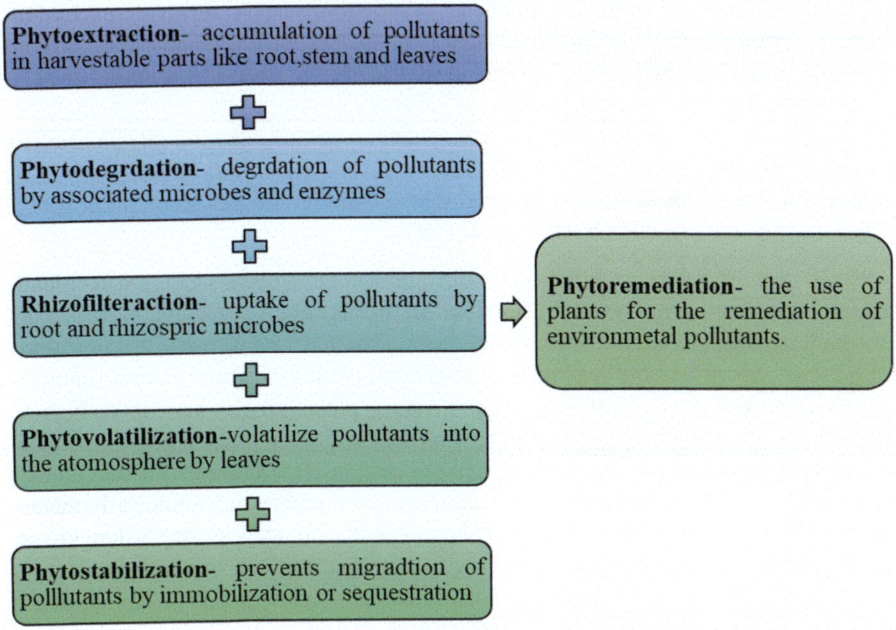

Fig. 12.1 Different forms of phytoremediation for the remediation of environmental pollutants

technology (Shehzadi et al. 2015). It is a good alternative option and can be applied to clean organic and inorganic pollutants in air, soil and water systems. Phytoremediation is a very broad application, which may be sub-divided into five branches depending upon the nature of pollutants and treatment methodologies. Major branches of phytoremediation are phytoextraction, phytodegradation, rhizofiltration phytostabilization and phytovolatilization (Salt et al. 1998; Glick 2003, 2010) (Fig. 12.1).

a. *Phytoextraction*: The use of pollutant-accumulating plants to remove metals or organics from the soil by concentrating them in some harvestable parts. This technique represents one of the largest economical opportunities for the remediation of metal contaminated and radionuclides polluted sites (Lee 2013).
b. *Phytodegradation*: The use of plants and associated microorganisms to degrade organic pollutants like PCBs, PAHs, pesticides, etc. Plants and their associated microbes produce various enzymes like catalase, peroxidase, laccase, reductase which help in the uptake and degradation of the pollutants. This technique is highly suitable for the degradation of moderately hydrophobic organic chemicals (Morikawa and Erkin 2003).
c. *Rhizofiltration*: In this techniques plant roots are involved in the absorption and adsorption of toxic pollutants, mainly metals from water and soil ecosystems. In this technique rhizospheric microorganism like PGPRs, endophytes and mycorrhizal fungi play a vital role in the removal of pollutants.

d. *Phytostabilization*: The use of plants to reduce the bioavailability of pollutants, by preventing their leaching and spread into water, soil and air. This process cannot degrade pollutants, but it can only reduce the mobility of pollutants by immobilization and sequestration.
e. *Phytovolatilization*: This technique uses plants to volatilize pollutants and mostly used to remove pollutants from air. Toxic heavy metals like As and Hg can be bio-methylated to volatile compounds, which can then be lost to the atmosphere by leaf surfaces of plant (Raskin et al. 1997).

Plants have the ability to accumulate substantial concentrations of toxic pollutants through their above (leaf and stem) and below (roots) plant parts. After accumulation, pollutants are degraded, metabolized, sequestrated or excreted into non-toxic substrate. In this process of pollutant degradation, plant-associated microorganisms play an important and viable role by promoting plant growth and detoxification mechanisms (Weyens et al. 2015). The root region of any plant remains in direct contact with soil microbes and the rhizospheric region. This includes an area around the root region which is abundant with microbial community and activity. Relationship between a healthy plant and microbes depends on various important factors including temperature, pH, microbial growth, tolerance ability of plant, repression or induction of catabolic genes and enzymes, nature of pollutant, potentiality of the strain, etc. (Alexander 2000; Diaz Martinez et al. 2018).

The rhizospheric region of the plant contains a variety of microorganisms including plant growth promoting rhizobacteria, endophytes, mycorrhizae fungus. These microbes get their nutrition from root exudates and in reverse, they help plants through their metabolic detoxification mechanisms which convert toxic compounds that are detrimental to plant growth and life into non-toxic substrate (Adieze et al. 2012). The synergistic relationship between plants roots and their associated microorganisms has been well studied and found effective for improved and enhanced remediation of contaminated environments and pollutants (Muller and Shann 2006; Gkorezis et al. 2016; Lal et al. 2018). Mostly PGPRs endorse plant growth directly by enabling resource procurement or by diminishing the inhibitory effects of several pathogens on plant growth in the form of biocontrol agent (Glick 2012). *Pseudomonas*, *Bacillus*, *Brevibacillus* are well-known bacterial genera which promote growth and yield in different hyperaccumulator plants indirectly by acting as biocontrol agents. Biocontrol agents not only inhibit pathogens but also increase plant growth through improved recycling of nutrients and minerals such as nitrogen, phosphates and other nutrients (Ryan et al. 2008).

"*Mycorrhiza*" is the symbiotic relationship between the fungal species and roots of vascular plants. These fungi help in improving plant growth by increasing the surface area of the plant's roots which enhance nutrient and water uptake in roots from the soil. Mycorrhizal fungi can protect plants from abiotic (e.g., increased heavy metal concentrations) and biotic (e.g., soil-borne pathogens) stress through the enhanced nutrient supply (Kohler et al. 2007; Chibuike 2013).

12.5 Mechanisms Involved in Plant and Microbe Interaction

Soil-borne microorganisms like bacteria, fungi, algae, yeast and cyanobacteria show different functional activities and interactions with the rhizospheric region of the plants. Understanding the mechanisms between plants and microbes is a crucial topic of scientific research in reference to agriculture and environmental sustainability. The plants and microbes interact with each other for their existence into the stressful environment via competition or coexistence to reduce toxic pressure and adaptations through active secretions of substrates that are known to vary between plant species. Plant root exudates are a useful source of energy and nutrients for microbes (Ma et al. 2016). Microorganisms are benefitted from plant root exudates and can thus alleviate phytotoxicity of plants and help to stimulate their growth against higher concentration of toxic pollutants by direct or indirect mechanisms through the induction of defence enzymes such as catalase, SOD, dehydrogenase, urease, etc. (Bhaduri et al. 2015).

Microbes can also stimulate plant growth by solubilization of essential nutrients in soil like nitrogen, phosphate, potassium by secreting siderophores, plant growth regulating hormones and fixation of nitrogen. Microbes can also evolve various resistance mechanisms for the degradation of pollutants such as detoxification, redox reaction, acidification, chelation, precipitation. Siderophore production by bacterial cells enhances plant growth by increasing protein content and chlorophyll content and decreases the heavy metal accumulation (Burd et al. 2000). Rhizospheric bacteria enhance the ability to take up iron by plant roots from the complex ligand siderophore compounds, produced by them to increase the growth and development of plants and to remediate metal contaminated environment (Rashmi et al. 2013). Plant–microbe interaction plays an important role in the exploration and enhancement of phytoremediation (Mishra et al. 2019).

Plant–microbe partnership needs more attention and advanced molecular studies to treat contaminated sites, as microbes provide resistance ability, metabolic degradation pathways for decreasing phytotoxicity and improving phytoremediation effectiveness. Microbes play a vital role in the biogeochemical cycle of minerals/nutrients and energy flow. The potentiality of biological nitrogen fixation as a low cost source of nitrogen and phosphorus solubilization by phosphate solubilizing bacteria and mycorrhizal fungi is very well studied and reported earlier (Bhaduri et al. 2015; Ozyigit and Dogan 2015).

12.6 Applications of Phytoremediation

12.6.1 In Metal Removal

Increasing contamination of toxic heavy metal pollution is becoming a global challenge for the entire environment and living beings. Heavy metals are highly toxic, hazardous and persistent in nature due to their chemical reactivity and bioavailability (Lotfy and Mostafa 2014). Therefore, these can easily accumulate in soil and water ecosystems and can persist in these matrices for many years and affect the physicochemical and biological properties of soil and water. These metals can also accumulate inside the living bodies and cause severe respiratory, developmental, mutagenic and carcinogenic related problems. Excessive deposition of toxic metals has an inhibitory action on enzymatic and metabolic activities, which sometimes leads to death also.

According to previous studies various hyperaccumulator plant species like *Allium schoenoprasum* (Goland 2006), *Brassica juncea* (Singh and Fulekar 2012), *Brassica napus* (Sheng and Xia 2006), *Cucumis sativus* (Takeda et al. 2006), *Cicer arietinum* (Garg et al. 2014), *Zea maize* (Abhilash et al. 2016), etc. are reported for the uptake of higher concentration of Ni, Co, Ca, Cd, Zn, Pd, As and Cr toxic metals from soil (Lee 2013). These plants can accumulate >100 mg/kg Cd, 1000 mg/kg Ni, Cu and Pb and ~10,000 mg/kg Zn and Mn in their leaves and stems under metal contaminated soils (Baker and Brooks 1989). Aquatic plants species like *Eichhornia crassipes* (water hyacinth), *Pistia stratiotes* (water lettuce) are efficient accumulator of toxic metals and capable to transform into their less toxic form (Saleh 2012; Das et al. 2014).

12.6.2 In Hydrocarbon Degradation

Hydrocarbons are organic, xenobiotic chemical compounds, whose excess concentration in the environment poses a serious matter of concern due to their ominously detrimental and deleterious impact on human, plants and on animals as well. Phytoremediation of hydrocarbons offers a safe and eco-friendly remediation process to degrade organic pollutants. The mutualistic interaction between plant roots and their associated microbes is more potent to endorse effective and efficient degradation of persistent organic pollutants (Asemoloye et al. 2017). PAHs, PCBs, 2, 4, 6-TNT and other organic pollutants can be degraded into stable (CO_2 and H_2O) and non-toxic intermediates by phytoremediation. Plants and their associated microbes produce various degradative enzymes that are capable in the complete mineralization of such pollutants.

Various plant species like *Helianthus annuus*, *Brassica juncea*, *Festuca arundinacea and Lolium perenne* are reported for the remediation of organic contaminants. These can degrade organic substances by means of releasing root exudates that promote microbial activity, certain degradative enzymatic activity in the vicinity for the potential biotransformation of organic compounds (Siciliano et al. 1998). Significant phytoremediation study of phenanthrene and pyrene was reported by Cheema et al. (2009) in acidic and spiked soil by single and combined plant cultivation system using *Zea maize*, *Medico sativa*, *Brassica napus*. These plants significantly enhance the adsorption of PAHs from polluted soils. *Cannabis sativa* (Campbell et al. 2002), *Oryza sativa* (Du et al. 2011) are also reported to have the potentiality for the reduction and degradation of benzopyrene and chrysene.

Meanwhile, application of grasses for the purpose of phytoremediation of organic pollutants has also been a promising approach to enhance this process. *Schizachyrium scoparium* (little bluestem grass) is capable to reduce the efficient amount of PAHs in combination with *P. virgatum*. *Lolium multiflorum* and *Cynodon dactylon* have been reported for the proficient degradation of naphthalenes (White et al. 2006).

12.6.3 In Pesticide Degradation

Phytoremediation is the most efficient technique to prevent the migration of pesticide and their hazardous residues into the soil and water ecosystems via leaching and surface runoff. Aquatic plants have extraordinary potential for the degradation of pesticides due to their great functionality as biosinks and biofilters of the contaminants (Gao et al. 2000). Tondon et al. (2016) reported the degradation of N-methyl carbamate pesticide (carbofuran) degradation by three aquatic plants like *Chrysopogon zizanioides*, *Acorus gramineus* and *Scirpus cyperinus* at different concentrations and found that *Acorus gramineus* was the most efficient and potent plant for the phytoremediation. *Cyperus rotundus* and *Scirpus schoenoplectus* were found as a good candidate for the treatment of water, contaminated with highly toxic pesticide monocrotophos by Tondon et al. (2017).

Aquatic plants having a very rich root system and microbial diversity help in the absorption or uptake of pollutant and their residues. *Lemna minor*, *Elodea canadensis*, *Cabomba aquatic* and *Eichhornia crassipes* are free-floating, widespread aquatic plants that are useful in the phytoremediation of pesticides (Olette et al. 2008).

12.7 Economic Importance of Phytoremediation

Phytoremediation is not only an environmentally friendly technology but also a very efficient, feasible cost-effective technology from economical point of view because of its easy availability and less expansive methodology as plants can be easily grown without much effort and can be monitored easily. The use of plants and their associated microbes for the mitigation and reclamation of contaminated sites is a viable approach in cleaning and restoring our natural environment in a more healthy and green way (Sumiahadi and Acar 2018). Phytoremediation is not only beneficial for the environment but also for our society, farmers and agriculture. Phytoremediation contributes to biodiversity improvement, soil restoration, carbon sequestration, protection of environment stability and in reducing hazardous chemical and pollutants (Adams et al. 2013). Hyperaccumulator plants are highly functional and valuable in phytoremediation technique because of their potentiality to tolerate and grow at higher concentration of environmental pollutants. They are capable of accumulation and degradation of toxic pollutants into non-toxic or less toxic substance.

12.8 Future Perspectives

Phytoremediation is an undoubtedly tremendous approach for the remediation of contaminated environments, but the success of this technology depends upon its proper implementation at different environmental parameters with suitable and potent plant species that survive and tolerate such harsh stress conditions to a variety of contaminants. Selection of most efficient plant species to degrade a specific pollutant is one of the basic and most important concepts of this technology (Dhir 2013). It is a good treatment strategy for the degradation of a variety of toxic environmental pollutants but more advanced and molecular mechanistic information are required to understand the relationships between plants and microbes for accurate expected outcomes of this technology. Use of omics techniques like metagenomics, meta-transcriptomic, proteomic and metabolomics can provide deeper knowledge and understanding of the synergistic relationship between plant–microbe, microbe–metal and microbe–microbe (Bell et al. 2014).

Applications of GMOs along with greenhouse experiments in agronomic soil conditions ensure a better knowledge of plant biology and their capability to enhance phytoremediation at large scale projects. Furthermore, monitoring vegetation of native plants of contaminated sites provides a good summary about the environmental conditions, survival and management strategies under stress environment for better implementation of phytoremediating plants (Dushenkov 2003; Asemoloye et al. 2019).

12.9 Conclusion

The exploitation of phytoremediation technology for the mitigation and removal of toxic and hazardous pollutants from the environment is a very pragmatic and eco-friendly approach. It is a highly valuable and much easier technology for that does not require any expensive or huge equipment. Common peoples and farmers are easily connected to this technology, as it is more convenient than conventional physicochemical treatment approaches. Although plants are capable of removal or degradation of toxic pollutants but in this degradation process, soil-borne microorganism also plays a very important role in their growth promotion under stressful environment. Microbial associations can promote plant growth by siderophore formation, chelation, phosphate solubilization, acidification and redox reactions, which affects the bioavailability and solubility of pollutants. Therefore, the plant–microbe interaction is a very interesting and imperative relationship, which needs more attention and research in respect to understanding the deeper knowledge of their genetic and molecular level mechanism for such mutualistic relation in the remediation of contaminated environments and betterment of phytoremediation in the near future in a more sustainable way.

Acknowledgements The senior author is highly thankful to all the contributors for their suggestions in writing, editing and proofreading of this chapter.

References

Abhilash MR, Srikantaswamy S, Shiva Kumar D, Jagadish K, Shruthi L (2016) Phytoremediation of heavy metal industrial contaminated soil by *Spinacia oleracea* L. and *Zea mays* L. Int J Appl Sci 4(1):192–199

Adams A, Raman A, Hodgkins D (2013) How do plants used in phytoremediation in constructed wetlands, a sustainable remediation wetlands, perform in heavymetal-contaminated mine sites? Water Environ. https://doi.org/10.1111/j.1747-6593.2012.00357.x

Adieze IE, Orji JC, Nwabueze RN, Onyeze GOC (2012) Hydrocarbon stress response of four tropical plants in weathered crude oil contaminated soil in microcosms. Int J Environ Stud 69 (3):490–500

Alexander M (2000) Aging, bioavailability and over stimulation of risk from environmental pollutants. Environ Sci Technol 34:4259–4265. https://doi.org/10.1021/es001069t

Ascarrunz ME, Tirado N, Gonzáles AR, Cuti M, Cervantes R, Huici O, Jors E (2006) Evaluación de riesgo genotóxico: biomonitorización de trabajadores agrícolas de Caranavi, Guanay, Palca y Mecapaca, expuestos a plaguicidas. Cuad Hosp Clín 51(1):7–18

Asemoloye MD, Ahmad R, Jonathan SG (2017) Synergistic action of rhizospheric fungi with *Megathyrsus maximus* root speeds up hydrocarbon degradation kinetics in oil-polluted soil. Chemosphere 187:1–12

Asemoloye MD, Jonathan SG, Ahmad R (2019) Synergistic plant-microbes interactions in the rhizosphere: a potential headway for the remediation of hydrocarbon polluted soils. Int J Phytol 21(2):71–83

ATSDR (2008) Agency for toxic substances and disease registry. U.S. Department of Health and Human Services. Public Health Service, Atlanta

Baker AJM, Brooks RR (1989) Terrestrial higher plants which hyper accumulate metallic elements-review of the distribution, ecology, and phytochemistry. Biorecovery 1:81–126

Bansal V, Kim KH (2015) Review of PAH contamination in food products and their health hazards. Environ Int 84:26–38

Barlow F (1985) Chemistry and formulation. In: Haskel PT (ed) Pesticide application: principles and practice. Oxford University Press, Oxford, pp 1–34

Bell TH, Joly S, Pitre FE, Yergeau E (2014) Increasing phytoremediation efficiency and reliability using novel omics approaches. Trends Biotechnol 32:271–280

Bhaduri D, Pal S, Purakayastha TJ, Chakraborthy K, Yadav RS, Akhtar MS (2015) Soil quality and plant microbe interactions in rhizosphere. Sust Agric Rev 17:307–335

Bharagava RN, Mishra S (2018) Hexavalent chromium reduction potential of *Cellulosimicrobium* sp. isolated from common effluent treatment plant of tannery industries. Ecotoxicol Environ Saf 147:102–109

Bolognesi C, Merlo FD (2011) Pesticides: human health effects. In: Nriagu JO (ed) Encyclopedia of environmental health. Elsevier, Burlington, pp 438–453

Burd GI, Dixon DG, Glick BR (2000) Plant growth-promoting bacteria that decrease heavy metal toxicity in plants. Can J Microbiol 46:237–245

Campbell S, Paquin D, Awaya JD, Li QX (2002) Remediation of benzo (a) pyrene and chrysene-contaminated soil with industrial hemp (*Cannabis sativa*). Int J Phytol 4:157–168

Cheema SA, Khan MI, Shen C, Tang X, Farooq M, Chen L, Chen Y (2009) Degradation of phenanthrene and pyrene in spiked soils by single and combined plants cultivation. J Hazard Mater 16:207–211

Chibuike GU (2013) Use of mycorrhiza in soil remediation. Sci Res Essays 835:1679–1687

Das S, Goswami S, Talukdar AD (2014) A study on cadmium phytoremediation potential of water lettuce, *Pistia stratiotes* L. Bull Environ Contam Toxicol 92:169–174

Dhir B (2013) Phytoremediation: role of aquatic plants in environmental clean-up. https://doi.org/10.1007/978-81-322-1307-9_3

Diaz Martinez ME, Argumedo-Delira RS, Trejo-Tellez LI (2018) Lead phytoextraction from printed circuit computer boards by *Lolium Perenne* L. and *Medicago sativa* L. Int J Phytorem 20(5):432

Dixit R, Wasiullah D, Malaviya K (2015) Bioremediation of heavy metals from soil and aquatic environment: an overview of principles and criteria of fundamental processes. Sustainability 7:2189–2212

Du W, Sun Y, Cao L, Huang J, Ji R, Wang X, Wu J, Zhu J, Guo H (2011) Environmental fate of phenanthrene in lysimeter planted with wheat and rice in rotation. J Hazard Mater 188:408–413

Dushenkov S (2003) Trends in phytoremediation. Plant Soil 249:167–175

Flora SJS, Saxena G, Gautam KP, Gill KD (2007) Lead induced oxidative stress and alterations in biogenic amines in different rat brain regions and their response to combined administration of DMSA and MiADMSA. Chem Biol Interact 170:209–220

Gadd GM (2010) Metals, minerals and microbes: geomicrobiology and bioremediation. Microbiology 156:609–643

Galloway T, Handy R (2003) Immunotoxicity of organophosphorous pesticides. Ecotoxicology 12:345–363

Gao J, Garrison AW, Hoehamer C, Mazur CS, Wolfe NL (2000) Uptake and phytotransformation of o,p'-DDT and p,p-DDT by axenically cultivated aquatic plants. J Agric Food Chem 481 (1):6121–6127

Garg N, Singla P, Bhandari P (2014) Metal uptake, oxidative metabolism, and mycorrhization in pigeon pea and pea under arsenic and cadmium stress Turk. J Agric For 39:234–250

Gkorezis P, Daghio M, Franzetti A, Van Hamme JD, Sillen W, Vangronsveld J (2016) The interaction between plant and bacteria in the remediation of petroleum hydrocarbons: an environmental perspective. Front Microbiol 7:1836. https://doi.org/10.3389/fmicb.2016.01836

Glick BR (2003) Phytoremediation: synergistic use of plants and bacteria to clean up the environment. Biotechnol Adv 21(5):383–393

Glick BR (2010) Using soil bacteria to facilitate phytoremediation. Biotechnol Adv 28:367–374

Glick BR (2012) Plant growth promoting bacteria: mechanisms and applications. Scientifica. https://doi.org/10.6064/2012/963401

Goland GA (2006) Plant tolerance to heavy metals, a risk for food toxicity or a means for food fortification with essential metals: the *Allium schoenoprasum* model. In: Twardowska I, Allen HE, Haggblom MM (eds) Soil and water pollution monitoring, protection and remediation. Springer, Cham, pp 479–478

IARC (1990) IARC monographs on the evaluation of carcinogenic risks to humans, vol 49. IARC Scientific Publications, Lyon

Irfan M, Hayat S, Ahmad A, Alyemeni MN (2013) Soil cadmium enrichment: allocation and plant physiological manifestations. Saudi J Biol Sci 20(1):1–10

Jayaraj R, Megha P, Sreedev P (2016) Organochlorine pesticide, their toxic effects on living organisms and their fate in the environment. Interdiscip Toxicol 9:90–100. https://doi.org/10.1515/intox-2016-0012

Kohler J, Caravaca F, Carrasco L, Roldan A (2007) Interactions between a plant growth-promoting rhizobacterium, an AM fungus and a phosphate-solubilising fungus in the rhizosphere of *Lactuca sativa*. Appl Soil Ecol 35:480–487

Lal S, Ratna S, Said OB, Kumar R (2018) Biosurfactant and exopolysaccharaide-assisted rhizobacterial techniques for the remediation of heavy metal contaminated soil: an advancement in metal phytoremediation technology. Environ Technol Innov 10:243–263

Landrum P, Giesy J, Oris J, Allred P (1987) Photoinduced toxicity of polycyclic aromatic hydrocarbons to aquatic organisms. In: Oil in freshwater: chemistry, biology, countermeasure technology. Pergamon, Elmsford, pp 304–318

Lee JH (2013) An overview of phytoremediation as a potentially promising technology for environmental pollution control. Biotechnol Bioprocess Eng 18:431–439. https://doi.org/10.1007/s12257-013-0193-8

Lee BK, Vu VT (2010) Sources, distribution and toxicity of polycyclic aromatic hydrocarbons (PAHs) in particulate matter. In: Villanyi V (ed) Air pollution, pp 99–120

Lotfy SM, Mostafa AZ (2014) Phytoremediation of contaminated soil with cobalt and chromium. J Geochem Explor 144:367–373

Lundstedt S, White PA, Lemieux CL (2007) Sources, fate, and toxic hazards of oxygenated polycyclic aromatic hydrocarbons (PAHs) at PAH-contaminated sites. Ambio 36(6):475–485

Lushchaka VI, Matyiishyna TM, Husaka VV, Storeyb JM, Storeyb KB (2018) Pesticide toxicity: a mechanistic approach. https://doi.org/10.17179/excli2018-1710

Ma Y, Oliviera RS, Freitas H, Zhang C (2016) Biochemical and molecular mechanism of plant-microbe-metal interactions: relevance for phytoremediation. Front Plant Sci. https://doi.org/10.3389/fpls.2016.00918

Marzooghi SM, DI Toro DM (2017) A critical review of polycyclic aromatic hydrocarbon phototoxicities models. Environ Toxicol Chem 36(5):1138–1148

Mishra S, Bharagava RN (2016) Toxic and genotoxic effects of hexavalent chromium in environment and its bioremediation strategies. J Environ Sci Health 34(1):1–32. https://doi.org/10.1080/10590501.2015.1096883

Mishra S, Sartele GD, Ferreira LFR and Bharagava RN (2019) Plant-microbe interactions: an ecofriendly approach for the remediation of metal contaminated environments. Mater Sci Mater Eng. https://doi.org/10.1016/B987-0-12-803581-8.11508-5.

Morikawa H, Erkin OC (2003) Basic processes in phytoremediation and some applications to air pollution control. Chemosphere 52:15533–11558

Muller KE, Shann JR (2006) PAH dissipation in spiked soil: impacts of bioavailability, microbial activity, and trees. Chemosphere 64(6):1006–1014

Nagajyoti PC, Lee KD, Sreekanth TVM (2010) Heavy metals, occurrence and toxicity for plants: a review. Environ Chem Lett 8:199–216

Olette R, Couderchet M, Biaginati S, Eullaffroy P (2008) Toxicity and removal of pesticides by selected aquatic plants. Chemosphere 70:1414–1421

Ozyigit II and Dogan I (2015) Plant microbe interactions in phytoremediation. Soil Remediat Plants. https://doi.org/10.1016/B978-0-12-799937-1.00009-7

Ramos KS, Moorthy B (2005) Bioactivation of polycyclic aromatic hydrocarbon carcinogens within the vascular wall: implications for human atherogenesis. Drug Metab Rev 37(4):595–610. https://doi.org/10.1080/03602530500251253

Rashmi V, Shylaja NR, Rajalaksmi S, D'Souza F, Prabaharan D, Uma L (2013) Siderophore mediated uranium sequestration by marine cyanobacterium *Synechococcus elongates*. Bioresour Technol 130:204–210

Raskin I, Smith RD, Salt DE (1997) Phytoremediation of metals: using plants to remove pollutants from the environment. Curr Opin Biotechnol 8:221–226

Rusiecki JK, Baccarelli A, Bollati V, Tarantini L, Mooore LE, Bonefeld-Jorgenson EC (2008) Global DNA hypomethylation is associated with high serum-persistent organic pollutants in Greenlandic unit. Environ Health Perspect 116:1547–1552

Ryan RP, Germaine K, Franks A, David RJ, Dowling DN (2008) Bacterial endophytes: recent developments and applications. FEMS Microbiol Lett 278:1–9

Saleh HM (2012) Water hyacinth for phytoremediation of radioactive waste simulate contaminated with cesium and cobalt radionuclides. Nucl Eng Des 242:425–432

Salt DE, Smith RD, Raskin I (1998) Phytoremediation. Annu Rev Plant Physiol Mol Biol 49:643–668

Shehzadi M, Fatima K, Imran A, Mirza MS, Khan QM, Afzal M (2015) Ecology of bacterial endophytes with wetland plants growing in textile effluent for pollutant-degradation and plant growth promotion potential. Plant Biosyst. https://doi.org/10.1080/11263504.2015.1022238

Sheng XF, Xia JJ (2006) Imporement of rape (Brassica napus) plant growth and cadmium uptake by cadmium-resistant bacteria. Chemosphere 64(6):1036–1042

Siciliano SD, Goldie H, Germida JJ (1998) Enzymatic activity in root exudates of Dahurian wild rye *(Elymus dauricus)* that degrades 2-chlorobenzoic acid. J Agric Food Chem 46:5–7

Singh A, Fulekar MH (2012) Phytoremediation of heavy metals by *Brassica juncea* in aquatic and terrestrial environment. In: The plant family brassicaceae: contribution towards phytoremediation. Springer, Cham, pp 153–169

Singh P, DeMarini DM, Dick CAJ, Tabor DG, Ryan JV, Linak WP (2004) Sample characterization of automobile and forklift diesel exhaust particles and comparative pulmonary toxicity in mice. Environ Health Perspect 112:820–825. https://doi.org/10.1289/ehp.6579

Sohail E, Waseem A, Chae WL, Jong JL, Imitiaz H (2004) Endocrine disrupting pesticides: a leading cause of cancer among rural people in Pakistan. Exp Oncol 26(2):98–105

Sumiahadi A, Acar R (2018) A review of phytoremediation technology: heavy metals uptake by plants. IOP Conf Ser 142:012023. https://doi.org/10.1088/1755-1315/142/1/012023

Takeda R, Sato Y, Yoshimura R, Komemushi S, Sawabe A (2006) Accumulation of heavy metals by cucumber and Brassica juncea under different cultivation conditions. In: Proc. Ann. Int. Conf. on soil sediments water energy (Massachusetts), vol 11. The Berkeley Electronic Press, Berkeley, pp 293–299

Tchounwou PB, Patlolla AK, Centeno JA (2003) Carcinogenic and systemic health effects associated with arsenic exposure-a critical review. Toxicol Pathol 31(6):575–588

Tondon SA, Deore R, Parab A (2016) Removal of pesticide carbofuran using wetland plants. Int J Adv Biol Res 6(4):482–485

Tondon SA, Deore R (2017) Removal of pesticide monocrotophos using wetland plants. Global J Biosci Biotechnol 6(4):671–676

Unwin J, Cocker J, Scobbie E, Chambers H (2006) An assessment of occupational exposure to polycyclic aromatic hydrocarbons in the UK. Ann Occup Hyg 50:395–403

Weyens N, Popek R, Witters N, Przybysz A, Espenshade J, Gawronska H, Vangronsveld J, Gawronski SW (2015) The role of plant-microbe interaction and their exploitation for phytoremediation of air pollutants. Int J Mol Sci 16:25576–25604. https://doi.org/10.3390/ijms161025576

White PM, Wolf DC, Thoma GJ, Reynolds CM (2006) Phytoremediation of alkylated polycyclic aromatic hydrocarbons in a crude oil-contaminated soil. Water Air Soil Pollut 169:207–220

Wuana RA, Okieimen FE (2011) Heavy metals in contaminated soils: a review of sources, chemistry, risks and best available strategies for remediation. Int Sch Res Netw ISRN Ecol 20:402647. https://doi.org/10.5402/2011/402647

Zhang Y, Tao S (2009) Global atmospheric emission inventory of polycyclic aromatic hydrocarbons (PAHs) for 2004. Atmos Environ 43(4):812–819. https://doi.org/10.1016/j.atmosenv.2008.10.050

Chapter 13
Understanding the Role of Microbes and Plants in the Management of Heavy Metal Stress: A Current Perspective

Saima Saif, Almas Zaidi, and Mohd. Saghir Khan

Abstract Significant amount of heavy metals is regularly added to the soils globally due to various natural and anthropogenic activities. The heavy metal uptake and accumulation in crops lead to yield losses. A common consequence of heavy metal toxicity to plants is an excessive accumulation of reactive oxygen species (ROS) and methylglyoxal (MG). Both molecules may cause peroxidation of lipids, oxidation of protein, DNA damage, inactivation of enzymes, and/or affect other vital plant constituents. Higher plants have evolved a balanced antioxidant defence system and a glyoxalase system to scavenge ROS and MG. Besides plants, microbes can also be used to remove heavy metals from polluted soils. This chapter highlights the suitability of various strategies adopted by useful soil microbiota and plants to eradicate heavy metal toxicity and consequently to enhance crop production in metal stressed soils.

13.1 Introduction

13.1.1 Heavy Metal Pollution and Its Agro-Ecological Impact: An Overview

Heavy metal pollution has emerged as a global challenge in developed and developing countries, which limits the economic growth and causes human health problems via food chain (Tchounwou et al. 2012; Khalid et al. 2017). Classically, heavy metals which are biologically and industrially important refer to a group of toxic elements having densities greater than 5 g cm^{-3}. Recently the term has also been used for metals and semimetals with potential human or environmental toxicity (Saunders et al. 2013; Mohmand et al. 2015). In soils, metals may be added through transport of continental dusts, emissions from volcanoes, and by weathering of metal-enriched rocks (Algreen et al. 2012). Natural inputs are unlikely to add

S. Saif (✉) · A. Zaidi · M. S. Khan
Department of Agricultural Microbiology, Faculty of Agricultural Sciences, Aligarh Muslim University, Aligarh, Uttar Pradesh, India

A. Sharma (ed.), *Microbes and Signaling Biomolecules Against Plant Stress*, Rhizosphere Biology, https://doi.org/10.1007/978-981-15-7094-0_13

considerably higher amount of HM to soils but due to rapid industrialization, there has been massive addition of heavy metals emanating from anthropogenic activities which has resulted in a greater public concern (Shaheen et al. 2017). Arsenic, lead, cadmium, and mercury are some of the notable soil heavy metals. Despite metal pollution arising from sewage application, sewage has primarily served as an important supplementary source to combat agricultural irrigation crisis in areas deficient in good quality waters (Li et al. 2017). Irrigation by wastewater in agriculture practices is common in many countries including China, India, Pakistan, Mexico, and Iran. The use of wastewater for irrigation has been found economical and to a certain extent has also solved the problem of effluent disposal (Navarro et al. 2015). Recent findings suggest that 35.9 Mha of irrigated crop lands are located in the vicinity of wastewater treatment plants and out of 82% of the agricultural fields, 75% are irrigated by wastewater (Thebo et al. 2017). Despite these benefits, use of wastewater for irrigation can pose substantial risks to the plants (Marrugo-Negrete et al. 2017). Although soil has some ability to clear and degrade pollutants via microbial metabolism and transformation but it is not enough to control heavy metal pollutants that accumulate in groundwater or soil solution due to continuous release of untreated pollutants or changes in soil pH. Hence, build-up of heavy metals in soils can restrict soil functioning (fertility) and result in toxicity to plants and physiological activities of microbes which in turn affect the quality and safety of foods severely and cause eventually human health problems (Meng et al. 2016).

13.1.2 Metal Toxicity to Human Health

The non-biodegradability and lethal effect of heavy metals is a serious problem for human health worldwide (Azimi et al. 2017). Though metals like Zn, Cu, Ni, Co and Cr at small quantities play a pivotal role in metabolic and physiological processes of plants (Singh et al. 2016), humans (Yamada 2013), and microorganisms (Boer et al. 2014). For instance, they affect redox processes, regulate osmotic pressure, and stabilize molecules through electrostatic interactions and act as cofactors for numerous enzymes and electron transport systems (Emamverdian et al. 2015). In contrast, there are non-essential heavy metals like Ag, As, Cd, Pb, and Hg which do not have any biological importance to living organisms and are very toxic even at very low concentrations. Heavy metals usually enter the human body via different food chains, inhalation, and ingestion. Once inside the human body, they stimulate the immune system and may cause nausea, anorexia, vomiting, gastrointestinal abnormalities, and dermatitis (Megido et al. 2017). Metals like Cr, As, Zn, Ur, Se, Au, and Ni may also adversely affect the quality of soil, crop production as well as public health (Venkanna and Karthikeyan 2017). These pollutants in general are major cause of life-threatening degenerative diseases such as Alzheimer's disease, atherosclerosis, cancer, Parkinson's disease, etc. Due to the detrimental effects of heavy metals, there is urgent need to find strategies to effectively eradicate HM from the environment and stabilize the ecosystem.

13.1.3 Metal–Microbe–Plant Interactions: A General Perspective

13.1.3.1 Heavy Metals and Rhizospheric Microflora

Soil microorganisms play a key role in maintaining soil fertility through various activities such as organic matter disintegration, adsorption, and pH balance and hence optimize crop production (Shahbaz et al. 2017). However, soil microbiota are greatly influenced when exposed to stress factors like heavy metals (Gutierrez-Ginés et al. 2014), high pH (Wu et al. 2017), salinity (He et al. 2017), extreme temperature (Akkermans et al. 2017), and chemical pollution (Gianfreda and Rao 2017). Among microbial communities, the bacterial communities in general have been reported to be the most severely affected by high HM concentration as compared to fungal population (Rajapaksha et al. 2004). The beneficial or detrimental effect of HMs onto microbial cells however depends on concentration, speciation, and duration of exposure of metals.

13.1.3.2 Impact of HM on Soil Microbial Composition and Function

Soil microbial community structure serves as an important marker for polluted soil since long-term exposure to pollutants changes the microbial compositions and functions. On the contrary, the long-term exposure to metals may also help microorganisms to adapt to the stressed environment (Kaci et al. 2016). However, various in-depth analysis have shown that metal pollution had significant impact on bacterial community structure causing changes in the relative abundance of specific bacterial taxa, but not bacterial taxon richness and community composition revealing their resilient nature in the ecosystem. Therefore, various metagenomic studies have focused on identifying the microbial communities and their relationship to the changing soil properties (Azarbad et al. 2015). In a study, Zhang et al. (2017) using 16S rRNA gene sequencing observed a general reorganization of soil microbial communities persistently exposed to metals specifically Cd and Pb in rhizosphere. On the other hand, Azarbad et al. (2015) assessed functional and potential microbial diversity (using Geo Chip 4.2) along two different gradients of metal polluted sites in Southern Poland. It was found that metal pollution caused negative impact on the relative abundance of specific bacterial genera including *Acidobacteria*, *Bacteroidetes*, *Actinobacteria*, *Chloroflexi*, *Planctomycetes, Firmicutes,* and *Proteobacteria*. Also, there were significant correlations between a group of metal-resistance genes and among the bacterial genera with metal concentrations in soil. Due to prolonged exposure to high metal concentrations, majority of the microflora that were found sensitive and unable to tolerate high doses of metals became extinct, while certain community members with different functional roles such as denitrification and metal resistance adapted and survived to form the basis for the emergence of other new community. Another important marker that indicates normal soil functionality is the soil microbial biomass that shows changes in soil properties due to complex environmental modifications (Hornick and Buschmann 2018; Rai et al. 2018). Yuan et al. (2015) reported a negative correlation between microbial viability and extended exposure to Pb. Similarly, Yao et al. (2017) studied the biological attributes of heavy metal-contaminated soils and established relationships between environmental variables and community composition. The microbes that were already tolerant became more competitive and thus were more in number. In a similar study, reduction in bacterial community, fungi, and actinobacteria in a heavy metal poisoned site located in one of the municipality of Brazil under the influence of a Votorantim Metal Company was reported by Dos Santos et al. 2016. Apart from composition and density, HMs have also been found to negatively affect the soil enzymes activity like cellulase, alkaline phosphatase, invertase, arylsulfatase, dehydrogenase, β-glucosidase, etc. (Burges et al. 2015).

13.1.3.3 Heavy Metal Toxicity to Physiological Processes of Microbes: A Brief Account

In general, high concentrations of heavy metals or metals above certain threshold level cause discrete and apparent injuries to microbial cells due to oxidative stress, protein dysfunction, or membrane damage (Olaniran et al. 2013). Essentially, metal ions have variable targets within microbial cells and hence affect various microbial activities (Table 13.1).

13.1.3.4 HMs-Induced Phytotoxicity and Physicochemical Changes in Plants

Although plants possess several strategies to offset metal toxicity but beyond certain limits such mechanisms often become ineffective and hence the survival of plant is compromised (Clemens and Ma 2016) (Table 13.2). The toxicity of heavy metals to plants varies with plant genotypes, metals species and its concentration, and soil characteristics (Topcuoğlu 2016).

13.2 Plant Defence Mechanisms Against Heavy Metal

13.2.1 Antioxidant Defence System

Plants have developed a number of strategies to overcome the adverse impacts imposed by heavy metals. Heavy metal toxicity also leads to the over production of ROS, and in turn causes peroxidation of many vital cell constituents. In this way, plants have an efficient defence system comprising of a set of enzymatic and non-enzymatic antioxidants.

Table 13.1 Effect of heavy metals on microbial growth and activities

Metal	Effects on microorganisms	Reference
Cadmium	Damage nucleic acid, denature protein, inhibit cell division and transcription, inhibit C and N mineralization, oxidative damage	Thomas and Benov (2018)
Chromium	Elongation of lag phase, inhibit growth and oxygen uptake, damage DNA	Thorgersen et al. (2017), Fathima and Rao (2018)
Copper	Disrupt cellular function, inhibit growth and enzyme activities, oxidative stress	Warnes and Keevil (2011), Saphier et al. (2018), Águila-Clares et al. (2018)
Mercury	Decrease population size, denature protein, disrupt cell membrane, inhibit enzyme function	Mahbub et al. (2017), LaVoie and Summers (2018)
Lead	Growth inhibition, denature nucleic acid and protein, inhibit enzymes activities and transcription, membrane damage	Adam et al. (2014)
Nickel	Reduce lipid content, disrupt cell membrane, inhibit enzyme activities, oxidative stress	Gupta and Karthikeyan (2016)
Silver	Cell lysis, inhibit cell transduction and growth	Westersund (2018), Choi et al. (2018)
Zinc	Decrease biomass, inhibit growth, DNA damage, membrane destruction	Ishida (2018)

Table 13.2 Heavy metals affecting plant health

Metal	Effects on plants	Reference
Chromium	Chlorosis, delayed, senescence, wilting, biochemical lesions, reduced biosynthesis germination, stunted growth, oxidative stress	Kabir (2016), Anjum et al. (2017)
Copper	Chlorosis, oxidative stress, retarded growth	Li et al. (2018)
Mercury	Affect photosynthesis and antioxidative system, enhance lipid peroxidation, induced genotoxic effect, inhibit plant growth, nutrient uptake and homeostasis, oxidative stress	Mishra et al. (2016)
Lead	Reduce growth, affect photosynthesis chlorosis, inhibit enzyme activities and seed germination, oxidative stress	Venkatachalam et al. (2017), Silva et al. (2017); Ahmad et al. (2018)
Nickel	Inhibit growth, decrease chlorophyll content, inhibit enzyme activities and growth, reduced nutrient uptake	Mir et al. (2018)
Cadmium	Affect growth, decrease chlorophyll content, inhibit growth, oxidative stress	Andresen and Küpper (2013), Shanying et al. (2017)
Zinc	Affect photosynthesis, inhibit growth rate, reduce chlorophyll content, germination rate and plant biomass	Dotaniya et al. (2018)

A wide variety of antioxidants that include superoxide dismutase (SOD), peroxidase (POD), catalase (CAT), and glutathione-s-transferase (GST) may efficiently convert superoxide radicals into H_2O_2 and subsequently H_2O and O_2, whereas low molecular weight non-enzymatic antioxidants like proline, ascorbic acid, and glutathione may directly detoxify ROS. These two groups of antioxidants may professionally quench a wide range of toxic oxygen derivatives and prevent the cells from oxidative stress. Depending upon their localization at different compartments of the cell, their quenching mechanism also differs and acts in an organized manner. For example, SODs are a group of metalloenzymes that convert superoxide radical (SOR, $O_2^{\bullet-}$) into hydrogen peroxide, whereas CAT, guaiacol peroxidase (GPX), and a variety of general PODs are involved in breakdown of H_2O_2 (Gautam et al. 2018).

13.2.2 Cellular Homeostasis

Proline has been considered as one of the most important osmoticum found in the cellular system exposed to metal, water and saline stress, etc. Under different stresses including HM stress, proline accumulates in cytosol and helps in several ways: (1) maintain intracellular redox homeostasis potential, (2) protects enzymes, (3) sustains 3-D structure of proteins and vital organelles including cell membrane, and (4) reduces the risk of peroxidation of lipids and proteins (Aslam et al. 2017). In addition, proline prevents the disruption of membranes by forming clusters with H_2O molecules and stabilizes their structures (Slama et al. 2015). Under stress, proline provides variable benefits to plants such as (1) it chelates heavy metals in the cytoplasm (Sharma and Dietz 2006), (2) regulates the water potential which is often impaired by heavy metals (Kholodova et al. 2011), (3) maintains osmotic adjustment through cellular homeostasis and reduces metal uptake (Szabados and Savoure 2010). For instance, Hayat et al. (2013) showed that the exogenous application of proline alleviated the damaging effects of Cd in plants such as down regulation of water potential and thereby enhanced the growth and photosynthesis.

13.2.3 Role of Genes in Metal Uptake and Their Transportation

A number of genes are expressed differentially after the exposure to heavy metal stress which activates specific enzymes to overcome the negative impact of stress. For instance, genetically modified tobacco callus showed more resistance to methyl mercury (CH_3Hg^+) than the wild-type. MerB enzyme, a product of merB gene was found to dissociate CH_3Hg^+ to less toxic Hg^{2+} which accumulates as Hg-polyP complex in tobacco cells (Nagata et al. 2010). Mostly, the detoxification/sequestration process occurs in plant vacuoles and involves several transporters, namely ABC, CDF, HMA, and NRAMP (Singh et al. 2011). Moreover in another study, over expression of AtPCS1 and CePCS genes enhanced the remediation efficiency of tobacco plants under As and Cd stress by causing increase in phytochelatins level (Wojas et al. 2008). Besides, introduction of a gene encoding moth bean D1-pyrroline-5-carboxylate synthetase (P5CS) that initiates proline synthesis in green microalga *Chlamydomonas reinhardti* assists to increase HM tolerance. Accumulation of more than 80% higher proline level than the wild-type cells caused a genetically modified microalga grow more rapidly under higher Cd concentrations (Hassinen et al. 2009). The mechanism of heavy metal detoxification in hyperaccumulators that protects themselves by the overexpression of genes of reduced glutathione (GSH), cysteine, and o-acetylserine resulting in increase in the antioxidant activities (Anjum et al. 2014). Some genes are exclusively expressed in hyper accumulator phenotypes such as HMA4 gene under heavy metal stress. Similar to this, other studies have revealed that upstream regulation of salicylic acid, *NgSAT*, etc. results in increased serine acetyltransferase activity and higher GSH level and resulted in tolerance to Ni, Co, Zn, and to a small extent Cd (Freeman et al. 2004, 2005; Freeman and Salt 2007).

13.3 Metal Detoxification Approaches

13.3.1 Physico-Chemical Remediation of Heavy Metals

Taking into the account of the issues of metal toxicity, many techniques have been employed for the removal and/or recovery of heavy metals from polluted environments. Some established conventional procedures for heavy metal removal and/or recovery from solution include adsorption processes, chemical precipitation, electrochemical techniques, chemical oxidation or reduction reactions, ion exchange, evaporative recovery, reverse osmosis, and sludge filtration (Chen and Li 2010). However, these techniques have certain limitations as they are expensive, sometimes impractical, and not specific for metal-binding properties. Furthermore, high reagent requirement, generation of toxic waste, and unpredictable nature of metals are some of the disadvantages associated with these methods. Majority of these methods are ineffective when metal concentration in the solution is less than 100 mg/L (Ahluwalia and Goyal 2007). Separation by physical and chemical techniques is also challenging due to high solubility of heavy metal salts in solution. Thus, there is a need to develop inexpensive and suitable techniques which could be applicable under metal stressed conditions.

13.3.2 Biological Approaches to Combat Heavy Metal Stress

Bioremediation is an environment friendly innovative technique for the removal and recovery of heavy metal from the polluted areas. It involves living organisms (e.g. algae, bacteria, fungi, or plants) and/or their associated activities that reduce and/or recover heavy metal pollutants into less hazardous forms. It has been employed for the removal of heavy metals from contaminated wastewaters and soils. This method is considered a viable and appealing alternative to physical and chemical techniques since it involves the use of inexpensive microorganisms and provides long-term environmental benefits (Emenike et al. 2018). These organisms help to detoxify hazardous components in the environment. The bioremediation process can function naturally or can be improved through the addition of electron acceptors, nutrients, or other factors.

13.3.2.1 Microbial Resistance Strategies Towards Heavy Metals

The rhizospheric microbes play important role in HM detoxification in contaminated soils. According to Pires et al. (2017) the predominant bacterial populations in HM contaminated sites belong to Firmicutes, Proteobacteria, and Actinobacteria and the most common genera are *Bacillus*, *Pseudomonas*, and *Arthrobacter*. In order to survive under metal stress conditions, bacteria have developed certain strategies to regulate the intracellular levels of HMs. Microbial resistance to heavy metals occurs via acquisition of specific resistance systems such as efflux and uptake and extracellular precipitation. In a study, Karthik et al. (2017) isolated rhizobacterial strain AR6 from the rhizosphere of *Phaseolus vulgaris* which showed high Cr (VI) tolerance and multifarious plant growth promoting traits. Also, it was observed that the detoxification of toxic Cr (VI) occurred directly by enzymatic reduction to less toxic Cr (III) by chromate reductase or indirectly by making complexes with metabolites (Karthik et al. 2017). Different chromate reductases (e.g. ChrR, YieF, NemA, and LpDH) have been identified that are located either in cytoplasm or membrane bound in a bacterial cell (Huang et al. 2016). Similarly, other PGPR with improved metal remediation efficiency have been found to facilitate the growth of plants under adverse toxic conditions (Gopalakrishnan et al. 2018). Apart from traditional PGPR, both symbiotic and free living rhizobia have also been found capable of detoxifying HM and consequently upgrading the quality of contaminated soils (Checcucci et al. 2017; Rangel et al. 2017). Summarily, the metal removal mechanism can be grouped into five categories (1) extracellular precipitation, (2) intracellular accumulation, (3) oxidation and reduction reactions, (4) methylation and demethylation, and (5) extracellular binding and complexation (Ojuederie and Babalola 2017) as presented in Table 13.3.

13.3.2.2 Metal Extrusion Strategies

Mostly the resistance mechanisms in bacteria known till date are encoded on plasmids and transposons and consequently have high probability of gene transfer or spontaneous mutation that help bacteria to acquire resistance against heavy metals. For example, in gram-negative bacteria (e.g. *Ralstonia eutropha*), a czc system is found responsible for the resistance to Cd, Zn, and Co. The czc-genes

Table 13.3 Strategies adopted by metal-tolerant bacteria to overcome metal stress

Mechanism	Organism	Description and effectiveness	References
Bioaccumulation and biosorption	*Sinorhizobium* sp.	Improved growth and nodulation in *M. sativa* under cd, cu, Pb, and Zn stress	Zribi et al. (2012)
Bioaccumulation	*Delftia* sp. B9	Intracellular dissolution of cd Reduce cd accumulation in rice grain	Liu et al. (2018)
Biotransformation and bioaccumulation	*Micrococcus* KUMAs 15	Arsenite oxidation and accumulation	Paul et al. (2018)
Biosorption	*Bacillus* sp. MC3B-22 and *Microbacterium* MC3B-10	EPS mediated sorption of Cd^{2+}	Camacho-Chab et al. (2018)
Bioreduction	*Bacillus subtilis* MA13	Reduction of Cr(VI) via Cr reductases enhanced growth and photosynthetic pigments of soybean	Wani et al. (2018)

encode the cation-proton antiporter (CzcABC) that exports Cd, Zn, and Co. Similarly, ncc system found in *Alcaligenes xylosoxidans* displayed resistance to Ni, Cd, and Co. On the contrary, Cd resistance mechanism in *Staphylococcus, Bacillus,* or *Listeria* operates through Cd-efflux ATPase. Two most well studied Cu resistance systems (cop) are observed in *P.syringae* pv. tomato and pco in *E.coli.* The cop genes encode for different Cu-binding proteins, which sequester Cu in the periplasm or in the outer membrane. However, the pco system acts through an ion-dependent Cu antiporter (Kunito et al. 1997). Naturally occurring PGPR also show resistance to zinc which is mostly through efflux system, for example, a P-type ATPase efflux system transports Zn ions across the cytoplasmic membrane via ATP hydrolysis (Beard et al. 1997), while RND-driven transporter system moves Zn across the cell wall of gram-negative bacteria through a proton gradient (Nies 1999). Likewise, Ni resistance is inducible and depends on energy-dependent efflux system driven by chemiosmotic proton-antiporter system (Taghavi et al. 2001).

13.3.2.3 Biotransformation

Numerous microorganisms have ability to reduce/transform a wide variety of multivalent metals that pose major threat to the environment. Though, various PGPR strains possessing metal reducing ability have been identified (Mallick et al. 2018), reduction of chromium only by PGPR will be discussed in the following section.

Among different forms of chromium, the hexavalent chromium is the more toxic and carcinogenic owing to its high solubility in water, rapid permeability through biological membranes, and subsequent interaction with intracellular proteins and nucleic acids (Kamaludeen et al. 2003). Among various forms of Cr, Cr (III) does not migrate freely in natural systems because it tends to precipitate as Cr (III) minerals or is removed by adsorption. Hence, reduction of toxic Cr (VI) to Cr (III) is a useful approach to remediate Cr (VI) affected environments (Thatoi and Pradhan 2017). In this regard, numerous chromium resistant PGPR like *Pseudomonas* sp., *Bacillus, Stenotrophomonas*, *Serratia*, *Arthrobacter,* and rhizobia have been identified and applied in Cr (VI) contaminated soils (Baldiris et al. 2018; Dong et al. 2018). Detoxification of chromium by microbes may occur directly or indirectly and can be affected by pH, incubation period, chromate concentration, and types of microbes (aerobic/anaerobic) involved (Narayani and Shetty 2013). In the direct mode, the microbes absorb chromium and then enzymatically (chromium reductases) reduce it (Mala et al. 2015). While in the indirect mode, metabolic products (reductants or oxidants) of the microbes in soil, such as H_2S, chemically reduce chromium by redox reactions. Jin et al. (2017) observed that the removal of Cr (VI) by *Acinetobacter* strain WB-1 was due to surface immobilization along with intracellular and extracellular reduction. Enzymatic reduction of Cr (VI) to Cr (III) usually is accomplished by chromium reductases and occurs both anaerobically (Masaki et al. 2015) as well aerobically (He et al. 2015) and sometimes involves chemical reactions associated with compounds such as nucleotides, sugars, amino acids, vitamins, organic acids or glutathione.

13.3.2.4 Bioaccumulation

Bioaccumulation strategy involves the uptake of metal ions by an organism either directly following exposure to a contaminated medium or indirectly by consumption of contaminants. Intracellular accumulation of toxic metals is an energy-dependent transport system which depends on (1) intrinsic properties, (2) physiological and genetic adaptation, (3) metal speciation, availability, and toxicity (Sinha et al. 2013). Once taken up, toxic metals pass through biological membranes via carrier mediated transport, endocytosis, ion pumps, ion channels, complex permeation, and lipid permeation (Adriano 2017). Permeabilization of cell membranes to toxic elements can lead to further exposure of intracellular metal-binding sites resulting in enhanced passive accumulation. Several methods have been used for detecting the accumulation and localization of metals inside the bacterial cells. For example, using TEM, Podder and Majumder (2018) found that the growth and bioaccumulation of arsenic ions by *Corynebacterium glutamicum* MTCC 2745 varied with pH, inoculum size, contact time, temperature, and concentrations of peptone and As. In another study, AAS analysis of the culture products from *B. amyloliquefaciens* treated with Cr (VI) for 45 h showed the distribution of Cr(III) in pellet and culture supernatant in the range of 37.4 ± 1.7 and 62.6 ± 3.4 mg L^{-1}, respectively (Das et al. 2014). In SEM images, the Cr (VI) treated bacterial pellets looked rough, coagulated, and porous, whereas the untreated pellets appeared smooth, regular, and non-porous. Also, TEM–EDX study of the bacterial precipitates under Cr (VI) treatment had nanometric range of intracellular Cr (III). Bioaccumulation process has several advantages like it is a metabolically active process of living organisms that works through adsorption, intracellular accumulation, and bioprecipitation. However, bioaccumulation also has limitation because it is applied on live cells only. The living cells however have the potential of recombination or mutant formation which can change morphological and physiological features of the strain. Besides, high concentration of applied or already present metals significantly damage the surface of living cells and may lead to partial loss in cell-binding abilities and consequently release of accumulated metals back into solution (Kadukova and Vircikova 2005).

13.3.2.5 Biosorption

Biosorption of heavy metals by certain living or dead microbial biomass is a very effective solution to remediate even very dilute aqueous solutions (Dadrasnia et al. 2015). Biosorption consists of several mechanisms, such as ion exchange, adsorption, chelation, and diffusion through cell walls and membranes, which differs and depends on the species used, the origin and processing of biomass, and the chemistry of solution. Biosorption in fact is a non-enzymatic process wherein pollutants are adsorbed onto the cell surface (Sulaymon et al. 2012). The uptake of metal could be active or passive (Vijayaraghavan and Yun 2008) and both may occur independently or simultaneously. Of these, passive process is relatively nonspecific (Volesky 2007) and does not involve cellular metabolism. Here, metal binds to poly ionic cell walls through ion exchange. This process is not affected by physical conditions such as pH and ionic strength. It is a reversible and fairly rapid process requiring only 5–10 min. For complete biosorption of heavy metals. The active process is slow and depends on cellular metabolism and therefore is influenced by uncouplers, metabolic inhibitors, and temperature. In the active process, the metal complexes with specific proteins like metallothioneins which is found in vacuole. For example, biosorption capacity of live and dead cells of a novel *Bacillus* strain for chromium showed that both live and dead biomass followed the monolayer biosorption on the active surface sites. Scanning electron microscopy and FTIR indicated significant influence on the morphological features of the dead cells during biosorption of chromium. Approximately 92% and 70% desorption efficiencies were achieved using dead and live cells, respectively (Dadrasnia et al. 2015). However, whatever may be the mode of metal uptake, the adsorption occurs due to the nonspecific binding of ionic species to cell surface associated or extracellular polysaccharides and proteins (Cristani et al. 2012) of different bacterial cell organization, bacterial cell walls, and envelopes (Puyen et al. 2012). Among bacteria, the cell walls of gram-positive bacteria in general bind larger quantities of toxic metals than the envelopes of the gram-negative bacteria (Silver and Phung 1996). FTIR studies have revealed that various functional moieties such as hydroxyl, amino, carboxylate, phosphoryl, etc. were present on the surface of bacterial cell which participated in metal binding and hence assisted biosorption process (Patil and Unnikrishnan 2017).

13.4 Role of Bioactive Molecules Secreted by PGPR in HM Removal

Many PGPR colonizing plants have been found to play significant roles in mobilization or immobilization of heavy metals and consequently reducing the availability and/or toxicity of metals to plants (Table 13.4). However, this metal accumulating ability and plant colonizing potential of rhizospheric bacteria together could be of practical importance in alleviating metal toxicity when bioinoculated plants are grown in metal-contaminated soils (Kidd et al. 2017). Fatnassi et al. (2015) reported that co-inoculation of plants, treated with 1 mM Cu and 2 mM Cu (1) increased the dry weight as compared with Cu-treated and uninoculated plants, (2) decreased Cu uptake in the roots, (3) increased copper tolerance status of *Vicia faba* compared to uninoculated plants exposed to Cu stress.

In a similar study, Mohamed and Almaroai (2017) found that the phosphate solubilizer *Bacillus* sp., *Azotobacter* sp., and *Pseudomonas* sp. produced a substantial amount of IAA both in the absence and presence of heavy metals. Besides these merits, they significantly decreased the uptake of heavy metals in corn plants grown under metal stress. Similarly, Subrahmanyam et al. (2018) reported that green gram, inoculated by *Enterobacter* sp. C1D had significantly better length and dry biomass of shoot and root and chlorophyll content when grown in the presence of Cr (VI). Confocal laser scanning microscopy (CLSM) of the roots showed heavy bacterial loads on root surface specifically at the root tip and the point of root hair/lateral root formation. Moreover, the elevated IAA levels and ACC deaminase activity enabled *Enterobacter* sp. C1D to enhance green gram production in Cr (VI)-amended soils.

13.5 Role of EPS as Biosorbents in Heavy Metal Removal

Extracellular polysaccharides, the complex biomolecules are consisted of proteins, polysaccharides, uronic acid, humic-like substances, nucleic acid, lipids, and glycoproteins and surround the bacterial cells (Sheng et al. 2010). Metal sorption by EPS is believed to be an important self-protection strategy of microbial cells against toxic substances (Bhunia et al. 2018). The ionic nature of metal, its size, and charge density regulate its interaction with negatively charged EPS (Gupta and Diwan 2017). Since EPS are usually the first barrier of microbial cells that directly contact and interact with metals, they are of vital importance not because they protect the interior microbial cells but also play important role in remediation of metal-contaminated environments (Ayangbenro and Babalola 2018). Recently, numerous PGPR found to secrete EPS are involved in metal removal, for example, EPS of *B. subtilis* and *P. putida* strains significantly enhanced their Cu (II) adsorption capacity (Fang et al. 2013). The complex and diverse EPS properties make it usually difficult to understand the adsorption behavior of microbial biomass. In recent times, FTIR has frequently been used to obtain structural information on metal-binding

Table 13.4 Role of metal-tolerant PGPR secreting bioactive molecules in phytoremediation process

Bacteria	Bioactive molecules	Role of PGPR	References
Pseudomonas and rhizobium sullae	IAA, siderophores, P solubilization	Increased growth of Sulla coronaria under cd stress, enhanced antioxidant response and cd accumulation in roots	Chiboub et al. (2018)
Bacillus	IAA, Ni, ACC deaminase	Increased plant growth and facilitated Ni accumulation	Akhtar et al. (2018)
Mesorhizobium ciceri	N_2 fixation	Enhanced growth of chickpea and Cr concentration in roots favoring phytostabilization	Velez et al. (2017)
Gordonia alkanivorans, Cupriavidus necator, and Sporosarcina luteola	EPS, IAA, NH3, N2 fixation	Increased phytoextraction of As and Hg in B. juncea and L. albus	Franchi et al. (2017)
Pseudomonas sp.	Auxin, siderophore, P solubilization, ACC deaminase	Enhanced plant growth and increased nodule biomass Influenced phytostabilization	Soussou et al. (2017)
Bacillus sp.	Zn solubilization	Zn mobilization and accumulation, yield enhancement in soybean and wheat	Khande et al. (2017)
Enterobacter	Siderophore, IAA	Increased Fe uptake, immobilized Cd^{2+} in rhizosphere, influenced phytostabilization of Cd^{2+}	Chen et al. (2017)
Burkholderia cepacia	IAA, Siderophore	Enhanced growth of Brassica rapa, Zn uptake	Kang et al. (2017)
Bradyrhizobium liaoningensis	–	Increased Ni and Fe uptake in Pongamia pinnata from V Fe magnetite mine failing site	Yu et al. (2017)
Mesorhizobium	IAA, Siderophore, ACC, TCP solubilization	Enhanced multi element tolerance in Leucaena leucocephala	Rangel et al. (2017)
Variovorax paradoxus		Increased biomass of plant and Ni uptake in roots and shoots	Durand et al. (2016)
Brevibacterium casei	NH3, ACC, IAA, HCN	Increased biomass, enhanced cd, Zn, cu accumulation in shoots	Plociniczak et al. (2016)
Bacillus sp.	Zn solubilization, P and K solubilization, biocontrol	Enhanced Zn translocation	Shakeel et al. (2015)
P. aeruginosa	Siderophore	As accumulation in all plant	Jeong et al. (2014)

complex associated with various functional groups of EPS. Hence, the chemical state and the IR absorption spectrum of EPS functional groups change sensitively upon binding with metals. The FTIR spectral band in the region between 4000 and

400 cm^{-1}usually shows the major characteristic bands of various bonds present in functional groups of EPS. For example, the Cu^{2+} complexation by carboxyl functional group of EPS can be clearly reflected by the change in peak intensity of FTIR spectral band at both 1400 cm^{-1} (associated with the stretching vibration of COO^{2-} bond from carboxylic group and deformation vibration of -OH from alcohols and phenol groups) and 1080 cm^{-1} (attributed to the stretching vibration of -OH group) (Mohite et al. 2018). An interesting feature called *rescuing* was highlighted wherein EPS secreting (metal-tolerant) strains displayed a protective effect towards non-EPS secreting sensitive strain in co-culture experiments (Nocelli et al. 2016). Similarly, the non-symbiotic N_2 fixers such as *A. chroococcum* secreted EPS and formed complexes with Pb and Hg ions in pH dependent manner. At lower pH (between 4 and 5), the maximum adsorption was 40–47% of initial metal ion concentration in solution (Rasulov et al. 2013).

13.6 Role of Metallothioneins and Phytochelatins

Metallothioneins (MTs) are cysteine-rich, heavy metal-binding protein molecules synthesized due to mRNA translation (Guo et al. 2013) and play a crucial role in uptake, detoxification, and accumulation of metal. MTs are gene-encoded polypeptides. They show greater affinity for metals such as Cd, Cu, Zn, and As by cellular sequestration, homeostasis of intracellular metal ions as well as modification of metal transport. In addition to detoxification of heavy metals, plant MTs also play an important role in cell growth and proliferation, repair of plasma membrane, repair of damaged DNA, scavenge ROS, and maintenance of redox level (Emamverdian et al. 2015). Naik et al. (2012) explored the role of bacterial MTs in Pb-resistant bacterial isolates: *Salmonella choleraesuis* strain 4A, *Proteus penneri* strain GM10, *Bacillus subtilis* strain GM02, *Pseudomonas aeruginosa* strain 4EA, *Proteus penneri* strain GM03, and *Providencia rettgeri* strain GM04 which were quarantined from soil, polluted with car battery waste from Goa, India. All the isolates except *P. aeruginosa* strain 4EA had plasmids. Both bacterial MTs and intracellular bioaccumulation of Pb in *S.choleraesuis* strain 4A and *P.penneri* strain GM10 were responsible for the Pb resistance. Similarly Li et al. (2015) found that expression of PcPCS1gene from bean pear (*Pyrus calleryana* Dcne.) was induced after Cd and Cu treatments and *E.coli* with over-expressed PcPCS1 had enhanced tolerance to Cd, Cu, Na, and Hg. *E.coli* cells transformed with pPcPCS1 survived in solid M9 medium containing 2 mM Cd^{2+},4 mM Cu^{2+}, 4.5%(w/v)Na^{+}, or200 μMHg^{2+}. Moreover, the growth curve showed that1.5 mM Cd^{2+}, 2.5 mMCu^{2+}, 3.5% (w/v)Na^{+}, and100 μMHg^{2+}had no effect on growth of *E.coli* cells transformed with pPcPCS1. Also, the content of PCs and accumulation of Cd, Cu, Na, and Hg ions were enhanced in the recombinant *E.coli* strain Rosetta™(DE3).

Phytochelatins (PCs) are short-chain thiol-rich repetitions of peptides synthesized from sulfur-rich glutathione by the enzyme phytochelatin synthase which defend plants against environmental stresses such as salinity, drought, herbicide, and heavy metals (Emamverdian et al. 2015). They are used as biomarkers for an early detection of heavy metal stress in plants (Saba et al. 2013). Metallothioneins have much affinity for a wide range of metals such as Cu, Zn, Cd, and As by cellular sequestration, homeostasis of intracellular metal ions as well as adjustment of metal transport. Apart from detoxification of heavy metals, plant MTs also play a role in maintenance of the redox level, repair of plasma membrane, cell proliferation, and its growth, repair of damaged DNA, and scavenge ROS (Emamverdian et al. 2015). Glutathione (GSH), a precursor of phytochelatin, has been reported to play a vital role in metal detoxification (Hasanuzzaman et al. 2017) and in protecting plant cells from other environmental stresses including intrinsic oxidative stress reactions.

13.6.1 Phytoremediation Strategies

Some of the strategies mediated by plants often termed as phytoremediation include:

1. Phytoextraction (phytoaccumulation): Uptake of the contaminant by plant roots from the environment and its translocation into harvestable plant biomass.
2. Rhizofiltration: Use of plant roots to absorb or adsorb contaminants present in solution form in the surrounding of root zone.
3. Phytostabilization: Reduction of mobility and bioavailability of pollutants in environment either by physical or chemical effects.
4. Phytodegradation (phytotransformation): Phytodegradation is the use of plants and microorganisms to uptake, metabolize, and degrade the contaminant. In this approach, plant roots are used in association with microorganisms to detoxify soil contaminated with organic compounds. It is also known as phytotransformation. Some plants are able to decontaminate soil, sludge, sediment, and ground and surface water by producing various enzymes. This approach involves organic compounds, including herbicides, insecticides, chlorinated solvents, and inorganic contaminants (Pivetz 2001).
5. Rhizodegradation: The breakdown of contaminants in the soil through microbial activity that is enhanced by the presence of root zone is called rhizodegradation. This process uses microorganisms to consume and digest organic substances for nutrition and energy. Natural substances released by the plant roots, sugars, alcohols, and acids contain organic carbon that provides food for soil microorganisms and establishes a dense root mass that takes up large quantities of water. This process is used for the removal of organic substance (contaminants) in soil medium (Moreno et al. 2008).
6. Phytovolatilization: Phytovolatilization is the use of green plants to extract volatile contaminants, such as Hg and Se, from the polluted soils and to ascend them into the air from their foliage (Karami and Shamsuddin 2010).

Phytoremediation as a technique is inexpensive since it involves plants which can be grown and monitored easily (Saraswat and Rai 2012). Moreover, the recovery and re-use of valuable products of this method are easy, because it uses natural biological

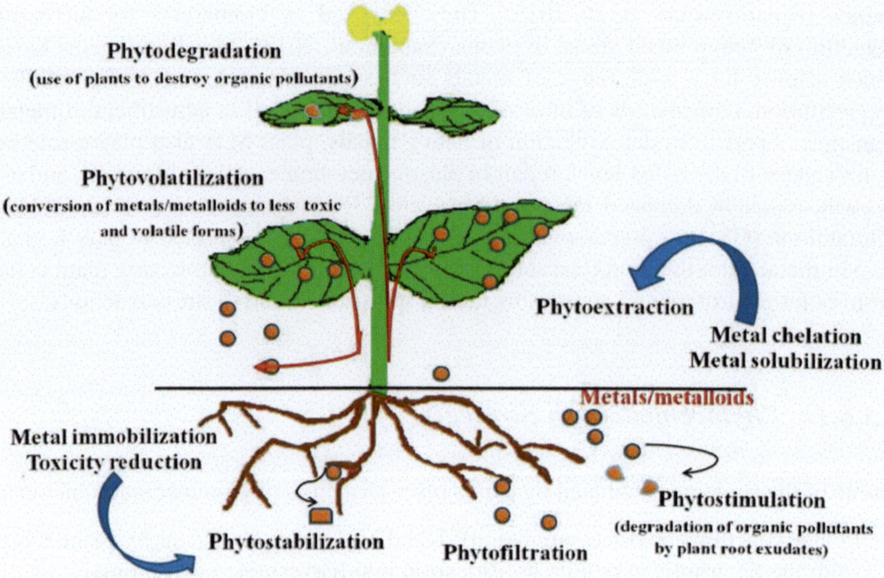

Fig. 13.1 Strategies employed for HM removal using plants. Adapted from Ojuederie and Babalola (2017)

materials. Additionally, plants can be modified for any target characteristics, and the original state of the environment could be restored.

It is relatively a recent technology and perceived as cost-effective, efficient, novel, eco-friendly, and solar-driven technology with good public acceptance. Phytoremediation is an area of active current research (Fig. 13.1). New efficient metal hyperaccumulators are being explored for their application in phytoremediation and phytomining.

Plant species for phytoremediation are selected based on their root depth, the nature of the contaminants and the soil, and regional climate. The root depth directly impacts the depth of soil that can be remediated. It varies greatly among different types of plants, and can also vary significantly for one species depending on local conditions such soil structure, depth of a hard pan, soil fertility, cropping pressure, contaminant concentration, or other conditions. The cleaning depths are approximately <3 feet for grasses, <10 feet for shrubs, and <20 feet for deep rooting trees. The nature of on-site contaminants is a principal factor in the selection of a plant for phytoremediation (Sharma and Reddy 2004). Grasses are the most commonly evaluated plants used for phytoremediation (Shu et al. 2002). They are preferably used for phytoremediation because as compared to trees and shrubs, herbaceous plants, especially grasses, rapid growth, large amount of biomass, strong resistance, effective stabilization to soils, and ability to remediate different types of soils (Elekes 2014). They are pioneers and usually adapted to adverse conditions such as low soil nutrient content, stress environment, and shallow soils (Malik et al. 2010). The large surface area of fibrous roots of grasses and their intensive penetration of soil reduces leaching, runoff, and erosion via stabilization of soil and offers advantages for phytoremediation. Molecular tools are being used to better understand the mechanisms of metal uptake, translocation, sequestration, and tolerance in plants.

13.6.2 PGPR: Assisted Phytoremediation

Degraded soils, following the harmful effects of emissions from non-ferrous metal smelters, are usually arid from organic material and devoid of proper microflora. Therefore, the poor condition of the soil environment makes it impossible to carry out an effective biological remediation of the degraded area. For this purpose, some organic substances, such as sewage sludge, are used to support the process of phytoremediation, because they are a source of biogenic elements and soil microorganisms (Kacprzak et al. 2014). Among the disadvantages, phytoremediation is a lengthy process and is affected greatly by the changing environmental conditions. Thus, the ability of plants to remove/sequester metals in contaminated sites can be improved by applying PGPR with various phytoremediation methods. When single or mixture of inoculants are applied along with phytoremediation process, PGPR affect the mobility and availability of metals to plants by releasing numerous chelating substances, acidification, phosphate solubilization, and redox changes (Zoomi et al. 2017). Chen et al. (2018a) reported that successful in situ phytoremediation depends on beneficial interactions between roots and microbes and plant cultivation. Soil amendment increases microbial diversity and restructures microbial communities. Rhizo-compartmentalization through selection of a specific core root microbiome by the metal-tolerant plant *H. cannabinus* with Enterobacteriaceae, Pseudomonadaceae, and Comamonadaceae included a large number of metal-tolerant and plant growth-promoting bacteria. The root-associated microbial community formed niche-assembled patterns and predominantly had *Proteobacteria, Actinobacteria,* and *Chloroflexi* under metal-contaminated conditions.

Złoch et al. (2017) evaluated the role of three metallotolerant siderophore-producing *Streptomyces* sp. B1–B3 strains in the phytoremediation of heavy metals using *S. dasyclados*. The bacterial inoculation significantly stimulated biomass and reduced oxidative stress. Moreover, the bacteria affected the speciation of heavy metals and finally their mobility, thereby enhancing the uptake and bioaccumulation of Zn, Cd, and Pb in the biomass. The best capacity for phytoextraction was noted for strain B1, which had the highest siderophore secretion ability. Five metal resistant PGPR (*Ralstonia eutropha* 1C2, *Chryseobacterium humi* ECP37, *Pseudomonas fluorescens* S3X, *Rhizobium radiobacter* EC1B, and *Pseudomonas reactans* EDP28) were investigated by Moreira et al. (2016) for their in vitro growth promoting traits and for their ability to induce growth of maize seedlings exposed to Zn and Cd. They showed that some bacteria only enhanced PGP traits when exposed to metals. The bacterial strains ECP37 and EDP28 were most efficient in improving seedling growth with increasing metal concentrations, followed by S3X. When inoculated in maize grown in mine soil, these strains also outperformed the others by increasing shoot biomass and elongation, metal accumulation, and by decreasing it in roots. The most evident effect of doubling the inoculum size was the increase in Cd accumulation, which was of 17% and 31% in roots and shoots, respectively. Other effects included a slight reduction in shoots' biomass (13%) and a general decrease in P content.

Sinorhizobium meliloti CCNWSX0020 was used to assess its effect on *Medicago sativa* seedlings under Cu stress (Chen et al. 2018b). This rhizobium inoculation alleviated Cu-induced growth inhibition, regulating antioxidant enzyme activities and increased nitrogen concentration in *M. sativa* seedlings. Moreover, the total amount of Cu uptake in inoculated plants was significantly increased compared with non-inoculated plants, and the increase in the roots was much higher than that in the shoots, thus decreasing the transfer coefficient and promoting Cu phytostabilization.

Bianucci et al. (2018) in a recent investigation observed no changes in growth variables (shoot and root dry weight) of soybean plants inoculated with *Bradyrhizobium* sp. grown in As (V) contaminated soils. Regarding As uptake by plants, metalloid accumulation followed the same distribution pattern among strains. Furthermore, at 6 μM As (V), *Bradyrhizobium* inoculated soybean revealed a significantly lower translocation factor (TF) in comparison to other inoculated strains and promoted As phytostabilization. At the highest As (V) concentration, only *B. diazoefficiens* USDA110 was able to nodulate the legume; however, a significant decrease in the number and dry weight of nodules and N content was observed. Similarly, Mallick et al. (2018) reported that two As-resistant halophilic bacterial strains *Kocuria flava* AB402 and *Bacillus vietnamensis* AB403 from mangrove rhizosphere of Sundarban could tolerate 35 mM and 20 mM of arsenite, respectively. Also, As had a variable impact on EPS synthesis, biofilm formation, and root association ability of both the bacterial strains. When used as inoculum they promoted the growth of rice seedlings by decreasing As uptake and accumulation in plants. Gupta et al. (2018) isolated Cr resistant plant growth promoting *Pseudomonas* sp. (strain CPSB21) from the tannery effluent contaminated agricultural soils and evaluated the plant growth promoting activities, oxidative stress tolerance, and Cr^{6+} bioremediation potential. Assessment of different plant growth promoting traits, such as P solubilization, IAA production, siderophores, ammonia, and HCN production, revealed that the strain CPSB21 could serve as an efficient plant growth promoter under laboratory conditions. Further, the plant growth, pigment content, N and P uptake, and Fe accumulation were reduced when sunflower and tomato were grown in Cr (VI) amended soils. However, inoculation of strain CPSB21 alleviated the Cr^{6+} toxicity and enhanced the plant growth parameters and nutrient uptake. Apart from these, Cr toxicity had varied response on oxidative stress tolerance at graded Cr^{6+}concentration on both plants, and increase in SOD and CAT activity and reduction in MDA were observed following inoculation of strain CPSB21. Additionally, inoculation of CPSB21 enhanced the uptake of Cr^{6+} in sunflower, while no substantial increment was observed in inoculated tomato plants.

In order to select the PGPR for phytoremediation of heavy metal contamination, Roman-Ponce et al. (2017) isolated bacterial strains from the rhizosphere of two endemic plants, *Prosopis laevigata* and *Sphaeralcea angustifolia,* grown in a heavy metal-contaminated zone in Mexico. These rhizobacterial strains were characterized for the growth at different pH and salinity, extracellular enzyme production, solubilization of phosphate, heavy metal resistance, and plant growth-promoting (PGP) traits, including production of siderophores and IAA. Overall, the rhizobacteria showed multiple PGP traits. These rhizobacteria were also resistant to high levels of heavy metals including As as a metalloid (up to 480 mmol L^{-1} As(V), 24 mmol L^{-1} Pb(II), 21 mmol L^{-1} Cu(II), and 4.5 mmol L^{-1} Zn(II)). Inoculation of *Brassica nigra* seeds with *Microbacterium* sp. CE3R2, *Microbacterium* sp. NE1R5, *Curtobacterium* sp. NM1R1, and *Microbacterium* sp. NM3E9 facilitated the root development and significantly improved the seed germination and root growth in the presence of 2.2 mmol L^{-1}Zn (II).

Additionally, siderophores released by PGPR including legume nodulating rhizobia into the rhizosphere serve as an Fe source for plants (Ivanov et al. 2012) and therefore help to fulfil the Fe deficiency of plants in Fe limiting soils. Considering this, it is generally suggested to use PGPR in soils deficient in Fe. To substantiate this, metal resistant PGPR such as *P. putida* strains and *P. fluorescens* strains (showing IAA, siderophores and ACC deaminase production) were reported to show significant improvement in growth attributes of inoculated canola and barley plants even when grown with various concentrations of $CdCl_2$ and $PbNO_3^{2-}$. Furthermore, the translocation factor indicated that inoculated canola and barley had abilities of Cd and Pb phytoextraction in soil contaminated with respective metal. Conclusively, the enhancement in the inoculated canola and barley plants occurred due to the protection against the inhibitory effects of Cd and Pb by PGPR in addition to their ability to provide IAA, siderophore, and ACCD to the developing plants (Yancheshmeh et al. 2011). Nayak et al. (2018) used *B. cereus* (T1B3) strain able to produce various bioactive compounds such as ACC deaminase, indole acetic acid, and siderophores, nitrogen fixation, and P solubilization. Removal capacity (mg L^{-1}) of T1B3 strain was 82% for Cr^{+6} (100), 92% for Fe (100), 67% for Mn (50), 36% for Zn (50), 31% for Cd (30), 25% for Cu (30), and 43% for Ni (50) during the active growth cycle in HM-amended soil.

Ma et al. (2015) isolated a PGPB strain SC2b from the rhizosphere of *Sedum plumbizincicola* grown in Pb/zinc (Zn) mine soils and characterized as *Bacillus* sp. using partial 16S ribosomal DNA sequencing analysis. Strain SC2b exhibited high levels of resistance to Cd (300 mg/L), Zn (730 mg/L), and Pb (1400 mg/L). Besides possessing various PGP features such as secretion of ACC deaminase, utilization of 1-aminocyclopropane-1-carboxylate, solubilization of P, and production of IAA and siderophore, the strain mobilized high concentration of heavy metals from the soils and exhibited different biosorption capacity toward the tested metal ions. Under pot trial, this metal resistant PGPB SC2b elevated the shoot and root biomass and leaf chlorophyll content of *S. Plumbizincicola* besides enhanced Cd and

Zn uptake through metal mobilization or plant-microbial mediated changes in chemical or biological soil properties.

A metal resistant bacterium *Enterobacter ludwigii* isolated from the rhizosphere of Kair "*Capparis decidua*" by Singh et al. (2018) was screened for its phytoextraction ability under gradient metal stress conditions. Among the PGP traits, isolate showed the production of ACC deaminase, produced IAA, and solubilized the inorganic P. The isolate was resistant to Zn, Ni, Cu, and Cd. Furthermore, inoculation of the test isolate significantly increased various growth parameters of wheat plants and also improved the photosynthetic pigments. In addition, inoculation resulted in significant increase in the Zn content in wheat plants under metal stress. Bacterial application significantly increased various compatible solutes such as proline content, total soluble sugar and decreased the malondialdehyde (MDA) content as compared to control, illustrating its protective effect under metal induced oxidative stress in wheat plants.

13.7 Conclusion

The toxic effect of heavy metals may result from the accumulation of HMs over time in vital parts of humans, plants as well as microbes. The direct and indirect mode of deleterious effect of HMs in plants results due to several reasons such as reduction in dry biomass, alteration in chlorophyll content and at molecular level due to over generation of ROS, damage to essential macromolecules, and thus constrain crop productivity. Undoubtedly the PGPR-assisted HM phytoremediation used in native or engineered forms is reported to have greater remediation potential but their impact on ecosystems needs to be elucidated before commercialization. Various steps of regulatory networks via plant-associated microbes including the synergistic action of plants and microbes and their mechanism for metal mobilization, transformation, and detoxification should be investigated for unravelling the dynamics of plant–microbe–metal interactions in the soils. Further monitoring and managing microbial heavy metal remediation require the characterization of the fate and behavior of the compounds of interest in the environment.

Acknowledgement Research work of various authors cited in this chapter and facilities provided by the University are thankfully acknowledged.

References

Adam V, Chudobova D, Tmejova K, Cihalova K, Krizkova S, Guran R, Kominkova M, Zurek M, Kremplova M, Jimenez AM, Konecna M (2014) An effect of cadmium and lead ions on *Escherichia coli* with the cloned gene for metallothionein (MT-3) revealed by electrochemistry. Electrochim Acta 140:11–19

Adriano DC (2017) Uptake and accumulation of metals in bacteria and fungi. In: Biogeochemistry of trace metals. CRC Press, Boca Raton, pp 289–310

Águila-Clares B, Castiblanco LF, Quesada JM, Penyalver R, Carbonell J, López MM, Marco-Noales E, Sundin GW (2018) Transcriptional response of *Erwinia amylovora* to copper shock: in vivo role of the copA gene. Mol Plant Pathol 19(1):169–179

Ahluwalia SS, Goyal D (2007) Review microbial and plant derived biomass for removal of heavy metals from wastewater. Bioresour Technol 98(12):2243–2257

Ahmad I, Akhtar MJ, Mehmood S, Akhter K, Tahir M, Saeed MF, Hussain MB, Hussain S (2018) Combined application of compost and *Bacillus* sp. CIK-512 ameliorated the lead toxicity in radish by regulating the homeostasis of antioxidants and lead. Ecotoxicol Environ Saf 148:805–812

Akhtar MJ, Ullah S, Ahmad I, Rauf A, Nadeem SM, Khan MY, Hussain S, Bulgariu L (2018) Nickel phytoextraction through bacterial inoculation in *Raphanus sativus*. Chemosphere 190:234–242

Akkermans S, Fernandez EN, Logist F, Van Impe JF (2017) Introducing a novel interaction model structure for the combined effect of temperature and pH on the microbial growth rate. Int J Food Microbiol 240:85–96

Algreen M, Rein A, Legind CN, Amundsen CE, Karlson UG, Trapp S (2012) Test of tree core sampling for screening of toxic elements in soils from a Norwegian site. Int J Phytoremediation 14:305–319

Andresen E, Küpper H (2013) Cadmium toxicity in plants. In: Cadmium: from toxicity to essentiality. Springer, Dordrecht, pp 395–413

Anjum NA, Gill SS, Gill R, Hasanuzzaman M, Duarte AC, Pereira E et al (2014) Metal/metalloid stress tolerance in plants: role of ascorbate, its redox couple, and associated enzymes. Protoplasma 251:1265–1283

Anjum SA, Ashraf U, Khan I, Tanveer M, Shahid M, Shakoor A et al (2017) Phyto-toxicity of chromium in maize: oxidative damage, osmolyte accumulation, anti-oxidative defense and chromium uptake. Pedosphere 27:262–273

Aslam M, Saeed MS, Sattar S, Sajad S, Sajjad M, Adnan M, Iqbal M, Sharif MT (2017) Specific role of proline against heavy metals toxicity in plants. Int J Pure Appl Biosci 5(6):27–34

Ayangbenro A, Babalola O (2018) Metal (loid) bioremediation: strategies employed by microbial polymers. Sustainability 10(9):3028

Azarbad H, Niklińska M, Laskowski R, Van Straalen NM, Van Gestel CAM, Zhou J, He Z, Wen C, Röling WF (2015) Microbial community composition and functions are resilient to metal pollution along two forest soil gradients. FEMS Microbiol Ecol 91:1–11

Azimi A, Azari A, Rezakazemi M, Ansarpour M (2017) Removal of heavy metals from industrial wastewaters: a review. Chem Bio Eng Rev 4(1):37–59

Baldiris R, Acosta-Tapia N, Montes A, Hernández J, Vivas-Reyes R (2018) Reduction of hexavalent chromium and detection of chromate reductase (ChrR) in *Stenotrophomonas maltophilia*. Molecules 23(2):406

Beard SJ, Hashim R, Hernandez J, Hughes M, Poole RK (1997) Zinc (II) tolerance in *Escherichia coli* K-12: evidence that the zntA gene (o732) encodes a cation transport ATPase. Mol Microbiol 25(5):883–891

Bhunia B, Uday US, Oinam G, Mondal A, Bandyopadhyay TK, Tiwari ON (2018) Characterization, genetic regulation and production of cyanobacterial exopolysaccharides and its applicability for heavy metal removal. Carbohydr Polym 179:228–243

Bianucci E, Godoy A, Furlan A, Peralta JM, Hernández LE, Carpena-Ruiz RO, Castro S (2018) Arsenic toxicity in soybean alleviated by a symbiotic species of *Bradyrhizobium*. Symbiosis 74 (3):167–176

Boer JL, Mulrooney SB, Hausinger RP (2014) Nickel-dependent metalloenzymes. Arch Biochem Biophys 544:142–152

Burges A, Epelde L, Garbisu C (2015) Impact of repeated single-metal and multi-metal pollution events on soil quality. Chemosphere 120:8–15

Camacho-Chab JC, Castañeda-Chávez MD, Chan-Bacab MJ, Aguila-Ramírez RN, Galaviz-Villa I, Bartolo-Pérez P, Lango-Reynoso F, Tabasco-Novelo C, Gaylarde C, Ortega-Morales BO (2018) Biosorption of cadmium by non-toxic extracellular polymeric substances (EPS) synthesized by Bacteria from marine intertidal biofilms. Int J Env Res Public Health 15(2):314

Checcucci A, Bazzicalupo M, Mengoni A (2017) Exploiting nitrogen-fixing rhizobial symbionts genetic resources for improving phytoremediation of contaminated soils. In: Enhancing Cleanup of Environmental Pollutants: Biological Approaches, vol 1. Springer International Publishing, Cham, pp 275–288

Chen YH, Li FA (2010) Kinetic study on removal of copper (II) using goethite and hematite nano-photocatalysts. J Colloid Interface Sci 347(2):277–281

Chen Y, Yang W, Chao Y, Wang S, Tang YT, Qiu RL (2017) Metal-tolerant *Enterobacter* sp. strain EG16 enhanced phytoremediation using *Hibiscus cannabinus* via siderophore-mediated plant growth promotion under metal contamination. Plant Soil 413(1–2):203–216

Chen Y, Ding Q, Chao Y, Wei X, Wang S, Qiu R (2018a) Structural development and assembly patterns of the root-associated microbiomes during phytoremediation. Sci Total Environ 644:1591–1601

Chen J, Liu YQ, Yan XW, Wei GH, Zhang JH, Fang LC (2018b) Rhizobium inoculation enhances copper tolerance by affecting copper uptake and regulating the ascorbate-glutathione cycle and phytochelatin biosynthesis-related gene expression in *Medicago sativa* seedlings. Ecotoxicol Environ Saf 162:312–323

Chiboub M, Jebara SH, Saadani O, Fatnassi IC, Abdelkerim S, Jebara M (2018) Physiological responses and antioxidant enzyme changes in *Sulla coronaria* inoculated by cadmium resistant bacteria. J Plant Res 131(1):99–110

Choi Y, Kim HA, Kim KW, Lee BT (2018) Comparative toxicity of silver nanoparticles and silver ions to *Escherichia coli*. J Environ Sci 66:50–60

Clemens S, Ma JF (2016) Toxic heavy metal and metalloid accumulation in crop plants and foods. Ann Rev Plant Biol 67:489–512

Cristani M, Naccari C, Nostro A, Pizzimenti A, Trombetta D, Pizzimenti F (2012) Possible use of *Serratia marcescens* in toxic metal biosorption (removal). Environ Sci Pollut Res Int 19:161–168

Dadrasnia A, Chuan Wei KS, Shahsavari N, Azirun MS, Ismail S (2015) Biosorption potential of *Bacillus salmalaya* strain 139SI for removal of Cr (VI) from aqueous solution. Int J Environ Res Public Health 12(12):15321–15338

Das S, Mishra J, Das SK, Pandey S, Rao DS, Chakraborty A, Sudarshan M, Das N, Thatoi H (2014) Investigation on mechanism of Cr (VI) reduction and removal by *Bacillus amyloliquefaciens*, a novel chromate tolerant bacterium isolated from chromite mine soil. Chemosphere 96:112–121

Dong L, Zhou S, He Y, Jia Y, Bai Q, Deng P, Gao J, Li Y, Xiao H (2018) Analysis of the genome and chromium metabolism-related genes of *Serratia* sp. S2. Appl Biochem Biotechnol 185 (1):140–152

Dos Santos JV, Varón-López M, Soares CRFS, Leal PL, Siqueira JO, De Souza Moreira FM (2016) Biological attributes of rehabilitated soils contaminated with heavy metals. Environ Sci Pollut Res 23:6735–6748

Dotaniya ML, Rajendiran S, Coumar MV, Meena VD, Saha JK, Kundu S, Kumar A, Patra AK (2018) Interactive effect of cadmium and zinc on chromium uptake in spinach grown in vertisol of Central India. Int J Environ Sci Technol 15(2):441–448

Durand A, Piutti S, Rue M, Morel JL, Echevarria G, Benizri E (2016) Improving nickel phytoextraction by co-cropping hyperaccumulator plants inoculated by plant growth promoting rhizobacteria. Plant Soil 399(1–2):179–192

Elekes CC (2014) Eco-technological solutions for the remediation of polluted soil and heavy metal recovery. In: Hernández-Soriano MC (ed) Environmental risk assessment of soil contamination. InTech, Rijeka, pp 309–335

Emamverdian A, Ding Y, Mokhberdoran F, Xie Y (2015) Heavy metal stress and some mechanisms of plant defense response. Sci World J 2015:756120

Emenike CU, Jayanthi B, Agamuthu P, Fauziah SH (2018) Biotransformation and removal of heavy metals: a review of phytoremediation and microbial remediation assessment on contaminated soil. Environ Rev 26(2):156–168

Fang L, Wei X, Cai P, Huang Q, Chen H, Liang W, Rong X (2013) Role of extracellular polymeric substances in cu(II) adsorption on *Bacillus subtilis* and *Pseudomonas putida*. Bioresour Technol 102(2):1137–1141

Fathima A, Rao JR (2018) Is Cr (III) toxic to bacteria: toxicity studies using *Bacillus subtilis* and *Escherichia coli*as model organism. Arch Microbiol 200:453–462

Fatnassi IC, Chiboub M, Saadani O, Jebara M, Jebara SH (2015) Impact of dual inoculation with rhizobium and PGPR on growth and antioxidant status of *Vicia faba* L. under copper stress. C R Biol 338(4):241–254

Franchi E, Rolli E, Marasco R, Agazzi G, Borin S, Cosmina P, Pedron F, Rosellini I, Barbafieri M, Petruzzelli G (2017) Phytoremediation of a multi contaminated soil: mercury and arsenic phytoextraction assisted by mobilizing agent and plant growth promoting bacteria. J Soils Sediments 17(5):1224–1236

Freeman JL, Salt DE (2007) The metal tolerance profile of *Thlaspi goesingense* is mimicked in *Arabidopsis thaliana* heterologously expressing serine acetyl-transferase. BMC Plant Biol 7:63. https://doi.org/10.1186/1471-2229-7-63

Freeman JL, Garcia D, Kim D, Hopf A, Salt DE (2005) Constitutively elevated salicylic acid signals glutathione-mediated nickel tolerance in *Thlaspi* nickel hyperaccumulators. Plant Physiol 137:1082–1091

Freeman JL, Persans MW, Nieman K, Albrecht C, Peer W, Pickering IJ et al (2004) Increased glutathione biosynthesis plays a role in nickel tolerance in *Thlaspi* nickel hyperaccumulators. Plant Cell 16:2176–2191

Gautam S, Bhagyawant SS, Srivastava N (2018) Antioxidant responses and isoenzyme activity of hydroponically grown safflower seedlings under copper stress. Indian J Plant Physiol 23 (2):342–351

Gianfreda L, Rao MA (2017) Soil microbial and enzymatic diversity as affected by the presence of xenobiotics. In: Xenobiotics in the soil environment. Springer, Cham, pp 153–169

Gopalakrishnan S, Srinivas V, Venula A, Samineni S, Rathore A (2018) Influence of diazotrophic bacteria on nodulation, nitrogen fixation, growth promotion and yield traits in five cultivars of chickpea. Biocatal Agric Biotechnol 15:35–42

Guo J, Xu L, Su Y, Wang H, Gao S, Xu J, Que Y (2013) ScMT2-1-3, a metallothionein gene of sugarcane, plays an important role in the regulation of heavy metal tolerance/accumulation. Biomed Res Int 2013:904769

Gupta AD, Karthikeyan S (2016) Individual and combined toxic effect of nickel and chromium on biochemical constituents in *E. coli* using FTIR spectroscopy and principle component analysis. Ecotoxicol Environ Saf 130:289–294

Gupta P, Diwan B (2017) Bacterial exopolysaccharide mediated heavy metal removal: a review on biosynthesis, mechanism and remediation strategies. Biotechnol Rep 13:58–71

Gupta P, Rani R, Chandra A, Kumar V (2018) Potential applications of *Pseudomonas* sp. (strain CPSB21) to ameliorate Cr^{6+} stress and phytoremediation of tannery effluent contaminated agricultural soils. Sci Rep 8:4860

Gutierrez-Ginés MJ, Hernández AJ, Pérez-Leblic MI, Pastor J, Vangronsveld J (2014) Phytoremediation of soils co-contaminated by organic compounds and heavy metals: bioassays with *Lupinus luteus* L. and associated endophytic bacteria. J Environ Manag 143:197–207

Hasanuzzaman M, Nahar K, Anee TI, Fujita M (2017) Glutathione in plants: biosynthesis and physiological role in environmental stress tolerance. Physiol Mol Biol Plants 23(2):249–268

Hassinen VH, Tuomainen M, Peräniemi S, Schat H, Kärenlampi SO et al (2009) Metallothioneins 2 and 3 contribute to the metal-adapted phenotype but are not directly linked to Zn accumulation in the metal hyperaccumulator, *Thlaspi caerulescens*. J Exp Bot 60:187–196

Hayat S, Hayat Q, Alyemeni MN, Ahmad A (2013) Proline enhances antioxidative enzyme activity, photosynthesis and yield of *Cicer arietinum* L. exposed to cadmium stress. Acta Bot Croat 2:323–335

He D, Zheng M, Ma T, Li C, Ni J (2015) Interaction of Cr (VI) reduction and denitrification by strain *Pseudomonas aeruginosa* PCN-2 under aerobic conditions. Bioresour Technol 185:346–352

He H, Chen Y, Li X, Cheng Y, Yang C, Zeng G (2017) Influence of salinity on microorganisms in activated sludge processes: a review. Int Biodeterior Biodegrad 119:520–527

Hornick KM, Buschmann AH (2018) Insights into the diversity and metabolic function of bacterial communities in sediments from Chilean salmon aquaculture sites. Ann Microbiol 68:63–77

Huang H, Wu K, Khan A, Jiang Y, Ling Z, Liu P, Chen Y, Tao X, Li X (2016) A novel *Pseudomonas gessardii* strain LZ-E simultaneously degrades naphthalene and reduces hexavalent chromium. Bioresour Technol 207:370–378

Ishida T (2018) Antibacterial mechanism of Bacteriolyses of bacterial cell walls by zinc (II) ion induced activations of PGN autolysins, and DNA damages. J Genes Protein 1:1–7

Ivanov R, Brumbarova T, Bauer P (2012) Fitting into the harsh reality: regulation of iron-deficiency responses in dicotyledonous plants. Mol Plant 5:27–42

Jeong S, Moon HS, Nam K (2014) Enhanced uptake and translocation of arsenic in Cretan brake fern (*Pteris cretica L.*) through siderophore-arsenic complex formation with an aid of rhizospheric bacterial activity. J Hazard Mater 280:536–543

Jin R, Wang B, Liu G, Wang Y, Zhou J, Wang J (2017) Bioreduction of Cr (VI) by *Acinetobacter* sp. WB-1 during simultaneous nitrification/denitrification process. J Chem Technol Biotechnol 92(3):649–656

Kabir AH (2016) Biochemical and molecular changes in rice seedlings (*Oryza* sativa L.) to cope with chromium stress. Plant Biol 18(4):710–719

Kaci A, Petit F, Fournier M et al (2016) Diversity of active microbial communities subjected to long-term exposure to chemical contaminants along a 40-year-old sediment core. Environ Sci Pollut Res Int 23:4095–4110

Kacprzak M, Grobelak A, Grosser A, Prasad MN (2014) Efficacy of biosolids in assisted phytostabilization of metalliferous acidic sandy soils with five grass species. Int J Phytoremediation 16(6):593–608

Kadukova J, Vircikova E (2005) Comparison of differences between copper bioaccumulation and biosorption. Environ Int 31:227–232

Kamaludeen SP, Megharaj M, Juhasz AL, Sethunathan N, Naidu R (2003) Chromium microorganism interactions in soil: remediation implications. Rev Environ Contaim Toxicol 178:93–164

Kang SM, Shahzad R, Bilal S, Khan AL, You YH, Lee WH, Ryu HL, Lee KE, Lee IJ (2017) Metabolism-mediated induction of zinc tolerance in *Brassica rapa* by *Burkholderia cepacia* CS2-1. J Microbiol 55(12):955–965

Karami A, Shamsuddin ZH (2010) Phytoremediation of heavy metals with several efficiency enhancer methods. Afr J Biotechnol 9(25):3689–3698

Karthik C, Ramkumar VS, Pugazhendhi A, Gopalakrishnan K, Arulselvi PI (2017) Biosorption and biotransformation of Cr (VI) by novel *Cellulosimicrobium funkei* strain AR6. J Taiwan Inst Chem Eng 70:282–290

Khalid S, Shahid M, Niazi NK, Murtaza B, Bibi I, Dumat C (2017) A comparison of technologies for remediation of heavy metal contaminated soils. J Geochem Explor 182:247–268

Khande R, Sharma SK, Ramesh A, Sharma MP (2017) Zinc solubilizing *Bacillus* strains that modulate growth, yield and zinc biofortification of soybean and wheat. Rhizosphere 4:126–138

Kholodova V, Volkov K, Abdeyeva A, Kuznetsov V (2011) Water status in *Mesembryanthemum crystallinum* under heavy metal stress. Environ Exp Bot 71:382–389

Kidd PS, Alvarez-Lopez V, Becerra-Castro C, Cabello-Conejo M, Prieto-Fernandez A (2017) Potential role of plant-associated bacteria in plant metal uptake and implications in phytotechnologies. In: Advances in botanical research, vol 83. Academic Press, Amsterdam, pp 87–126

Kunito T, Nagaoka K, Tada N, Saeki K, Senoo K, Oyaizu H, Matsumoto S (1997) Characterization of cu-resistant bacterial communities in cu-contaminated soils. Soil Biol Plant Nutr 43:709–717

LaVoie SP, Summers AO (2018) Transcriptional responses of *Escherichia coli* during recovery from inorganic or organic mercury exposure. BMC Genomics 19(1):52

Li H, Cong Y, Lin J, Chang Y (2015) Enhanced tolerance and accumulation of heavy metal ions by engineered *Escherichia coli* expressing *Pyrus calleryana* phytochelatin synthase. J Basic Microbiol 55(3):398–405

Li CF, Wang F, Cao WT, Pan J, Lü JS, Wu QY (2017) Source analysis, spatial distribution and pollution assessment of heavy metals in sewage irrigation area farmland soils of Longkou City. Huan jing ke xue= Huanjing kexue 38(3):1018–1027

Li L, Zhang K, Gill RA, Islam F, Farooq MA, Wang J, Zhou W (2018) Ecotoxicological and interactive effects of copper and chromium on physiochemical, ultrastructural, and molecular profiling in Brassica napus L. BioMed Res Int 2018:9248123

Liu Y, Tie B, Li Y, Lei M, Wei X, Liu X, Du H (2018) Inoculation of soil with cadmium-resistant bacterium *Delftia* sp. B9 reduces cadmium accumulation in rice (Oryza sativa L.) grains. Ecotoxicol Environ Saf 163:223–229

Ma Y, Oliveira RS, Wu L, Luo Y, Rajkumar M, Rocha I, Freitas H (2015) Inoculation with metal-mobilizing plant-growth-promoting rhizobacterium *Bacillus* sp. SC2b and its role in rhizoremediation. J Toxic Environ Health A 78:931–944

Mahbub KR, Krishnan K, Naidu R, Megharaj M (2017) Mercury remediation potential of a mercury resistant strain *Sphingopyxis* sp. SE2 isolated from contaminated soil. J Environ Sci 51:128–137

Mala JG, Sujatha D, Rose C (2015) Inducible chromate reductase exhibiting extracellular activity in *Bacillus methylotrophicus* for chromium bioremediation. Microbiol Res 170:235–241

Malik RN, Husain SZ, Nazir I (2010) Heavy metal contamination and accumulation in soil and wild plant species from industrial area of Islamabad, Pakistan. Pak J Bot 42:291–301

Mallick I, Bhattacharyya C, Mukherji S, Dey D, Sarkar SC, Mukhopadhyay UK, Ghosh A (2018) Effective rhizoinoculation and biofilm formation by arsenic immobilizing halophilic plant growth promoting bacteria (PGPB) isolated from mangrove rhizosphere: a step towards arsenic rhizoremediation. Sci Total Environ 610:1239–1250

Marrugo-Negrete J, Pinedo-Hernández J, Díez S (2017) Assessment of heavy metal pollution, spatial distribution and origin in agricultural soils along the Sinú River Basin. Colomb Environ Res 154:380–388

Masaki Y, Hirajima T, Sasaki K, Okibe N (2015) Bioreduction and immobilization of hexavalent chromium by the extremely acidophilic Fe (III)-reducing bacterium *Acidocella aromatica* strain PFBC. Extremophiles 19:495–503

Megido L, Suárez-Peña B, Negral L, Castrillón L, Fernández-Nava Y (2017) Suburban air quality: human health hazard assessment of potentially toxic elements in PM10. Chemosphere 177:284–291

Meng W, Wang Z, Hu B, Wang Z, Li H, Cole Goodman R (2016) Heavy metals in soil and plants after long-term sewage irrigation at Tianjin China: a case study assessment. Agric Water Manag 171:153–161

Mir MA, Sirhindi G, Alyemeni MN, Alam P, Ahmad P (2018) Jasmonic acid improves growth performance of soybean under nickel toxicity by regulating nickel uptake, redox balance, and oxidative stress metabolism. J Plant Growth Regul 37:1–5

Mishra S, Alfeld M, Sobotka R, Andresen E, Falkenberg G, Küpper H (2016) Analysis of sublethal arsenic toxicity to *Ceratophyllum demersum*: subcellular distribution of arsenic and inhibition of chlorophyll biosynthesis. J Exp Bot 67:4639–4646

Mohamed HM, Almaroai YA (2017) Effect of phosphate solubilizing bacteria on the uptake of heavy metals by corn plants in a long-term sewage wastewater treated soil. Int J Environ Sci Develop 8:366

Mohite BV, Koli SH, Patil SV (2018) Heavy metal stress and its consequences on exopolysaccharide (EPS)-producing *Pantoea agglomerans*. Appl Biochem Biotechnol 186 (1):199–216

Mohmand J, Eqani SA, Fasola M, Alamdar A, Mustafa I, Ali N, Liu L, Peng S, Shen H (2015) Human exposure to toxic metals via contaminated dust: bio-accumulation trends and their potential risk estimation. Chemosphere 132:142–151

Moreira H, Pereira SI, Marques AP, Rangel AO, Castro PM (2016) Selection of metal resistant plant growth promoting rhizobacteria for the growth and metal accumulation of energy maize in a mine soil—effect of the inoculum size. Geoderma 278:1–11

Moreno FN, Anderson CWN, Stewart RB, Robinson BH (2008) Phytofiltration of mercury-contaminated water: volatilisation and plant-accumulation aspects. Environ Exp Bot 62 (1):78–85

Nagata T, Morita H, Akizawa T, Pan-Hou H (2010) Development of a transgenic tobacco plant for phytoremediation of methylmercury pollution. Appl Microbiol Biotechnol 87:781–786

Naik MM, Shamim K, Dubey SK (2012) Biological characterization of lead-resistant bacteria to explore role of bacterial metallothionein in lead resistance. Curr Sci 103:426–429

Narayani M, Shetty KV (2013) Chromium-resistant bacteria and their environmental condition for hexavalent chromium removal: a review. Crit Rev Environ Sci Technol 43(9):955–1009

Navarro I, Chavez A, Barrios J, Maya C, Becerril E, Lucario S, Jiménez B (2015) Wastewater Reuse for Irrigation - Practices, Safe Reuse and Perspectives. In: Javaid MS (ed) Irrigation and Drainage—Sustainable Strategies and Systems. InTech, Rijeka. https://doi.org/10.5772/59361

Nayak AK, Panda SS, Basu A, Dhal NK (2018) Enhancement of toxic Cr(VI), Fe and other heavy metals phytoremediation by the synergistic combination of native Bacillus cereus strain and Vetiveria zizanioides L. Int J Phytoremediation 20(7):682–691

Nies DH (1999) Microbial heavy metal resistance. Appl Microbiol Biotechnol 51:730–750

Nocelli N, Bogino PC, Banchio E, Giordano W (2016) Roles of extracellular polysaccharides and biofilm formation in heavy metal resistance of rhizobia. Materials 9(6):418

Ojuederie OB, Babalola OO (2017) Microbial and plant-assisted bioremediation of heavy metal polluted environments: a review. Int J Environ Res Public Health 14(12):1504

Olaniran AO, Balgobind A, Pillay B (2013) Bioavailability of heavy metals in soil: impact on microbial biodegradation of organic compounds and possible improvement strategies. Int J Mol Sci 14:10197–10228

Patil S, Unnikrishnan G (2017) Fourier transform infrared spectroscopic characterization of heavy metal-induced metabolic changes in the thermophiles. World J Pharm Pharmaceut Sci 7:592–599

Paul T, Chakraborty A, Islam E, Mukherjee SK (2018) Arsenic bioremediation potential of arsenite-oxidizing *Micrococcus* sp. KUMAs15 isolated from contaminated soil. Pedosphere 28 (2):299–310

Pires C, Franco AR, Pereira SIA, Henriques I, Correia A, Magan N et al (2017) Metal (loid)-contaminated soils as a source of culturable heterotrophic aerobic bacteria for remediation applications. Geomicrobiol J 34:1–9

Pivetz P (2001) Phytoremediation of contaminated soil and ground water at hazardous waste sites. EPA/540/S-01/500. United States Environmental Protection Agency (EPA), Washington DC, 36 p

Plociniczak T, Sinkkonen A, Romantschuk M, Sułowicz S, Piotrowska-Seget Z (2016) Rhizospheric bacterial strain *Brevibacterium casei* MH8a colonizes plant tissues and enhances cd, Zn, cu phytoextraction by white mustard. Front Plant Sci 7:101

Podder MS, Majumder CB (2018) Optimization of environmental conditions for the growth and bioaccumulation of as (III) and as (V) ions by growing *Corynebacterium glutamicum* MTCC 2745: inhibition kinetic study. Sustain Water Resour Manag 4(1):23–44

Puyen ZM, Villagrasa E, Maldonado J, Diestra E, Esteve I, Sole A (2012) Biosorption of lead and copper by heavy-metal tolerant *Micrococcus luteus* DE2008. Bioresour Technol 126:233–237

Rajapaksha RM, Tobor-Kapłon MA, Bååth E (2004) Metal toxicity affects fungal and bacterial activities in soil differently. Appl Environ Microbiol 70(5):2966–2973

Rai PK, Rai A, Singh S (2018) Change in soil microbial biomass along a rural-urban gradient in Varanasi (U.P., India). In: Geology, ecology, and landscapes. Taylor & Francis Group, London, pp 1–8

Rangel WM, Thijs S, Janssen J, Oliveira Longatti SM, Bonaldi DS, Ribeiro PR, Jambon I, Eevers N, Weyens N, Vangronsveld J, Moreira FM (2017) Native rhizobia from Zn mining soil promote the growth *of Leucaena leucocephala* on contaminated soil. Int J Phytoremediation 19(2):142–156

Rasulov BA, Yili A, Aisa HA (2013) Biosorption of metal ions by exopolysaccharide produced by *Azotobacter chroococcum* XU1. J Environ Prot 4(09):989

Roman-Ponce B, Reza-Vázquez DM, Gutierrez-Paredes S, María de Jesús DE, Maldonado-Hernandez J, Bahena-Osorio Y, Estrada-De los Santos P, Wang ET, Vásquez-Murrieta MS (2017) Plant growth-promoting traits in rhizobacteria of heavy metal-resistant plants and their effects on *Brassica nigra* seed germination. Pedosphere 27(3):511–526

Saba H, Jyoti P, Neha S (2013) Mycorrhizae and phytochelators as remedy in heavy metal contaminated land remediation. Int Res J Environ Sci 2:74–78

Saphier M, Silberstein E, Shotland Y, Popov S, Saphier O (2018) Prevalence of monovalent copper over divalent in killing *Escherichia coli* and *Staphylococcus aureus*. Curr Microbiol 75 (4):426–430

Saraswat S, Rai JP (2012) Prospective application of *Leucaena leucocephala* for phytoextraction of Cd and Zn and nitrogen fixation in metal polluted soils. Int J Phytoremediation 13:271–288

Saunders JE, Jastrzembski BG, Buckey JC, Enriquez D, MacKenzie TA, Karagas MR (2013) Hearing loss and heavy metal toxicity in a Nicaraguan mining community: audiological results and case reports. Audiol Neurootol 18:101–113

Shahbaz M, Kuzyakov Y, Sanaullah M, Heitkamp F, Zelenev V, Kumar A, Blagodatskaya E (2017) Microbial decomposition of soil organic matter is mediated by quality and quantity of crop residues: mechanisms and thresholds. Biol Fertil Soils 53(3):287–301

Shaheen SM, Antoniadis V, Biswas J, Wang H, Ok YS, Rinklebe J (2017) Biosolids application affects the competitive sorption and lability of cadmium, copper, nickel, lead, and zinc in fluvial and calcareous soils. Environ Geochem Health 39(6):1365–1379

Shakeel M, Rais A, Hassan MN, Hafeez FY (2015) Root associated Bacillus sp. improves growth, yield and zinc translocation for basmati rice (Oryza sativa) varieties. Front Microbiol 6:1286

Shanying HE, Xiaoe YA, Zhenli HE, Baligar VC (2017) Morphological and physiological responses of plants to cadmium toxicity: a review. Pedosphere 27(3):421–438

Sharma HD, Reddy KR (2004) Geoenvironmental engineering: site remediation, waste containment and emerging waste management technologies. Wiley, New York

Sharma SS, Dietz KJ (2006) The significance of amino acids and amino acid-derived molecules in plant responses and adaptation to heavy metal stress. J Exp Bot 57:711–726

Sheng G-P, Yu H-Q, Li X-Y (2010) Extracellular polymeric substances (EPS) of microbial aggregates in biological wastewater treatment systems. Rev Biotechnol Adv 28(6):882–894

Shu WS, Xia HP, Zhang ZQ (2002) Use of vetiver and three other grasses for revegetation of Pb/Zn mine tailings: field experiment. Int J Phytoremediation 4:47–57

Silva S, Silva P, Oliveira H, Gaivão I, Matos M, Pinto-Carnide O et al (2017) Pb low doses induced genotoxicity in *Lactuca sativa* plants. Plant Physiol Biochem 112:109–116

Silver S, Phung LT (1996) Bacterial heavy metal resistance: new surprises. Annu Rev Microbiol 50:753–789

Singh A, Parihar P, Singh R, Prasad SM (2016) An assessment to show toxic nature of beneficial trace metals: too much of good thing can be bad. Int J Curr Multidisciplinary Stud 2:141–144

Singh P, Mishra R, Jha S, Raghuvanshi P, Jha S, Prabhat (2018) Effect of inoculation of zinc-resistant bacterium *Enterobacter ludwigii* CDP-14 on growth, biochemical parameters and zinc uptake in wheat (*Triticum aestivum* L.) plant. Ecol Eng 116:163–173

Singh RK, Anandhan S, Singh S, Patade VY, Ahmed Z, Pande V (2011) Metallothionein-like gene from *Cicer microphyllum* is regulated by multiple abiotic stress. Protoplasma 248:839–847
Sinha S, Mishra RK, Sinam G, Mallick S, Gupta AK (2013) Comparative evaluation of metal phytoremediation potential of trees, grasses and flowering plants from tannery wastewater contaminated soil in relation with physico-chemical properties. Soil Sediment Contam 22:958–983
Slama I, Abdelly C, Bouchereau A, Flowers T, Savoure A (2015) Diversity, distribution and roles of osmoprotective compounds accumulated in halophytes under abiotic stress. Ann Bot 115 (3):433–447
Soussou S, Brunel B, Pervent M, van Tuinen D, Cleyet-Marel JC, Baudoin E (2017) Rhizobacterial Pseudomonas spp. strains Harbouring acdS gene could enhance Metallicolous legume nodulation in Zn/Pb/cd mine tailings. Water Air Soil Pollut 228:142
Subrahmanyam G, Sharma RK, Kumar GN, Archana G (2018) *Vigna radiata* var. GM4 plant growth enhancement and root colonization by a multi-metal-resistant plant growth-promoting bacterium *Enterobacter* sp. C1D in Cr (VI)-amended soils. Pedosphere 28(1):144–156
Sulaymon AH, Ebrahim SE, Mohammed-Ridha MJ (2012) Equilibrium, kinetic, and thermodynamic biosorption of Pb (II), Cr (III), and Cd (II) ions by dead anaerobic biomass from synthetic wastewater. Environ Sci Pollut Res Int 20:175–187
Szabados L, Savoure A (2010) Proline: a multifunctional amino acid. Trends Plant Sci 15(2):89–97
Taghavi S, Delanghe H, Lodewyckx C, Mergeay M, Lellie DV (2001) Nickel resistance based minitransposons: new tools for genetic manipulation of environmental bacteria. Appl Environ Microbiol 67:1015
Tchounwou PB et al (2012) Heavy metals toxicity and the environment. PMC. EXS 101:133–164
Thatoi HN, Pradhan SK (2017) Detoxification and bioremediation of hexavalent chromium using microbes and their genes: an insight into genomic, proteomic and bioinformatics studies. In: Microb Biotechnol. Springer, Singapore, pp 287–306
Thebo AL, Drechsel P, Lambin EF, Nelson KL (2017) A global, spatially-explicit assessment of irrigated croplands influenced by urban wastewater flows. Environ Res Lett 12:7
Thomas M, Benov L (2018) The contribution of superoxide radical to cadmium toxicity in *E. coli*. Biol Trace Elem Res 181(2):361–368
Thorgersen MP, Lancaster WA, Ge X, Zane GM, Wetmore KM, Vaccaro BJ, Poole FL, Younkin AD, Deutschbauer AM, Arkin AP, Wall JD (2017) Mechanisms of chromium and uranium toxicity in *Pseudomonas stutzeri* RCH2 grown under anaerobic nitrate-reducing conditions. Front Microbiol 8:1529
Topcuoğlu B (2016) Heavy metal mobility and bioavailability on soil pollution and environmental risks in greenhouse areas. Int J Adv Agric Environ Eng 3(1):208
Velez PA, Talano MA, Paisio CE, Agostini E, González PS (2017) Synergistic effect of chickpea plants and *Mesorhizobium* as a natural system for chromium phytoremediation. Environ Technol 38(17):2164–2172
Venkanna RM, Karthikeyan JSK (2017) Soil quality and heavy metal contamination of soils in Mindi industrial area, Visakhapatnam, Andhra Pradesh, India. Indian J Ecol 44(4):774–777
Venkatachalam P, Jayalakshmi N, Geetha N, Sahi SV, Sharma NC, Rene ER et al (2017) Accumulation efficiency, genotoxicity and antioxidant defense mechanisms in medicinal plant *Acalypha indica* L. under lead stress. Chemosphere 171:544–553
Vijayaraghavan K, Yun YS (2008) Bacterial biosorbents and biosorption. Biotechnol Adv 26:266–291
Volesky B (2007) Biosorption and me. Water Res 41:4017–4029
Wani PA, Wahid S, Singh R, Kehinde AM (2018) Antioxidant and chromium reductase assisted chromium (VI) reduction and Cr (III) immobilization by the rhizospheric *Bacillus* helps in the remediation of Cr (VI) and growth promotion of soybean crop. Rhizosphere 6:23–30
Warnes SL, Keevil CW (2011) Mechanism of copper surface toxicity in vancomycin-resistant enterococci following wet or dry surface contact. Appl Environ Microb 77(17):6049–6059

Westersund C (2018) Investigating the role of the cell envelope in *E. coli* relating to silver sensitivity and resistance (Master's thesis, Science). University of Calgary, Alberta, Canada
Wojas S, Clemens S, Hennig J, Sklodowska A, Kopera E et al (2008) Overexpression of phytochelatin synthase in tobacco: distinctive effects of AtPCS1 and CePCS genes on plant response to cadmium. J Exp Bot 59:2205–2219
Wu W, Dong C, Wu J, Liu X, Wu Y, Chen X, Yu S (2017) Ecological effects of soil properties and metal concentrations on the composition and diversity of microbial communities associated with land use patterns in an electronic waste recycling region. Sci Total Environ 601:57–65
Yamada K (2013) Cobalt: its role in health and disease. In: Interrelations between essential metal ions and human diseases. Springer, Dordrecht, pp 295–320
Yancheshmeh JB, Khavazi K, Pazira E, Solhi M (2011) Evaluation of inoculation of plant growth-promoting rhizobacteria on cadmium and lead uptake by canola and barley. Afr J Microbiol Res 5:1747–1754
Yao XF, Zhang JM, Tian L, Guo JH (2017) The effect of heavy metal contamination on the bacterial community structure at Jiaozhou Bay, China. Brazilian J Microbiol 48:71–78
Yu X, Li Y, Li Y, Xu C, Cui Y, Xiang Q, Gu Y, Zhao K, Zhang X, Penttinen P, Chen Q (2017) *Pongamia pinnata* inoculated with *Bradyrhizobium liaoningense* PZHK1 shows potential for phytoremediation of mine tailings. Appl Microbiol Biotech 101(4):1739–1751
Yuan L, Zhi W, Liu Y, Karyala S, Vikesland PJ, Chen X et al (2015) Lead toxicity to the performance, viability, and community composition of activated sludge microorganisms. Environ Sci Technol 49:824–830
Zhang X, Yang H, Cui Z (2017) Assessment on cadmium and lead in soil based on a rhizosphere microbial community. Toxicol Res 6(5):671–677
Złoch M, Kowalkowski T, Tyburski J, Hrynkiewicz K (2017) Modeling of phytoextraction efficiency of microbially stimulated *Salix dasyclados* L. in the soils with different speciation of heavy metals. Int J Phytoremediation 19(12):1150–1164
Zoomi I, Narayan RP, Akhtar O, Srivastava P (2017) Role of plant growth promoting rhizobacteria in reclamation of wasteland. In: Microbial biotechnology. Springer, Singapore, pp 61–80
Zribi K, Djébali N, Mrabet M, Khayat N, Smaoui A, Mlayah A, Aouani ME (2012) Physiological responses to cadmium, copper, lead, and zinc of *Sinorhizobium* sp. strains nodulating *Medicago sativa* grown in Tunisian mining soils. Ann Microbiol 62(3):1181–1188

Chapter 14
ROS Signaling Under Oxidative Stress in Plants

Deepesh Bhatt, Saurabh C. Saxena, and Sandeep Arora

Abstract Plants being sessile have to withstand an array of adverse environmental challenges throughout their life span. The chances of survival of a plant, under sub-optimal growth conditions, depend on the way they perceive and transmit environmental signals. Plants transmit these signals and initiate appropriate response with the help of an array of secondary messenger molecules which help them to adapt towards the changing environment. Secondary messengers are among the most versatile molecules or intermediates that include calcium messenger systems, reactive nitrogen oxide species along with the most prevailing reactive oxygen species. Examples of these signaling molecules include superoxide radical, hydrogen peroxide, nitric oxide, and hydroxyl radicals. One of the most intriguing facts about cellular metabolism is that some of the secondary messengers, involved in transducing signals are also part of potentially damaging molecules that can destroy membrane integrity and negatively affect the redox balance of the cell. During cellular metabolism, minimal generation of ROS takes place constantly, though its levels are kept under strict control.

Abiotic as well as biotic stresses lead to the production of reactive oxygen species. These reactive oxygen species are responsible for the activation of cellular defence network and causing cellular damage due to oxidative stress. There is a fine line separating these two seemingly antagonistic roles of ROS. The role of ROS as secondary messenger molecules is intricately linked with the downstream signaling and transcriptional regulation during oxidative stress in the plants. A delicate equilibrium between the role of ROS as a secondary messenger and as causative agents for oxidative damage determines the growth characteristics of a plant under

D. Bhatt
Department of Biotechnology, Shree Ramkrishna Institute of Computer Education and Applied Sciences, Veer Narmad South Gujarat University, Surat, Gujarat, India

S. C. Saxena
Department of Biochemistry, Central University of Haryana, Mahendragarh, Haryana, India

S. Arora (✉)
Department of Molecular Biology and Genetic Engineering (CBSH), G. B. Pant University of Agriculture and Technology, Pantnagar, Uttarakhand, India

A. Sharma (ed.), *Microbes and Signaling Biomolecules Against Plant Stress*, Rhizosphere Biology, https://doi.org/10.1007/978-981-15-7094-0_14

stress. Several enzymatic as well as non-enzymatic antioxidants present in the cells act as agents for effective scavenging of ROS, produced during different kinds of stressful conditions. The current chapter underlines the dynamics of ROS production and their role as a secondary messenger as well as inducer of oxidative damage.

14.1 Introduction

All the living organisms usually come across a combination of many abiotic stresses in their natural habitats. Plants routinely face sudden and extreme changes in environmental conditions but being sessile in nature, they cannot move away in order to weaken the ill effect of those stresses. It is a well known fact that environmental stress affects the plant growth and its development in a pessimistic way. Plant exhibits several defence responses to mitigate the deleterious effects of the stresses. During the adverse conditions, plant activates various cellular processes for the synthesis of signaling molecules that trigger the protective mechanisms to protect the plant. Under such extreme conditions, plants shift their physiological and biochemical eminence through molecular shifting or molecular rearrangement which is said to impart protection to the plants (Formentin et al. 2018; Ferrer et al. 2018). These adverse conditions may range from cold, drought, heat, salinity, heavy metal, UV radiation, and attack of pathogens. In various cell organelles, such unfavourable surroundings lead to generation of Reactive Oxygen Species (ROS) which interact with various macromolecules and cause severe damage to proteins, DNA, and lipids which subsequently affect normal functioning of the cell that ultimately put the plant in the oxidative damage condition (Tripathi et al. 2017; Banerjee et al. 2018). ROS are produced due to reduction of molecular oxygen by high energy exposure or electron transfer reactions. Sequential transfer of one, two or three electrons to oxygen leads to the formation of a superoxide radical ($O_2^{\bullet -}$), or a hydroxyl radical ($^{\bullet}OH$), super oxide anions (O_2^{-}), Hydrogen peroxide (H_2O_2), and singlet oxygen ($^1O_2$), respectively (Mittler 2017; Waszczak et al. 2018). ROS may further be generated via involvement of RBOH (respiratory burst oxidase homologs) and few similar classes of oxidases and peroxidases, via interconnected pathways (Suzuki et al. 2011; Gilroy et al. 2016; Mignolet-Spruyt et al. 2016). Plants produce them during metabolic activities in different cellular compartments or through leakage of electrons from electron transport chain in chloroplast and mitochondria where their increased concentration leads to peroxidation of lipids, damage to nucleic acid, oxidation of proteins, activation of programmed cell death (PCD) which ultimately leads to cell death (Sharma et al. 2012). When concentration of ROS, via any sort of cellular or biochemical perturbation exceeds the defence mechanism, the cell is said to be in state of oxidative stress and ion toxicity.

In spite of destructive activities, ROS are also demarcated as important secondary messengers in various cellular processes during stressful conditions (Desikan et al. 2001). There exists a dynamic balance between ROS production and scavenging which ultimately decides the impact of ROS in the cell. Due to their dual role, plant cell has to take extra cautions to control the concentration of ROS inside the cell in order to avoid any lethal injury. ROS generation (metabolically or for signaling purposes) or ROS scavenging is a stimulated process which occurs in most of the cellular compartments simultaneously. To control the concentration of ROS, plant has to detoxify or scavenge surplus ROS by means of its efficient antioxidative system which consists of enzymatic and non-enzymatic antioxidants (Noctor and Foyer 1998). In non-enzymatic category, phenols, ascorbate, carotenoid, glutathione, and tocopherols act as potent antioxidants within the cell, whereas superoxide dismutase (SOD) and enzymes of ascorbate-glutathione (AsA-GSH) cycle such as Ascorbate peroxidase (APX), Dehydroascorbate reductase (DHAR), Monodehydroascorbate reductase (MDHAR), Catalase (CAT), Guaiacol peroxidase (GPX), and Glutathione reductase (GR) serve as a part of enzymatic antioxidant system (Noctor and Foyer 1998). Many researchers have reported the increased concentration of these enzymatic and non-enzymatic antioxidant molecules during adverse environmental conditions (Bhatt et al. 2011).

In addition, ROS produced during stressful conditions are recognized by the sensors which further translate the signal to nucleus via a relay of redox reactions. Signal once reaches nucleus, initiates the translation of new proteins and modification in the existing proteins through generation of stress related transcription factors and their binding to corresponding cis-acting elements in the nucleus (Foyer and Noctor 2016). TFs have one DNA-binding domain that assists its binding to cis regulatory element (CREs) which activates or deactivates the gene expression and allow TFs to work as molecular switch. Minimal levels of ROS can be sensed by NADPH oxidase, also termed as respiratory burst homolog (RBOH) which gets activated through calcium mediated signaling leading to its own phosphorylation to further generate ROS (Gilroy et al. 2016). ROS can further be recognized by ROS recognizing receptors, like Mitogen Activated Protein Kinases (MAPK), phosphatases, and several categories of transcription factors. Through genomic and transcriptomic studies, it has been revealed that expression levels of MAP Kinases and Transcription factors (TFs) alter in response to stress. Therefore, plant shows a high level regulatory mechanism consisting of TFs, CREs that control the expression of stress related genes.

14.1.1 ROS and Its Types

Plants are known to synthesize ROS even under seemingly normal environmental conditions which are unwanted by-product of various metabolic processes in the cell. These free radicals or reactive molecules are produced through stepwise reduction of oxygen. In various organelles of plant cell such as chloroplasts, mitochondria, peroxisomes, production of ROS takes place where 1% of total oxygen absorbed by the plant is consumed. These ROS have dual role as secondary messenger in small quantities as well as deleterious for cell when produced in excess.

We have summarized generation of various types of ROS stepwise and their site of synthesis in detail.

1. Singlet oxygen ($^1O_2$)

Molecular Oxygen in its ground state has two electrons in its outer shell which has parallel spin. This makes molecular oxygen as an inactive molecule which does not react with organic molecule unless gets activated. There are two different activation mechanisms of oxygen: first, absorption of energy to reverse the spin of one of the unpaired electron and stepwise reduction. In the former mechanism, $^1O_2$ is formed while in later $O_2^{\bullet-}$, H_2O_2, and •OH are synthesized via reduction of molecular oxygen. Singlet oxygen is a typical ROS which is not formed through reduction of oxygen or electron transfer process rather it is produced through the reaction of oxygen with triplet form of chlorophyll in the antenna system of plant cell.

$$\text{Chl} \rightarrow 3\text{Chl}$$

$$3\text{Chl} + 3O_2 \rightarrow \text{Chl} + {}^1O_2$$

Environmental hardship favours deficiency of carbon dioxide (CO_2) inside the cell which is responsible for the formation of singlet oxygen. This can cause severe harm to both the photosystems (PS I & PS II). It can alter the metabolism and cause serious damage to protein, nucleic acid, and lipids in spite having the short life of around three microseconds (Krieger-Liszkay et al. 2008).

2. Superoxide radical ($O^{\bullet-\,2}$)

When oxygen absorbs sufficient energy and reverses the spin of one of the electrons, it allows oxygen to participate in divalent reduction reaction where simultaneous transfer of two electrons occurs (Apel and Hirt 2004). There is a spin restriction in molecular oxygen, following that oxygen cannot take four electrons at a time to produce H_2O. Therefore it accepts electrons sequentially, due to which reduction of O_2 leads to production of stable intermediates in a stepwise manner (Halliwell and Gutteridge 1984). The primary ROS is $O_2^{\bullet}$in the cell which starts cascade of reaction producing secondary ROS either directly or through enzyme driven processes that further depend upon the site of ROS formation (Valko et al. 2005). This radical is moderately reactive with a short half-life of 2–4 microseconds. It itself is not so deleterious for the cell but it gets transformed into $OH^{\bullet}$ and $^1O_2$ which are more reactive and can damage the cell, largely causing membrane lipid peroxidation (Halliwell 2006).

3. Hydrogen peroxide (H_2O_2)

$O_2^{\bullet}$ can be further reduced to H_2O_2 through non-enzymatic dismutation or SOD catalysed dismutation. Major source of H_2O_2 generation is electron transport chain of mitochondria, endoplasmic reticulum, chloroplast, and photorespiration and β oxidation of fatty acid in the plant cell. H_2O_2 is moderately active molecule and can readily cross cellular membranes and can cause oxidative damage far from the site of its synthesis. It is the only molecule which can pass through aquaporins in the membranes so it has gained wide spread attention as a crucial ROS and as regular signaling molecule over other ROS species (Bienert et al. 2007).

4. Hydroxyl radical ($OH^•$)

Among all the ROS family members, hydroxyl radical is the most reactive and toxic ROS known till now. It is formed through Fenton Reaction between H_2O_2 and $O_2^•$ catalysed by transition metals like Fe^{2+}& Fe^{3+} at neutral pH.

$$H_2O_2 + O_2^{•-} \rightarrow OH^- + O_2 + OH^•$$

Once formed, it can damage cellular components by lipid peroxidation and invariably causes membrane damage. Interestingly, there is no enzymatic defence system against this ROS, so it can lead to cell death when accumulated in high concentration (Pinto et al. 2003).

14.1.2 Sites of ROS Generation

ROS are produced under normal conditions as well as in stressful environmental conditions at several locations inside the cell such as mitochondria, cell wall, endoplasmic reticulum, chloroplast, peroxisome depending upon the availability of light. In dark, major site of production is mitochondria, while in light, chloroplast and peroxisomes are major site of production of ROS in the cell (Choudhury et al. 2013).

1. Mitochondria

Mitochondria are the major site for the production of ROS like H_2O_2 and $O_2^•$ (Navrot et al. 2007). Mitochondrial electron transport chain (ETC) is majorly responsible for reduction of oxygen through energized electrons which form ROS. Complex I and II are the major counter parts of mitochondrial ETC, which are responsible for the production of ROS (Moller et al. 2007; Noctor et al. 2007). Complex I or NADH dehydrogenase directly reduces oxygen within its flavoprotein region. When in complex I, there is a limited concentration of NAD^+ substrate; the electron flow becomes reversed from complex II to complex I. This process is responsible for increasing ROS production in complex I and it is regulated by ATP hydrolysis. Complex III or Ubiquinone-cytochrome region is also responsible for production of $O_2^•$. Many enzymes which are present in mitochondria can produce ROS as well. Some enzymes donate electrons to ETC and reduce oxygen and form the ROS indirectly, e.g. 1-galactono-γ lactone dehydrogenase (GAL), while some enzymes directly participate in the process of ROS formation like aconitase (Andreyev et al. 2005; Rasmusson et al. 2008).

2. Cell wall

Cell wall acts as a potent site for generation of ROS like $OH^{\bullet}$, $O^{\bullet-2}$, H_2O_2, and $^1O_2$ during adverse environmental conditions. Cell wall contains an enzyme lipoxygenase (LOX) which is responsible for hyper oxidation of poly unsaturated fatty acid (PUFA), which further acts as an active site for ROS generation. Diamine oxidase situated in cell wall uses diamines or polyamines to produce ROS during stress conditions. During biotic stress, when a pathogen attacks the cell then the lignin precursors cross link extensively with each other through H_2O_2 mediated pathway which in turn makes cell wall strong and rigid to inhibit the entry of the pathogen inside the cell (Sharma et al. 2012).

3. Endoplasmic Reticulum

Cytochrome P450 (Cyt P_{450}) in endoplasmic reticulum is responsible for ROS generation through NADPH-mediated Electron Transport System (Mittler 2002). Cyt P_{450} reacts with organic substrate (Rh) and further get reduced by a flavoprotein resulting into a free radical intermediate (Cyt P450 R^-). Immediately after production of this intermediate, it reacts with triplet oxygen ($_3O^2$) and oxygenated complex is formed (Cyt P450-ROO−). During stress condition, this oxygenated complex is converted into Cyt P450-Rh and $O^{\bullet-2}$ as a by-product.

4. Chloroplast

It is a well-organized organ made up of thylakoid membranes which contains light capturing and well efficient light harvesting machinery (Pfannschmidt 2003). The major source of ROS production in the chloroplast is core light harvesting system, i.e. Photosystems, PSI & PSII. During abiotic conditions like drought, salinity, cold, and heat, etc., the mehler reaction gets activated and water level gets limited and CO_2 concentration along with increased light results into the formation of $O^{\bullet-2}$

$$2O_2 + 2Fd\ (\text{reduced}) \rightarrow 2O\bullet\ -2 + 2\ Fd\ (\text{oxidized})$$

Afterward at PSI, resulted $2O^{\bullet-2}$ gets converted into H_2O_2 through a membrane bound Cu/Zn SOD (Miller et al. 2010). Other sources of electron leakage in PSI are 2Fe-2S and 4Fe-4S clusters. In PSII system, leakage of electrons occurs through Q^A and Q^B electron acceptor which results into the production of $2O^{\bullet-2}$, which further converts itself into a more toxic form like $OH^{\bullet}$ via Fenton reaction with H_2O_2 as an intermediate product. Chloroplast is also involved in oxidative stress induced programmed cell death as revealed in the study performed in transgenic tobacco where expression of anti-apoptotic Bcl-2 was studied (Chen and Dickman 2004). In this way, chloroplast acts as a potent site of ROS generation.

5. Peroxisomes

These are single membrane bound organelles and probably act as a major site for generation of H_2O_2 as they have an oxidative type of metabolism. Many metabolic processes like β-oxidation of fatty acids, glycolate oxidase reaction, the disproportionation of $O_2^{\bullet -}$, and the enzymatic reaction of flavin oxidases are responsible for the production of H_2O_2 (Baker and Graham 2013). Along with H_2O_2, $O_2^{\bullet -}$ radicals are also generated in peroxisome. In the peroxisomal matrix, Xanthine oxidase is situated which is responsible for conversion of anthine and hypoxanthine into uric acid which generates $O^{\bullet -}$ 2 as a by-product. Second site of $O^{\bullet -}$ 2 is NADPH dependent small ETC which is localized in peroxisomal membrane. This small ETC is consisted of NADH and cyt b, and releases $O^{\bullet -}$ 2 into cytosol after utilizing O_2 as an electron acceptor.

In addition to these sites, three peroxisomal membrane polypeptides having molecular weight of 18, 29, and 32 kDa are also responsible for $O^{\bullet -}$ 2 generation. To reduce cytochrome c, 18 and 32 kDa membrane polypeptides use NADH as the electron donor while 29 kDa protein uses NADPH as an electron donor in the reaction.

14.2 Antioxidative Systems in Plants: ROS Mitigation

Stress faced by plants usually falls in two broad categories: one is abiotic stress that is caused via non-living environmental factors like water deficit, salinity, high or low altitude, UV radiation, high wind velocity, etc., while the other category is biotic stress where living organisms are responsible for imposing stressful conditions namely caused by pathogen, bacteria, and fungi, parasites, animals, etc. Repercussions of any type of stress leads to an overall change in the internal stress regulatory machinery and subsequent generation of ROS, if ROS generation exceeds the level of mitigation, then it leads to a condition termed as oxidative stress which triggers cell toxicity. Furthermore, ROS production is linked to stomatal closure and dehydration, therefore disturbing the osmotic homeostasis of the plant Fig. 14.1.

14.2.1 Antioxidative Defence

The antioxidative defence system consists of two parts of defence machinery, one is enzymatic and other one is non enzymatic. This ROS defence system overall protects the plant from the hazardous effects of abiotic and biotic stresses. Various enzymes such as Catalase (CAT), Superoxide dismutase (SOD), Ascorbate peroxidase (APX), Dehydroascorbate reductase (DHAR), Monodehydroascorbate reductase (MDHAR), Guaiacol peroxidase (GPX), and Glutathione reductase (GR) make strong enzymatic antioxidative defence system in the plants to reduce ROS level and mitigate the effect of oxidative stress (Fig. 14.1). It also maintains the reduced glutathione (GSH) pool to balance redox state of a plant cell under stress (Gill and Tuteja 2010; Nahar et al. 2015). In addition, non-enzymatic antioxidants like

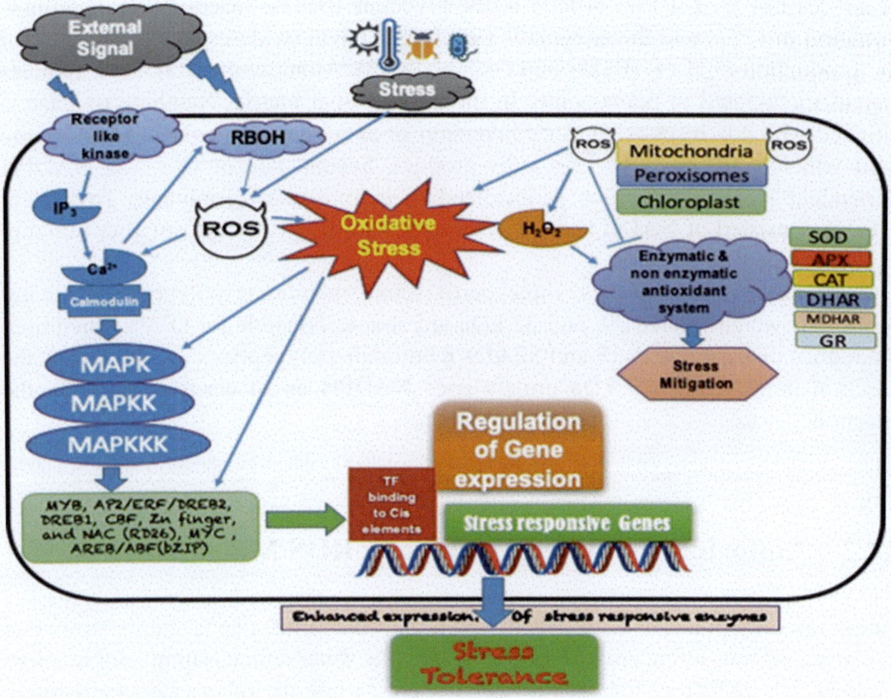

Fig. 14.1 Drought stress tolerance: an insight to resistance mechanism and adaptation in plants

ascorbic acid and GSH also play an important role in ROS detoxification and they also act as photo protectant (Khan et al. 2012).

14.3 Enzymatic Antioxidative Defence System

Plant cell has got a very efficient and dynamic antioxidative defence system which comprises of various enzymes such as Superoxide dismutase (SOD), Ascorbate peroxidase (APX), Catalase (CAT), Dehydroascorbate reductase (DHAR), Monodehydroascorbate reductase (MDHAR), Guaiacol peroxidase (GPX), and Glutathione reductase (GR).

14.3.1 Superoxide Dismutase (SOD)

This enzyme makes a first line of defence against oxidative stress conditions. It belongs to the family of metalloenzymes which are universally found in all the aerobic cells. It catalyses the conversion of $O^{\bullet -2}$ by dismutating it into O_2 and H_2O_2. Various studies have emphasized their role under stress and studies further indicated the upregulation of these enzymes during stress conditions (Boguszewska and Zagdańska 2012).

$$O^{\bullet -2} + O^{\bullet -2} + 2H^+ \rightarrow 2H_2O_2 + O_2$$

SOD can be classified into three main classes depending upon the type of metal it binds with, Mn SOD (found in mitochondria), Cu/Zn SOD (found in peroxisome, cytosol, and chloroplast), and Fe SOD (found in chloroplast).

14.3.2 Ascorbate Peroxidase (APX)

APX is a vital part of Ascorbate-Glutathione cycle. It performs the same role as that of catalase, while catalase scavenges H_2O_2 in peroxisome, APX scavenges H_2O_2 in cytosol and chloroplast. It uses ascorbic acid as a reducing agent and reduces H_2O_2 to H_2O and DHA

$$H_2O_2 + AA \rightarrow 2\ H_2O + DHA$$

APX is present in isoforms and depending upon the locations and amino acids, it can be found in cytosol, mitochondria, peroxisome, and chloroplast (Sharma and Dubey 2004). APX has a better affinity towards H_2O_2 than CAT and it is widely distributed inside the cell, so it effectively scavenges H_2O_2 at the time of prolonged stress conditions.

14.3.3 Catalase (CAT)

It is a tetrameric heme containing enzyme and shows high affinity towards H_2O_2. It does not have greater affinity towards organic peroxide (R-O-O-R) and does not need any reducing equivalent; this property makes it unique among all antioxidant enzymes. Catalase has a very high turnover rate (6×10^6 molecules of H_2O_2 to H_2O and O_2 min^{-1}). It is found predominantly in peroxisomes as these are the prime site of H_2O_2 production due to various metabolic activities like photorespiration, purine catabolism, fatty acid oxidation, and oxidative stress (Mittler 2002). According to some recent studies, CAT is also found in other cell organelles like cytosol, mitochondria, chloroplast but CAT activity in these organelles is questionable (Mhamdi et al. 2010). Angiosperm plants are known to have three types of CAT gene; CAT1 gene is expressed in pollens and seeds, while CAT2 genes are reported in photosynthetic parts of the plant and CAT3 gene is expressed in vascular tissues and leaves. Catalase enzyme is known to scavenge H_2O_2 in thermodynamically efficient way during adverse conditions.

$$H_2O_2 \rightarrow H_2O + (1/2)\ O_2$$

14.3.4 Dehydroascorbate Reductase (DHAR)

It uses reduced Glutathione (GSH) as an electron donor and reduces dehydroascorbate (DHA) to Ascorbic Acid (AA) (Eltayeb et al. 2007). It maintains the redox state of the cell by regulating ascorbic acid concentration in apoplast and symplast and thus regenerates ascorbic acid pool inside the cell (Chen and Gallie 2006). It is majorly found in seeds, root, and green and non-green shoots.

$$DHA + 2GSH \rightarrow AA + GSSG$$

14.3.5 Monodehydroascorbate Reductase (MDHAR)

It uses NADPH as a reducing agent for regenerating ascorbic acid from MDHA which has a very short lifespan and therefore regulates AA pool inside the cell. It has several forms which are found in different cell organelles such as peroxisomes, cytosols, mitochondria, chloroplast, and glyoxysomes.

$$MDHA + NADPH \rightarrow AA + NADP^+$$

It is localized in peroxisomes and mitochondria along with APX as it supplies AA to APX by maintaining its pool inside the cell, which further oxidizes AA and scavenges H_2O_2 in the process (Mittler 2002).

14.3.6 Glutathione Reductase (GR)

It is a flavoprotein oxidoreductase and reduces GSSG to GSH by using NADPH as a reducing agent. Further MDHA and DHA regenerate ascorbic acid by using reduced Glutathione (GSH), which itself gets converted into oxidized form, i.e. GSSG. It is majorly situated in chloroplast with small presence in cytosol and mitochondria. It is an integral component of ASC-GSH cycle and maintains high ratio of GSH/GSSG by formation of disulphide bond in Glutathione disulphide (Asada 1999). Reduced

Glutathione (GSH) is a low molecular weight compound which acts as a reducing agent to prevent the oxidation of thiol group and further reacts with potent ROS like $^1O_2$ and $OH^•$.

$$GSSG + NADPH \rightarrow 2GSH + NADP^+$$

14.3.7 *Guaiacol Peroxidase (GPX)*

This enzyme contains heme group and throws out excess H_2O_2 in both normal and stressed conditions. It utilizes H_2O_2 during degradation of Indole Acetic Acid (IAA) at the time of stress and plays important role in lignin biosynthesis. It uses guaiacol and pyrogallol as electron donor (Asada 1999). It is active intracellularly as well as extracellularly so it is recognized as a key enzyme for the removal of H_2O_2.

$$H_2O_2 + GSH \rightarrow H_2O + GSSG$$

Apart from enzymatic protectants similar protective strategy employed by plants for stress mitigation is making use of osmolytes which are nontoxic, highly polar, and low molecular weight compounds and help to maintain membrane integrity by stabilizing protein function of plant under stress (Saxena et al. 2013; Slama et al. 2015). Proline, sucrose, trehalose, glycine betaine are few examples of osmolytes whose levels is frequently known to elevate under stressful conditions (Slama et al. 2015; Masood et al. 2016; Per et al. 2017). Another large group of compounds like amino acid, sugar, and polyhydric alcohols also acts as osmoprotectants and ROS scavengers under oxidative stress.

ROS can further be recognized by various ROS sensors that translate the signal to nucleus via a relay of redox reactions for translation of new proteins via activating downstream transcription factors and upregulating numerous stress-responsive genes via binding to corresponding cis acting elements present in their promotor region thereby imparting stress mitigation.

14.4 ROS Mediated Downstream Signaling and Transcriptional Regulation

Unfavourable conditions during life cycle of a plant compels plant cell to cope with harsh environmental conditions. Therefore, plants have evolved key signaling enzymes/molecules that can perceive external environmental stimuli and transduce them across membranes and cytoplasm which result in an interaction with regulatory elements present in the genome, thus eliciting an appropriate adaptive response for protection and stress mitigation. This intricate mechanism of signal transduction is mediated via alterations in the levels of indigenous ROS molecules. In normal course of homeostasis, ROS are countered by various enzymatic or non-enzymatic antioxidants, however, under oxidative stress these may further be perceived by various downstream receptors which help transmitting the signal for eliciting the appropriate response thereafter. This signal transmittance is mediated via alteration key intermediates namely Ca^{2+}, Ca^{2+} binding protein mainly calmodulins, G protein activation and alteration in phospholipid signaling. Among all forms of ROS, H_2O_2 is supposed to be the most common form leading to the generation of calcium intermediates that modulate kinases like CDPKs and CBLs-CIPKs which are finally responsible for activating RBOH (respiratory burst oxidase homologues). RBOHs are membrane bound proteins having an oxidase domain, localized extracellularly and found responsible for generation of superoxide radical while an N-terminal calcium-binding domain, responsible for phosphorylation and contains EF hands, is localized towards the cytoplasmic side. (Ogasawara et al. 2008; Takeda et al. 2008). ROS usually have short life span therefore both sensing machinery and transducing mechanism need to be quick, efficient, and robust. Several intercellular and intracellular signaling cascades are activated as soon as ROS is generated and sensed and lead to a concomitant alteration in several downstream partners. The mechanisms that mostly prevail for ROS sensing are namely protein kinase and transcription factor based sensing and protein phosphatase inhibition based sensing which has been discussed in detail in the later part of the topic. These protein kinases are known to be the largest group of functional enzymes present in eukaryotes. Reports indicate that the plant genome consists of serine/threonine and histidine type protein kinase, however, in addition to above mentioned kinases the animal genome consists of Tyrosine kinase (Liang and Zhou 2018).

According to the plant diversity, protein kinases are categorized into four major clusters namely receptor-like kinases, Mitogen Activated Protein kinases, Calmodulin-like domain protein kinases, and Calcium dependent kinases. Apart from these, few less characterized kinases like SNF1/AMPK, PDK1, HKs, DGKs are also reported which have specific role in stress signaling (Chakradhar et al. 2019). Calmodulin-like domain protein kinases (CDPKs) and receptor-like kinases (RLKs) are most specific to plants but absent in other eukaryotic forms. Gauthier et al. (2011) reported two ROS sensors in *Arabidopsis thaliana*, namely LRR (leucine-rich repeat) and (RLKs) cysteine-rich receptor-like kinases. Domains of both these receptors are located extracellularly and known to modify while sensing ROS thus were found helpful in timely mitigation of oxidative stress. Further, authors reported that knockout lines for both these receptors were found susceptible against oxidative stress which signified their utility. Similarly COLD 1 is reported to be a sensor for cold stress and known to interact with some integral proteins thus regulating calcium gates in *Oryza sativa* (Zhu 2016). A sudden change in temperature following fluctuation in environmental conditions is found responsible for altering the fluidity of the membrane that is recognized by integral proteins namely receptor-like kinases (RLKs). Another kinase having an important role in ROS receptivity is OXI 1 (an oxidative signal inducible 1 kinase) and a serine/threonine protein kinase is known to trigger mitogen activated proteins and downstream transcription factors. MAPK cascade involves MAPKK kinase (MAPKKK), MAPK kinase (MAPKK), and MAPK where each one is activated via phosphorylation of adjacent kinase thus transferring the signal that further relays stress mediated responses downstream (Fig. 14.1). Apart from enzymes like kinases, other intermediates that play a key role in modulation of gene/protein expression and protective osmolytes are the stress inducible transcription factors (TF). TFs possess a DNA-binding domain and a transcriptional activation/repression domain that facilitate its binding to cis regulatory elements (CREs) present in the promoter region of numerous stress related genes, which allow activation/deactivation of those genes thus allowing TFs to serve as a molecular switch. Transcription factors, the key factors against oxidative stress tolerance, are mostly characterized predominantly into following subfamilies viz. MYB, AP2/ERF, bZIP, Zn finger, and NAC (Riechmann et al. 2000).

MYB transcription factors bears a conserved MYB repeats (R) towards N-terminus to perform DNA-binding and protein–protein interactions and a variable region towards C-terminal responsible for the regulatory activity present in the protein. According to Chen et al. (2015), *GbMYB5* gene was able to impart drought tolerance in cotton species as well as in transgenic tobacco by regulating the genes related to ROS scavenging namely genes for polyamine biosynthesis, late embryogenic abundant proteins LEA, and other genes responsible for drought response like RD26. Another class of transcription factors belonging to a subfamily APETALA2/ethylene response element-binding factors (AP2/ERF) type is majorly responsible for promoting plant development. It is also involved in providing stress tolerance against both biotic and abiotic stress. (Xu et al. 2011; Sharoni et al. 2010). AP2/ERF class has further been characterized into four subfamilies viz. AP2 (Apetala 2), RAV (related to ABI3/VP1), DREB (dehydration-responsive element-binding protein), and ERF (Rashid et al. 2012; Sharoni et al. 2010). Next class of TF (bZIP type transcription factors) that belongs to basic leucine zipper family, bears conserved N terminal domain was responsible for DNA binding and a C terminal leucine-rich motif was responsible for protein dimerization (Wang et al. 2015). This class of transcription factors is also known to interact with phytohormone ABA for providing drought tolerance in Rice and Arabidopsis. Studies have indicated that over expression of OsbZIP23, OsbZIP46, OsbZIP16 in rice and TabZIP60 in Arabidopsis was able to impart an enhanced tolerance against stresses like drought, salinity, and freezing as well as increased sensitivity to ABA (Zhang et al. 2015). Next class belongs to Zinc-finger proteins (ZFPs) which are consisted of 23 different subfamilies. Most characterized class among this family is WRKY transcription factor, which has a proven role in biotic and abiotic stresses alleviation (Li et al. 2014). This protection is said to be mediated via maintaining equilibrium between ROS generation and scavenging process (Baek et al. 2015). This fine-tuning helps in maintaining the levels of metabolites namely proline, chlorophyll, soluble sugars, as well as ABA (Wang et al. 2016a, b). Zinc-finger structure is diversified on the basis of presence of C and H residues which are present in secondary structure of these transcription factors. On the basis of C and H repeats these DNA-binding Zn fingers are classified into nine diverse groups: C2H2, C8, C6, C3HC4, C2HC, C2HC5, C4, C4HC3, and C3H (Luo et al. 2012; Gupta et al. 2012). Next type (NAC gene family) is the largest gene family and has most diverse role from development to leaf senescence, and from developing lateral roots to stress acclimation (Nuruzzaman et al. 2013). Majorly NAC gene family is named by the presence NAC domain that is formed collectively by NAM (for no apical meristem), ATAF 1 and 2 (Arabidopsis transcription activation factor), and CUC2 (for cup-shaped cotyledon). This family consists of a DNA-binding conserved N terminal nuclear localizing DNA-binding domain and a variable C terminal region responsible for regulating transcription (Olsen et al. 2005). Several researchers highlighted the important role of NAC variant's viz. OsNAC2/6, OsNAC10, SNAC1 genes/transcription factors, when over expressed in rice under stress, conferred enhanced stress tolerance, and enhanced grain yield, respectively. (Hu et al. 2006). Thus it could be concluded that as plants routinely face severe stressful regimes so they synchronize ROS level via activating/repressing several stress inducible transcription factors, protein kinases, and phosphatases. These factors work in tandem to regulate the expression of stress-responsive genes either positively or negatively for mitigating the lethal effect of oxidative stress.

14.5 Conclusion

Environmental stress namely biotic and abiotic, inadvertently disturbs the internal homeostasis of plants resulting in compromised growth and development, thus affecting the final yield potential. These stressful conditions lead to overproduction of ROS which are highly reactive molecules and alter the membrane properties via lipid peroxidation. In addition, it also makes irreversible damages to nucleic acids, lipids, and proteins thus hampering its proper functioning which is termed as oxidative stress. Plant possesses an efficient antioxidative defence system that counters the deleterious effect of ROS and oxidative stress with the help of enzymatic and non-enzymatic antioxidative defence pathways. Interestingly at lower level, ROS is also known to function as secondary messenger having several receptors for downstream relay of stress signal. ROS influences various proteins and enzymes namely protein kinases, which are categorized as receptor-like kinases, Mitogen Activated Protein kinases, Calmodulin-like domain protein kinases, and calcium dependent kinases. ROS also regulates levels of various stress inducible genes and transcription factors for stress abatement. While functioning as signaling molecule, a cross-talk exists between ROS and various regulatory genes/transcription factors, namely MYB, WRKY, DREB/ERF, and NAC which imply that stress tolerance is a complex interplay and regulated by an intricate regulatory mechanism. As every year we lose a significant portion of our crop yield due to these stresses therefore in order to achieve global food security and to feed burgeoning human population it becomes imperative to further decipher the role of ROS and its types, and signaling for stress mitigation and its complex interplay with defence related pathways in plants.

Acknowledgement Work of all the researchers cited in this chapter is sincerely acknowledged.

References

Andreyev AY, Kushnareva YE, Starkov AA (2005) Mitochondrial metabolism of reactive oxygen species. Biochem Mosc 70(2):200–214

Apel K, Hirt H (2004) Reactive oxygen species: metabolism, oxidative stress, and signal transduction. Annu Rev Plant Biol 55:373–399

Asada K (1999) The water-water cycle in chloroplasts: scavenging of active oxygens and dissipation of excess photons. Annu Rev Plant Biol 50(1):601–639

Baek D, Cha JY, Kang S, Park B, Lee HJ, Hong H, Chun HJ, Kim DH, Kim MC, Lee SY, Yun DJ (2015) The Arabidopsis a zinc finger domain protein ARS1 is essential for seed germination and ROS homeostasis in response to ABA and oxidative stress. Front Plant Sci 6:963

Baker A, Graham IA (eds) (2013) Plant peroxisomes: biochemistry, cell biology and biotechnological applications. Springer, Dordrecht

Banerjee A, Tripathi DK, Roychoudhury A (2018) Hydrogen sulphide trapeze: environmental stress amelioration and phytohormone crosstalk. Plant Physiol Biochem 132:46–53

Bhatt D, Negi M, Sharma P, Saxena SC, Dobriyal AK, Arora S (2011) Responses to drought induced oxidative stress in five finger millet varieties differing in their geographical distribution. Physiol Mol Biol Plants 17(4):347

Bienert GP, Møller AL, Kristiansen KA, Schulz A, Møller IM, Schjoerring JK, Jahn TP (2007) Specific aquaporins facilitate the diffusion of hydrogen peroxide across membranes. J Biol Chem 282(2):1183–1192

Boguszewska D, Zagdańska B (2012) ROS as signaling molecules and enzymes of plant response to unfavourable environmental conditions. In: Volodymyr L, Halyna S (eds) Oxidative stress-molecular mechanisms and biological effects. Intech Open, Rijeka, Croatia, pp 341–362

Chakradhar T, Reddy RA, Chandrasekhar T (2019) Protein kinases and phosphatases in stress transduction: role in crop improvement. In: Iqbal M, Khan R, Reddy PS, Khan NA (eds) Plant signaling molecules. Woodhead Publishing, Cambridge, UK, pp 533–547

Chen S, Dickman MB (2004) Bcl-2 family members localize to tobacco chloroplasts and inhibit programmed cell death induced by chloroplast-targeted herbicides. J Exp Bot 55 (408):2617–2623

Chen Z, Gallie DR (2006) Dehydroascorbate reductase affects leaf growth, development, and function. Plant Physiol 142(2):775–787

Chen T, Li W, Hu X, Guo J, Liu A, Zhang B (2015) A cotton MYB transcription factor, GbMYB5, is positively involved in plant adaptive response to drought stress. Plant Cell Physiol 56 (5):917–929

Choudhury S, Panda P, Sahoo L, Panda SK (2013) Reactive oxygen species signaling in plants under abiotic stress. Plant Signal Behav 8(4):23681–23686

Desikan R, Soheila AH, Hancock JT, Neill SJ (2001) Regulation of the Arabidopsis transcriptome by oxidative stress. Plant Physiol 127(1):159–172

Eltayeb AE, Kawano N, Badawi GH, Kaminaka H, Sanekata T, Shibahara T, Inanaga S, Tanaka K (2007) Overexpression of monodehydroascorbate reductase in transgenic tobacco confers enhanced tolerance to ozone, salt and polyethylene glycol stresses. Planta 225(5):1255–1264

Ferrer MA, Cimini S, López-Orenes A, Calderón AA, De Gara L (2018) Differential Pb tolerance in metallicolous and non-metallicolous *Zygophyllum fabago* populations involves the strengthening of the antioxidative pathways. Environ Exp Bot 150:141–151

Formentin E, Sudiro C, Perin G, Riccadonna S, Barizza E, Baldoni E, Lavezzo E, Stevanato P, Sacchi GA, Fontana P, Toppo S (2018) Transcriptome and cell physiological analyses in different rice cultivars provide new insights into adaptive and salinity stress responses. Front Plant Sci 9:204

Foyer CH, Noctor G (2016) Stress-triggered redox signalling: what's in pROSpect? Plant Cell Environ 39(5):951–964

Gauthier A, Idänheimo N, Brosché M (2011) Characterization of RLSs in Arabidopsis thaliana. In: Proceedings of the 10th international conference on reactive oxygen and Nitrogen species in plants, p 5

Gill SS, Tuteja N (2010) Reactive oxygen species and antioxidant machinery in abiotic stress tolerance in crop plants. Plant Physiol Biochem 48(12):909–930

Gilroy S, Białasek M, Suzuki N, Górecka M, Devireddy AR, Karpiński S, Mittler R (2016) ROS, calcium, and electric signals: key mediators of rapid systemic signaling in plants. Plant Physiol 171(3):1606–1615

Gupta SK, Rai AK, Kanwar SS, Sharma TR (2012) Comparative analysis of zinc finger proteins involved in plant disease resistance. PLoS One 7(8):e42578

Halliwell B (2006) Reactive species and antioxidants. Redox biology is a fundamental theme of aerobic life. Plant Physiol 141(2):312–322

Halliwell B, Gutteridge J (1984) Oxygen toxicity, oxygen radicals, transition metals and disease. Biochem J 219(1):1–14

Hu H, Dai M, Yao J, Xiao B, Li X, Zhang Q, Xiong L (2006) Overexpressing a NAM, ATAF, and CUC (NAC) transcription factor enhances drought resistance and salt tolerance in rice. Proc Nat Acad Sci USA 103(35):12987–12992

Khan TA, Mazid M, Mohammad F (2012) Potential of ascorbic acid against oxidative burst in plants under biotic stress: a review. J Ind Res Technol 2(2):72–80

Krieger-Liszkay A, Fufezan C, Trebst A (2008) Singlet oxygen production in photosystem II and related protection mechanism. Photosynth Res 98(1-3):551–564

Li WT, Chen WL, Yang C, Wang J, Yang L, He M, Wang JC, Qin P, Wang YP, Ma BT, Li SG (2014) Identification and network construction of zinc finger protein (ZFP) genes involved in the rice-'*Magnaporthe oryzae*' interaction. Plant Omics 7(6):540–548

Liang X, Zhou JM (2018) Receptor-like cytoplasmic kinases: central players in plant receptor kinase–mediated signaling. Annu Rev Plant Biol 69:267–299

Luo X, Bai X, Zhu D, Li Y, Ji W, Cai H, Wu J, Liu B, Zhu Y (2012) GsZFP1, a new Cys2/His2-type zinc-finger protein, is a positive regulator of plant tolerance to cold and drought stress. Planta 235(6):1141–1155

Masood A, Per TS, Asgher M, Fatma M, Khan MIR, Rasheed F, Hussain SJ, Khan NA (2016) Glycine betaine: role in shifting plants toward adaptation under extreme environments. In: Iqbal N, Nazar R, Khan NA (eds) Osmolytes and plants acclimation to changing environment: emerging omics technologies. Springer, New Delhi, pp 69–82

Mhamdi A, Queval G, Chaouch S, Vanderauwera S, Van Breusegem F, Noctor G (2010) Catalase function in plants: a focus on Arabidopsis mutants as stress-mimic models. J Exp Bot 61 (15):4197–4220

Mignolet-Spruyt L, Xu E, Idänheimo N, Hoeberichts FA, Mühlenbock P, Brosché M, Van Breusegem F, Kangasjärvi J (2016) Spreading the news: subcellular and organellar reactive oxygen species production and signalling. J Exp Bot 67(13):3831–3844

Miller GAD, Suzuki N, Ciftci-Yilmaz SULTAN, Mittler RON (2010) Reactive oxygen species homeostasis and signalling during drought and salinity stresses. Plant Cell Environ 33 (4):453–467

Mittler R (2002) Oxidative stress, antioxidants and stress tolerance. Trends Plant Sci 7(9):405–410

Mittler R (2017) ROS are good. Trends Plant Sci 22(1):11–19

Moller N, Christov C, Weare J (2007) Thermodynamic model for predicting interactions of geothermal brines with hydrothermal aluminum silicate minerals. In: Proceedings 32th workshop on geothermal reservoir engineering. Stanford University, Stanford, CA, pp 22–24

Nahar K, Hasanuzzaman M, Alam MM, Fujita M (2015) Roles of exogenous glutathione in antioxidant defense system and methylglyoxal detoxification during salt stress in mung bean. Biol Plant 59(4):745–756

Navrot N, Rouhier N, Gelhaye E, Jacquot JP (2007) Reactive oxygen species generation and antioxidant systems in plant mitochondria. Physiol Plant 129(1):185–195

Noctor G, Foyer CH (1998) Ascorbate and glutathione: keeping active oxygen under control. Annu Rev Plant Biol 49(1):249–279

Noctor G, De Paepe R, Foyer CH (2007) Mitochondrial redox biology and homeostasis in plants. Trends Plant Sci 12(3):125–134

Nuruzzaman M, Sharoni AM, Kikuchi S (2013) Roles of NAC transcription factors in the regulation of biotic and abiotic stress responses in plants. Front Microbiol 4:248

Ogasawara Y, Kaya H, Hiraoka G, Yumoto F, Kimura S, Kadota Y, Hishinuma H, Senzaki E, Yamagoe S, Nagata K, Nara M (2008) Synergistic activation of the Arabidopsis NADPH oxidase AtrbohD by Ca2+ and phosphorylation. J Biol Chem 283(14):8885–8892

Olsen AN, Ernst HA, Leggio LL, Skriver K (2005) NAC transcription factors: structurally distinct, functionally diverse. Trends Plant Sci 10(2):79–87

Per TS, Khan NA, Reddy PS, Masood A, Hasanuzzaman M, Khan MIR, Anjum NA (2017) Approaches in modulating proline metabolism in plants for salt and drought stress tolerance: Phytohormones, mineral nutrients and transgenics. Plant Physiol Biochem 115:126–140

Pfannschmidt T (2003) Chloroplast redox signals: how photosynthesis controls its own genes. Trends Plant Sci 8(1):33–41

Pinto E, Sigaud-kutner TC, Leitao MA, Okamoto OK, Morse D, Colepicolo P (2003) Heavy metal–induced oxidative stress in algae. J Phycol 39(6):1008–1018

Rashid M, Guangyuan H, Guangxiao Y, Hussain J, Xu Y (2012) AP2/ERF transcription factor in rice: genome-wide canvas and syntenic relationships between monocots and eudicots. Evol Bioinform Online 8:321–355

Rasmusson AG, Geisler DA, Møller IM (2008) The multiplicity of dehydrogenases in the electron transport chain of plant mitochondria. Mitochondrion 8(1):47–60

Riechmann JL, Heard J, Martin G, Reuber L, Jiang CZ, Keddie J, Adam L, Pineda O, Ratcliffe OJ, Samaha RR, Creelman R (2000) Arabidopsis transcription factors: genome-wide comparative analysis among eukaryotes. Science 290(5499):2105–2110

Saxena SC, Kaur H, Verma P, Petla BP, Andugula VR, Majee M (2013) Osmoprotectants: potential for crop improvement under adverse conditions. In: Tuteja N, Gill S (eds) Plant acclimation to environmental stress. Springer, New York, pp 197–232

Sharma P, Dubey RS (2004) Ascorbate peroxidase from rice seedlings: properties of enzyme isoforms, effects of stresses and protective roles of osmolytes. Plant Sci 167(3):541–550

Sharma P, Jha AB, Dubey RS, Pessarakli M (2012) Reactive oxygen species, oxidative damage, and antioxidative defense mechanism in plants under stressful conditions. J Bot 2012:1–26

Sharoni AM, Nuruzzaman M, Satoh K, Shimizu T, Kondoh H, Sasaya T, Choi IR, Omura T, Kikuchi S (2010) Gene structures, classification and expression models of the AP2/EREBP transcription factor family in rice. Plant Cell Physiol 52(2):344–360

Slama I, Abdelly C, Bouchereau A, Flowers T, Savoure A (2015) Diversity, distribution and roles of osmoprotective compounds accumulated in halophytes under abiotic stress. Ann Bot 115 (3):433–447

Suzuki N, Miller G, Morales J, Shulaev V, Torres MA, Mittler R (2011) Respiratory burst oxidases: the engines of ROS signaling. Curr Opin Plant Biol 14(6):691–699

Takeda S, Gapper C, Kaya H, Bell E, Kuchitsu K, Dolan L (2008) Local positive feedback regulation determines cell shape in root hair cells. Science 319(5867):1241–1244

Tripathi DK, Singh S, Singh S, Srivastava PK, Singh VP, Singh S, Prasad SM, Singh PK, Dubey NK, Pandey AC, Chauhan DK (2017) Nitric oxide alleviates silver nanoparticles (AgNps)-induced phytotoxicity in *Pisum sativum* seedlings. Plant Physiol Biochem 110:167–177

Valko MMHCM, Morris H, Cronin MTD (2005) Metals, toxicity and oxidative stress. Curr Med Chem 12(10):1161–1208

Wang Z, Cheng K, Wan L, Yan L, Jiang H, Liu S, Lei Y, Liao B (2015) Genome-wide analysis of the basic leucine zipper (bZIP) transcription factor gene family in six legume genomes. BMC Genomics 16(1):1–15

Wang C, Lu W, He X, Wang F, Zhou Y, Guo X, Guo X (2016a) The cotton mitogen-activated protein kinase kinase 3 functions in drought tolerance by regulating stomatal responses and root growth. Plant Cell Physiol 57(8):1629–1642

Wang Y, Sun T, Li T, Wang M, Yang G, He G (2016b) A CBL-interacting protein kinase TaCIPK2 confers drought tolerance in transgenic tobacco plants through regulating the stomatal movement. PLoS One 11(12):e0167962

Waszczak C, Carmody M, Kangasjärvi J (2018) Reactive oxygen species in plant signaling. Annu Rev Plant Biol 69:209–236

Xu ZS, Chen M, Li LC, Ma YZ (2011) Functions and application of the AP2/ERF transcription factor family in crop improvement. J Int Plant Biol 53(7):570–585

Zhang L, Zhang L, Xia C, Zhao G, Liu J, Jia J, Kong X (2015) A novel wheat bZIP transcription factor, TabZIP60, confers multiple abiotic stress tolerances in transgenic Arabidopsis. Physiol Plant 153(4):538–554

Zhu JK (2016) Abiotic stress signaling and responses in plants. Cell 167(2):313–324

Chapter 15
Bacterial Antagonists Effective Against Soil Borne and Foliar Pathogens

Stanzin Idong and Anil Kumar Sharma

Abstract Diseases affect annual and perennial woody plants severely, hence lead to major food losses and damage to natural ecosystem. Management of plant diseases is urgently required to revive economic and ecological losses. Chemical pesticides are heavily exploited on large scale to overcome the hurdle of these losses. But as the world is facing the problems of human health hazard and ground water pollution due to injudicious use of agrochemicals (especially pesticides) in agriculture and public sector therefore a large group of agriculture scientists is looking forward to develop ecofriendly, harmless, and sustainable system of bioinoculants, in the form of antagonistic bacteria. Bacterial antagonist acts as potential and effective biocontrol agent for most of the plant diseases occurring by soil borne pathogens or air borne pathogens. Plant pathogens can be suppressed with the induction of bacterial antagonist as they produce a variety of secondary metabolites with different biological functions such as inhibitors, antimicrobial, and organic matter degrading enzymes. These bacteria also activate defense mechanisms of plants against various diseases.

15.1 Introduction

Microorganisms are generally found associated with the plant rhizosphere, which is considered as a hot spot of microbial diversity (Santoyo et al. 2016). These microorganisms are involved in competition among themselves which could be synergistic or antagonistic with plants. On the basis of the location on plants these bacteria can be categorized as epiphytic and endophytic where they live sharing numerous characteristics for host plant growth promotion. Depending on the physiochemical properties of soil, per gram of soil contains several hundred million of bacteria

S. Idong
Department of Plant Pathology, College of Agriculture, Pantnagar, Uttarakhand, India

A. K. Sharma (✉)
Department of Biological Sciences, CBSH, G.B. Pant University of Agriculture and Technology, Pantnagar, Uttarakhand, India

A. Sharma (ed.), *Microbes and Signaling Biomolecules Against Plant Stress*, Rhizosphere Biology, https://doi.org/10.1007/978-981-15-7094-0_15

(Toppo and Naik 2015). Different rhizospheric bacteria have been reported to possess antagonistic activities against plant fungal and bacterial plant pathogens such as *Pseudomonas fluorescens*, *Agrobacterium radiobacter*, *Bacillus subtilis*, *B. amyloliquefaciens*, *Trichoderma virens*, *Burkholderia cepacia*, and *Gliocladium* spp. (Kloepper et al. 1989).

Plant growth promoting rhizobacteria (PGPR) constitute a group of microorganisms, which is used as biofertilizer and found to promote plant growth by increasing nutrient availability, phytostimulation, and biocontrol properties (Pandey and Malviya 2014). Population of microbes can increase or decrease depending on the changes in soil condition, i.e., moisture, temperature, or substrates like carbon. Bacteria/fungi generally release different antibiotics to suppress the growth of other microbes in the soil. The root exudates in the rhizospheric region have numerous essential nutrients and growth factors to support the growth of important bacteria. The wide effect of rhizosphere is usually observed with bacteria rather than Actinomycetes and fungi. Groups of gram negative, rod shaped, and nonsporulating bacteria which respond to root exudates are dominant bacteria (*Pseudomonas*, *Agrobacterium*).

The bacteria isolated from rhizosphere soils are beneficial to the plants by directly having an effect on nitrogen fixation, solubilization of nutrients, production of siderophores, β 1,3-glucanase, antibiotics, fluorescent pigments, hydrolytic enzymes, and stimulation of phytoalexin or flavonoids like compounds. Genus like *Azotobacter*, *Actinomycetes*, *Bacillus*, *Azospirillum*, and *Pseudomonas* produce auxin that stimulate plant growth. It has been estimated that 80% bacteria isolated from rhizosphere can produce plant growth regulators (Patten and Glick 1996). These strains must be competent, able to survive and colonize in the rhizospheric soil.

In agriculture many phytopathogenic fungi, bacteria, and viruses affect the crops and lead to develop different diseases, poor crop health, and ultimately yield loss. To overcome the losses pesticides are invariably used in agriculture practices. However population is growing at a faster pace and in order to feed huge population the chemicals are being used blindly and un-judiciously on large scale. Unfortunately the use of agrochemicals causes environmental pollution and health problems in human. Thus the alternative of costly agrochemicals is the antagonistic microorganisms that are found in association with host plants.

15.2 Bacterial Antagonists Effective Against Soil Borne Pathogens

Tripathi and Johti (2002) studied some potential bacterial isolates from the rhizosphere of pea and wheat against maize sheath blight caused by *Rhizoctonia solani*. Few isolates were effective against multiple diseases and some were effective against only one pathogen showing their diversity. Tiwari and Thrimurthy (2007)

Table 15.1 Bacterial antagonist effective against different soil borne diseases

Crop	Antagonist	Disease	Pathogen	References
Tomato	*Bacillus spp.2P2*	Collar rot	*Sclerotium rolfsii*	Sahu et al. (2019)
Rice	*Bacillus amyloliquefaciens* spp.	Sheath blight	*Rhizoctonia solani*	Abbas et al. (2019)
Faba bean	*Rahnella aquatilis B16C*	Root rot	*Fusarium solani*	Bahroun et al. (2018)
Tomato	*Bacillus amyloliquefaciens FBZ24*	Wilt	*Fusarium oxysporum f. sp.lycopersici*	Elanchezhiyan et al. (2018)
Cucumber	*Paenibacillus polymyxa*	Wilt	*Fusarium oxysporum sp. cucumerinum*	Du et al. (2017)
Apple	*Bacillus amyloliquefaciens GB1*	Valsa Canker	*Valsa mali*	Zhang et al. (2015)
Rapeseed	*Bacillus subtilis EDR4*	Sclerotinia rot	*Sclerotinia sclerotiorum*	Chen et al. (2014)
Rice	*Bacillus subtilis MBI 600*	Sheath blight	*Rhizoctonia solani*	Kumar et al. (2012)
Tomato	*Bacillus subtilis BSF 4*	Southern blight	*Sclerotium rolfsii*	De Curtis et al. (2010)
Banana	*Pseudomonas, Burkholderia*	Fusarium wilt	*Fusarium oxysporum f. sp. cubense*	Fishal et al. (2010)
Cotton	*Pseudomonas* spp.	Verticillium spp.	*Verticillium dahlia*	Erdogan and Benlioglu (2010)

isolated 21 isolates of *Pseudomonas fluorescens* from the rhizosphere of maize, rice, wheat, and chickpea from Raipur and Bastar region, where seven isolates showed bright fluorescence and also showed positive response of siderophore production. Among the isolates, PFR 1 and PFR 2 were significantly superior in increasing shoot and root length of plants. In vitro evaluation has confirmed the antagonistic ability of the two isolates against *Pyricularia grisea* and *Rhizoctonia solani* in dual culture test.

More than 10,000 species of fungi can cause disease in plants (Agrios 2005). A common fungal pathogen in agriculture crops is *Rhizoctonia* spp. existing in soil and attacks more than 200 species of crops. *Fusarium* has long been known for causing wilts in many species and considered as an important plant pathogen (Toppo and Naik 2015). Other important plant diseases are root rot, wilt, and sheath blight for which many antagonistic bacteria have been isolated and tested. Various *Pseudomonas sp.*, evaluated for antifungal activity against soil borne fungal plant pathogen, showed reduction in mycelium growth of *Fusarium oxysporum* (Table 15.1).

15.3 Bacterial Antagonists Effective Against Foliar Pathogens

Different biocontrol agents isolated so far are mainly found responsible for the management of soil borne pathogens as destruction due to the pathogen is more severe than foliar pathogens. The effective control of *A. alternata* of citrus, litchi, and muskmelon was observed with *Bacillus subtilis* (Jiang et al. 2001). In this study, rhizobacterial isolates were tested against the growth of *Alternaria brassicae* and mustard plants. Mates et al. (2019) reported that out of 29 isolates, *Bacillus velezensis* GF267 was considered to be a multi-site antagonist and controlled tomato bacterial spot. When *Bacillus velezensis* GF267 was applied on soil it increased the activity of peroxidase and polyphenoloxidase, amount of chlorophyll content, and reduced disease intensity (Table 15.2).

15.4 Mechanism

Various mechanisms involved in the inhibition of plant pathogens include secretion of lytic enzymes, siderophores, antibiotics, and volatile organic compounds (VOCs) emitted from bacterial antagonists which influence the mycelial growth of the soil borne phytopathogenic fungus *Rhizoctonia solani* Kühn. The biological control of pathogens can result from a combination of mechanisms, including competition for nutrients and space, antibiotic production, and induced systemic resistance. Iron depletion by the *M. pulcherrima* in the medium inhibited the mycelial growth and conidial germination of *B. cinerea*, *A. alternata*, and *P. expansum*. *Pseudomonas* known to produce the antibiotic 2, 4-diacetylphloroglucinol (DAPG) may also induce host defenses. Additionally, DAPG-producing bacterial antagonists can aggressively colonize root (Heydari and Pessarakli 2010). 2-phenylethanol isolated from *K. apiculata* (an antifungal volatile) has inhibitory effect on green and blue mold of citrus (Liu et al. 2014). Exo-β-1, 3-glucanase secreted by *P. membranifaciens* had a role in the biocontrol activity against *B. cinerea* on grapevine (Masih and Paul 2002).

15.5 Conclusion

Antagonistic bacteria can be effective in controlling the plant diseases through different mechanisms. These bacteria not only reduce the disease symptoms of plants by pathogens but also enhance plant growth and development. Farmers, in general still rely on the use of synthetic fungicides to control plant diseases. However, the misuse of these chemicals may cause serious environmental and health problems. Microbial antagonists are potential agents that can be explored to provide

Table 15.2 Bacterial antagonists effective against foliar diseases

Crop	Antagonist	disease	Pathogen	References
Tomato	*Bacillus subtilis*	Early blight	*Alternaria solani*	Awan and Shoaib (2019)
Rice	*Phosphorus solubilizing bacteria (PSB)*	Bacterial leaf blight	*Xanthomonas oryzae pv. oryzae*	Rasul et al. (2019)
Tomato	*Bacillus velezensis GF 267*	Bacterial leaf spot	*Xanthomonas perforans* *Xanthomonas gardneri*	Mates et al. (2019)
Rice	*Serratia nematodiphila CT-78*	Bacterial leaf blight	*Xanthomonas oryzae pv oryzae*	Khoa et al. (2016)
Grapevine	*Bacillus pumilus* *Paenibacillus* spp.	Grapevine trunk disease	*Phaeomoniella chlamydospora*	Haidar et al. (2016)
Potato	*Bacillus subtilis*	Late blight	*Phytophthora infestans*	Kumbar et al. (2019)
Rice	*Bacillus amyloliquefaciens* *BS5*	Brown spot	*Helminthosporium oryzae*	Prabhukarthikeyan et al. (2019)
Tomato	*Pseudomonas fluorescence SF4c*	Bacterial spot	*Xanthomonas vesicatoria*	Principe et al. (2018)
Cotton	*Bacillus licheniformis OE-04*	Anthracnose	*Colletotrichum gossypii*	Nawaz et al. (2018)
Mustard	*Bacillus amyloliquefaciens + earthworm*	Leaf blight	*Alternaria brassicae*	Ayuke et al. (2017)
Zucchini	*Serratia marcescens*	Powdery mildew	*Podosphaera xanthii*	Tesfagiorgis et al. (2014)
Tomato	*Streptomyces griseoviridisK61*	Early blight	*Alternaria alternata*	Cuppels et al. (2013)
Tomato	*Streptomyces lydicus WYEC108+* *Pseudomonas fluorescens A506*	Anthracnose	*Colletotrichum phomoides*	Cuppels et al. (2013)
Sweet orange	*Bacillus subtilis ACB-69*	Postbloom fruit drop	*Colletotrichum acutatum*	Kupper et al. (2012)
Chili	*Pseudomonas fluorescens*	Powdery mildew	*Leveillula taurica*	Anand et al. (2010)

effective and safe means to manage plant diseases. Several microorganisms have been tested and proven to possess antagonistic properties against plant pathogenic fungi to develop sustainable agriculture. Control of plant disease is very crucial and use of biocontrol agents is an alternative approach against chemical pesticide. Further exploitation and work are required to carry out on antagonistic bacteria to produce a commercial biocontrol agent for controlling phytopathogens.

Acknowledgements Authors acknowledge the facilities provided by the Department of Biological Sciences and Department of Plant Pathology GBPUA&T, Pantnagar.

References

Abbas A, Khan SU, Khan WU, Saleh TA, Khan MHU, Ullah S, Ikram M (2019) Antagonist effects of strains of Bacillus spp. against *Rhizoctonia solani* for their protection against several plant diseases: alternatives to chemical pesticides. C R Biol 342:124–135

Agrios GN (2005) Plant diseases caused by fungi. In: Plant pathology. Elsevier/Academic Press, Burlington, MA, p 4

Anand T, Chandrasekaran A, Kuttalam S, Senthilraja G, Samiyappan R (2010) Integrated control of fruit rot and powdery mildew of chilli using the biocontrol agent *Pseudomonas fluorescens* and a chemical fungicide. Biol Control 52:1–7

Awan ZA, Shoaib A (2019) Combating early blight infection by employing *Bacillus subtilis* in combination with plant fertilizers. Curr Plant Biol 20:100125

Ayuke FO, Lagerlöf J, Jorge G, Söderlund S, Muturi JJ, Sarosh BR, Meijer J (2017) Effects of biocontrol bacteria and earthworms on the severity of *Alternaria brassicae* disease and the growth of oilseed rape plants (*Brassica napus*). Appl Soil Ecol 117:63–69

Bahroun A, Jousset A, Mhamdi R, Mrabet M, Mhadhbi H (2018) Anti-fungal activity of bacterial endophytes associated with legumes against *Fusarium solani*: assessment of fungi soil suppressiveness and plant protection induction. Appl Soil Ecol 124:131–140

Chen Y, Gao X, Chen Y, Qin H, Huang L, Han Q (2014) Inhibitory efficacy of endophytic *Bacillus subtilis* EDR4 against *Sclerotinia sclerotiorum* on rapeseed. Biol Control 78:67–76

Cuppels DA, Higham J, Traquair JA (2013) Efficacy of selected streptomycetes and a streptomycete+ pseudomonad combination in the management of selected bacterial and fungal diseases of field tomatoes. Biol Control 67:361–372

De Curtis F, Lima G, Vitullo D, De Cicco V (2010) Biocontrol of *Rhizoctonia solani* and *Sclerotium rolfsii* on tomato by delivering antagonistic bacteria through a drip irrigation system. Crop Prot 29:663–670

Du N, Shi L, Yuan Y, Sun J, Shu S, Guo S (2017) Isolation of a potential biocontrol agent *Paenibacillus polymyxa* NSY50 from vinegar waste compost and its induction of host defense responses against *Fusarium* wilt of cucumber. Microbiol Res 202:1–10

Elanchezhiyan K, Keerthana U, Nagendran K, Prabhukarthikeyan SR, Prabakar K, Raguchander T, Karthikeyan G (2018) Multifaceted benefits of *Bacillus amyloliquefaciens* strain FBZ24 in the management of wilt disease in tomato caused by *Fusarium oxysporum* f. sp. *lycopersici*. Physiol Mol Plant Pathol 103:92–101

Erdogan O, Benlioglu K (2010) Biological control of *Verticillium* wilt on cotton by the use of fluorescent *Pseudomonas* spp. under field conditions. Biol Control 53:39–45

Fishal EMM, Meon S, Yun WM (2010) Induction of tolerance to *Fusarium* wilt and defense-related mechanisms in the plantlets of susceptible berangan banana pre-inoculated with *Pseudomonas* sp. (UPMP3) and *Burkholderia* sp. (UPMB3). Agric Sci China 9:1140–1149

Haidar R, Roudet J, Bonnard O, Dufour MC, Corio-Costet MF, Fert M, Fermaud M (2016) Screening and modes of action of antagonistic bacteria to control the fungal pathogen *Phaeomoniella chlamydospora* involved in grapevine trunk diseases. Microbiol Res 192:172–184

Heydari A, Pessarakli M (2010) A review on biological control of fungal plant pathogens using microbial antagonists. J Biol Sci 10:273–290

Jiang YM, Zhu XR, Li YB (2001) Postharvest control of litchi fruit rot by *Bacillus subtilis*. LWT-Food Sci Technol 34:430–436

Khoa ND, Giàu NDN, Tuấn TQ (2016) Effects of *Serratia nematodiphila* CT-78 on rice bacterial leaf blight caused by *Xanthomonas oryzae* pv. *oryzae*. Biol Control 103:1–10

Kloepper JW, Lifshitz R, Zablotowicz RM (1989) Free-living bacterial inocula for enhancing crop productivity. Trends Biotechnol 7:39–44

Kumar KVK, Yellareddygari SK, Reddy MS, Kloepper JW, Lawrence KS, Zhou XG, Miller ME (2012) Efficacy of *Bacillus subtilis* MBI 600 against sheath blight caused by *Rhizoctonia solani* and on growth and yield of rice. Rice Sci 19:55–63

Kumbar B, Mahmood R, Nagesha SN, Nagaraja MS, Prashant DG, Kerima OZ, Chavan M (2019) Field application of *Bacillus subtilis* isolates for controlling late blight disease of potato caused by *Phytophthora infestans*. Biocatal Agric Biotechnol 22:101366

Kupper KC, Corrêa FE, de Azevedo FA, Da Silva AC (2012) *Bacillus subtilis* to biological control of postbloom fruit drop caused by *Colletotrichum acutatum* under field conditions. Sci Hortic 134:139–143

Liu P, Cheng Y, Yang M, Liu Y, Chen K, Long CA, Deng X (2014) Mechanisms of action for 2-phenylethanol isolated from *Kloeckera apiculata* in control of *Penicillium* molds of citrus fruits. BMC Microbiol 14:242

Masih EI, Paul B (2002) Secretion of β-1, 3-glucanases by the yeast *Pichia membranifaciens* and its possible role in the biocontrol of *Botrytis cinerea* causing grey mold disease of the grapevine. Curr Microbiol 44:391–395

Mates ADPK, de Carvalho Pontes N, de Almeida Halfeld-Vieira B (2019) *Bacillus velezensis* GF267 as a multi-site antagonist for the control of tomato bacterial spot. Biol Control 137:104013

Nawaz HH, Rajaofera MN, He Q, Anam U, Lin C, Miao W (2018) Evaluation of antifungal metabolites activity from *Bacillus licheniformis* OE-04 against *Colletotrichum gossypii*. Pestic. Biochem Phys 146:33–42

Pandey A, Malviya T (2014) Production of antibiotics isolated from soil bacteria from rhizospheric and non-rhizospheric region of medicinal plants. Indian J Appl Res 4:25–32

Patten CL, Glick BR (1996) Bacterial biosynthesis of indole-3-acetic acid. Can J Microbiol 42:207–220

Prabhukarthikeyan SR, Yadav MK, Anandan A, Aravindan S, Keerthana U, Raghu S, Rath PC (2019) Bio-protection of brown spot disease of rice and insight into the molecular basis of interaction between *Oryza sativa*, *Bipolaris oryzae* and *Bacillus amyloliquefaciens*. Biol Control 137:104018

Principe A, Fernandez M, Torasso M, Godino A, Fischer S (2018) Effectiveness of tailocins produced by *Pseudomonas fluorescens* SF4c in controlling the bacterial-spot disease in tomatoes caused by *Xanthomonas vesicatoria*. Microbiol Res 212:94–102

Rasul M, Yasmin S, Zubair M, Mahreen N, Yousaf S, Arif M, Mirza MS (2019) Phosphate solubilizers as antagonists for bacterial leaf blight with improved rice growth in phosphorus deficit soil. Biol Control 136:103997

Sahu PK, Singh S, Gupta A, Singh UB, Brahmaprakash GP, Saxena AK (2019) Antagonistic potential of bacterial endophytes and induction of systemic resistance against collar rot pathogen *Sclerotium rolfsii* in tomato. Biol Control 137:104014

Santoyo G, Moreno-Hagelsieb G, del Carmen Orozco-Mosqueda M, Glick BR (2016) Plant growth-promoting bacterial endophytes. Microbiol Res 183:92–99

Tesfagiorgis HB, Laing MD, Annegarn HJ (2014) Evaluation of biocontrol agents and potassium silicate for the management of powdery mildew of zucchini. Biol Control 73:8–15
Tiwari PK, Thrimurthy VS (2007) Isolation and characterization of the *Pseudomonas fluorescens* from rhizosphere of different crops. J Mycol Plant Pathol 37:231–234
Toppo SR, Naik UC (2015) Isolation and characterization of bacterial antagonist to plant pathogenic fungi (*Fusarium* Spp.) from agro based area of Bilaspur. Int J Res Stud Biosci 8:6–14
Tripathi M, Johti B (2002) *In vitro* antagonistic potential of fluorescent pseudomonads and control of sheath blight of maize caused by *Rhizoctonia solani*. Ind J Microbiol 42:207–214
Zhang JX, Gu YB, Chi FM, Ji ZR, Wu JY, Dong QL, Zhou ZS (2015) *Bacillus amyloliquefaciens* GB1 can effectively control apple valsa canker. Biol Control 88:1–7

The manufacturer's authorised representative in the EU is Springer Nature Customer Service Centre GmbH, Europaplatz 3, 69115 Heidelberg, Germany. If you have any concerns regarding our products, please contact ProductSafety@springernature.com

Printed and bound by CPI Group (UK) Ltd, Croydon, CR0 4YY
30/09/2026
02225328-0002